# HUBBLE'S UNIVERSE

## Greatest Discoveries and Latest Images

# HUBBLE'S UNIVERSE

## Greatest Discoveries and Latest Images

### TERENCE DICKINSON

FIREFLY BOOKS

# A Firefly Book

Published by Firefly Books Ltd. 2017

Second printing, 2018

**Publisher Cataloging-in-Publication Data (U.S.)**

Names: Dickinson, Terence, author.
Title: Hubble's Universe : Greatest Discoveries and Latest Images / Terence Dickinson.
Description: Richmond Hill, Ontario, Canada : Firefly Books, 2017. | Second edition, revised and updated. | Includes bibliographical references and index. | Summary: "The book features 330 high resolution celestial portraits selected by the author, a four-page fold-out of the Andromeda Galaxy and an illuminating narrative that brings to life Hubble's journey and the fascinating forces at work in the universe." – Provided by publisher.
Identifiers: ISBN 978-1-77085-997-5 (hardcover)
Subjects: LCSH: Outer space – Exploration. | Hubble Space Telescope
   (Spacecraft) | Deep space – Pictorial works.
Classification: LCC QB500.268D535 |DDC 522.2919 – dc23

**Library and Archives Canada Cataloguing in Publication**

A CIP record for this title is available from Library and Archives Canada

Published in the United States by
Firefly Books (U.S.) Inc.
P.O. Box 1338, Ellicott Station
Buffalo, New York 14205

Published in Canada by
Firefly Books Ltd.
50 Staples Avenue, Unit 1
Richmond Hill, Ontario  L4B 0A7

Cover and interior design:
Janice McLean/Bookmakers Press Inc.

Printed in China

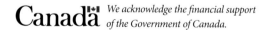
We acknowledge the financial support of the Government of Canada.

**Front cover:** NASA, ESA, the Hubble Heritage Team (STScI)
**Back cover:** NASA

# ACKNOWLEDGMENTS

This is a book that required an expert team, and I was fortunate to have one from day one. First on the list is Ray Villard, News Chief at the Space Telescope Science Institute, in Baltimore, Maryland, and a longtime friend and colleague. Ray's experience at space telescope headquarters dates back to 1986—four years before the launch of Hubble. This immersion of more than half an adult lifetime in all things Hubble has made him a walking encyclopedia of the history and accomplishments of the orbiting observatory. Ray has been invaluable in answering at least several hundred questions of mine during the preparation of the book you are now reading. For me to say he has been a stellar resource is to vastly understate his contributions.

Another essential part of the process is my publisher, Lionel Koffler, owner of Firefly Books. Lionel has partnered with me for 10 books over a span of three decades. I appreciate Lionel's insistence on uncompromising quality throughout the production of these books, allowing the delicate hues of astronomical images to be fully rendered on the printed page. With well over a million copies of my Firefly books in print, it has been a rewarding collaboration.

The handsome appearance of the book is entirely due to the graphic design talents of Janice McLean, a whirlwind of creativity and inventiveness that is a pleasure to witness. Editor Tracy C. Read, as always, has asked the incisive questions and offered many helpful suggestions, all of which improved the final work.

Finally, my deepest thanks are reserved for my wife, Susan, who has been my copyeditor and assistant for all 10 Firefly titles. The books couldn't have happened without her. A great copyeditor, which Susan is, breathes new life into ordinary prose—she makes my words sing.

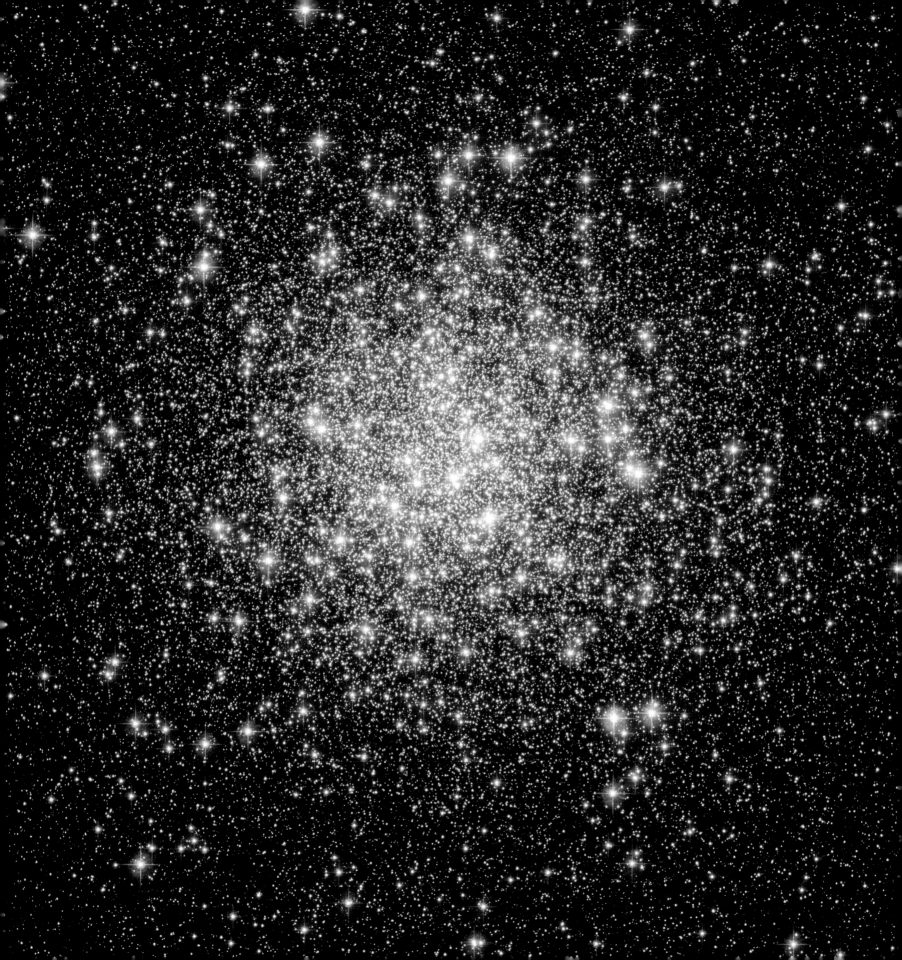

# CONTENTS

The dazzling globular star cluster Messier 9, or simply M9, contains hordes of stars swarming in a spherical cloud about 25,000 light-years from Earth. It is too faint to be seen with the naked eye, and when it was discovered by French astronomer Charles Messier in 1764, he observed it only as a faint smudge in his small telescope. He classified the cluster as a nebula ("cloud" in Latin). This Hubble Space Telescope portrait, the best image yet of M9, reveals 250,000 individual stars.

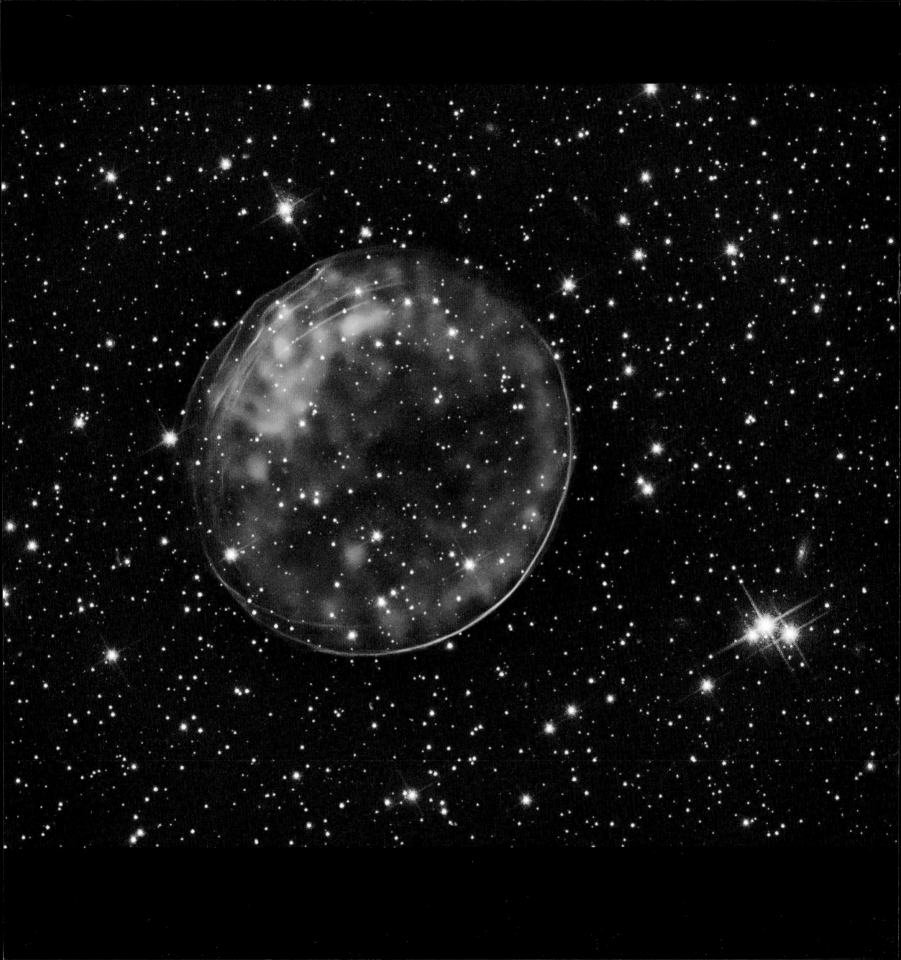

# INTRODUCTION

In addition to being one of the greatest scientific instruments of all time, the Hubble Space Telescope has given humanity a spectacular legacy of beautiful images of the universe. The best of these are displayed—and explained—in this book.

As a teenager in the 1950s, I was captivated by the science fiction of the brilliant visionary Arthur C. Clarke. Browsing the local library, I stumbled upon Clarke's early nonfiction work *The Exploration of Space*, published in 1951. Half a century later, *The New York Times* described this classic text as "a seamless blend of scientific expertise and poetic imagination that helped usher in the space age."

It was in the pages of Clarke's book that I first encountered the concept of a telescope in orbit around our planet. This telescope would peer at the universe from well above the interference of the Earth's ever turbulent atmosphere, which relentlessly blurs the view in ground-based telescopes and makes stars twinkle. Ahead of his time, Clarke outlined the advantages of an orbiting telescope compared with a telescope that might, at some future point, be installed on the Moon's surface, as had been suggested decades earlier. "Even the Moon's extremely tenuous atmosphere might affect certain delicate observations," he wrote. "[Moreover,] an observatory in space would be able to survey the complete sphere of the sky."

The orbiting scope should even be able to detect planets of nearby stars, enthused Clarke, "something quite out of the question with Earth-based equipment." I couldn't wait! During breaks at my first summer job in the shipping department of a publishing house, I made endless pencil sketches on large sheets of brown paper. I imagined just what the photos from that great eye-in-the-sky would look like—images that would show surface details on the moons of Jupiter, views deep within the core of the globular cluster M13, and so on—until my boss saw what I was up to and cautioned me not to waste any more shipping paper.

Today, the orbiting telescope Clarke envisioned is known as the Hubble Space Telescope, and it has been in service since 1990. That telescope has captured stupendous full-color images that depict the subjects of my crude brown-paper sketches and hundreds more of objects I hadn't yet conjured. What a pleasure it was to select more than 300 of Hubble's best cosmic portraits for the first edition of this book in 2012. Now, five years later, the Hubble Space Telescope continues to perform at its best, capturing dozens of new images—some as recent as 2017—which are collected in Chapter 11 in this expanded edition. The book is now an up-to-date collector's edition of Hubble's best work, more than 350 images, all accompanied by captions and text that will serve as navigational tools as you explore with one of the greatest scientific tools ever built.

*Hubble's Universe* is a celebration of the astonishing achievements of a remarkable discovery machine. Enjoy the excursion!

—*Terence Dickinson*

A celestial shell of interstellar gas being shocked by the blast wave from a supernova, the Ornament Nebula was imaged by the Hubble Space Telescope and combined with X-ray images from NASA's Chandra X-ray Observatory. The supernova—the explosive destruction of a star—occurred nearly 400 years ago and is 23 light-years across. The nebula is expanding at the rate of the Earth-to-Moon distance every minute.

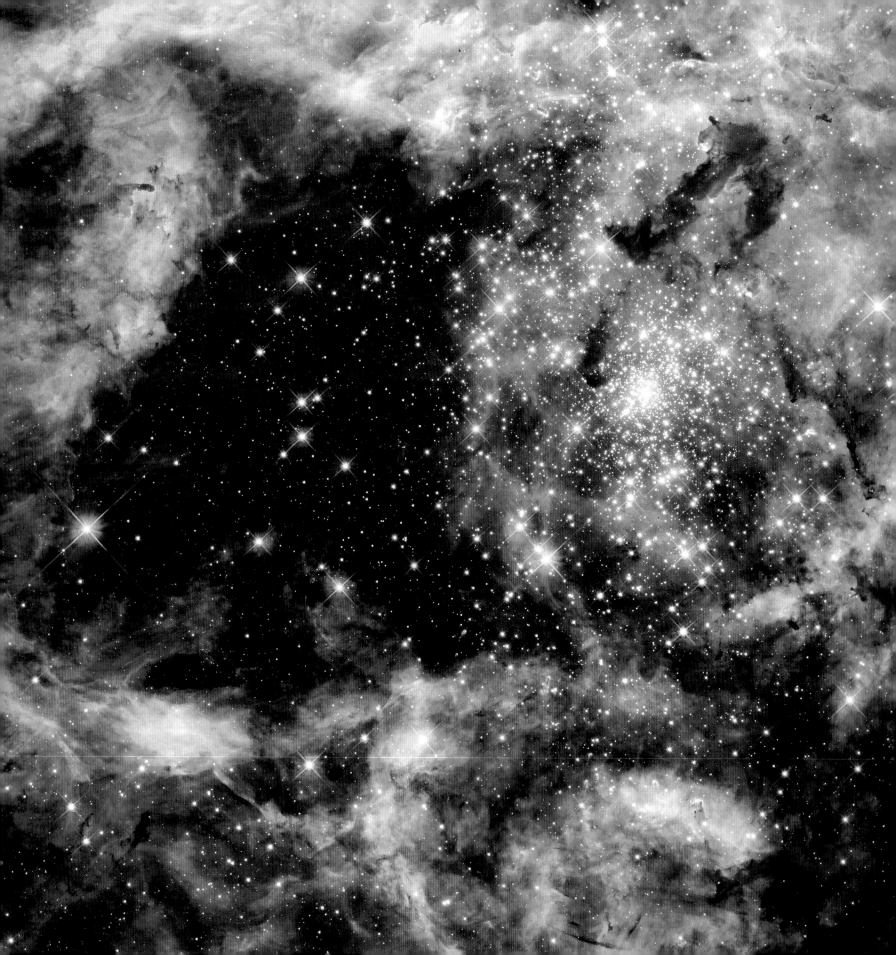

# HUBBLE'S UNIVERSE

The flagship of NASA's Great Observatory program, the Hubble Space Telescope is one of the most ambitious, legendary and nail-biting science endeavors in human history. The payoff has been immeasurable: Hubble has given us the universe.

Before Hubble was launched into orbit around Earth in 1990, there was lots of scientific discussion about what a space telescope might find. Hubble scientists agreed that it would be anticlimactic if, in fact, the orbiting telescope found simply what was predicted.

The top to-do-list items for Hubble were to measure the expansion rate of the universe, to find distant galaxies and to determine the chemical content of the space between the galaxies. But everyone expected Hubble's most important discoveries would provide answers to questions that astronomers do not know how to ask and find objects that were not yet even conceived.

Still, no one imagined that Hubble's photographs of deep space would be so utterly, jaw-droppingly beautiful. Or that the pictures would have a purely visceral appeal to an entire generation, making Hubble a household word synonymous with spectacular images of the cosmos. It was as though we were looking at the universe with a new eye. The crystal clarity of the images gave us a subliminal three-dimensional universe. The details were so clear and crisp that viewers were drawn into a fantasy landscape of unfathomable shapes and uncommonly vibrant colors.

Hubble's incredible sharpness is due to the fact that the telescope is located just above the Earth's turbulent atmosphere, which smears starlight (this is why stars seen in a dark, rural night sky appear to twinkle). If your eye were as sensitive as Hubble's, you could look from New York City and see the glow of a pair of fireflies in Tokyo. If the fireflies were six feet apart, you could resolve that there were two of them side by side. Hubble's super vision has shown us places in the cosmos that no generation of humans before us has ever seen.

The Hubble images have unveiled that beyond the starry sky we see on a clear night, there is another universe, an "undiscovered country" of discordant objects, violent explosions and cataclysmic collisions. Galaxies plow into one another; stars erupt from firestorms of gas and dust and light; newly forming stars shoot out saberlike jets of gas as birth announcements to the universe. All this drama plays out against a pitch-black backdrop that is unimaginably infinite.

A single Hubble snapshot can portray the universe as awesome, mysterious and breathtaking—and, at the same time, chaotic, overwhelming and foreboding. These pictures have become iconic, seminal and timeless. They can be found gracing everything from coffee cups to Times Square billboards and making cameo appearances in science fiction movies.

Hubble photographs know no national, political or ideological boundaries. There are no language or cultural barriers to being completely entranced by Hubble's vision. The images are so overpowering and humbling, they have touched people of all ages, a reminder that we inhabit a tiny planet in a vast universe with many remaining unknowns. If anything interesting is happening in the cosmos, people expect that the Hubble Space Telescope will take a look at it.

Hubble has been the biggest game changer in astronomy since Galileo first turned his

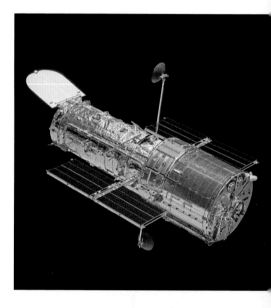

Orbiting 550 kilometers above the Earth's surface, the Hubble Space Telescope probes the universe in the pristine darkness of space. The Hubble image on the facing page is the most detailed portrait yet of the largest star nursery in our local galactic neighborhood. Known as R136, the massive stellar grouping is a turbulent star-birth region in the Large Magellanic Cloud.

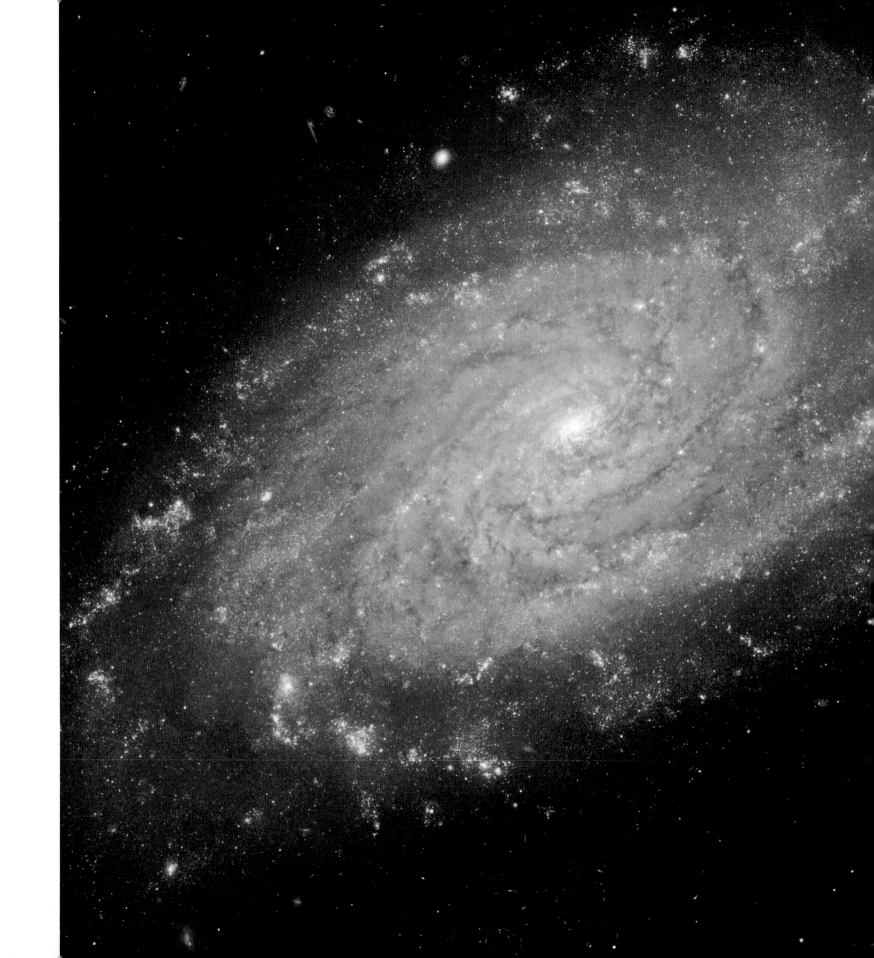

## CHAPTER 1

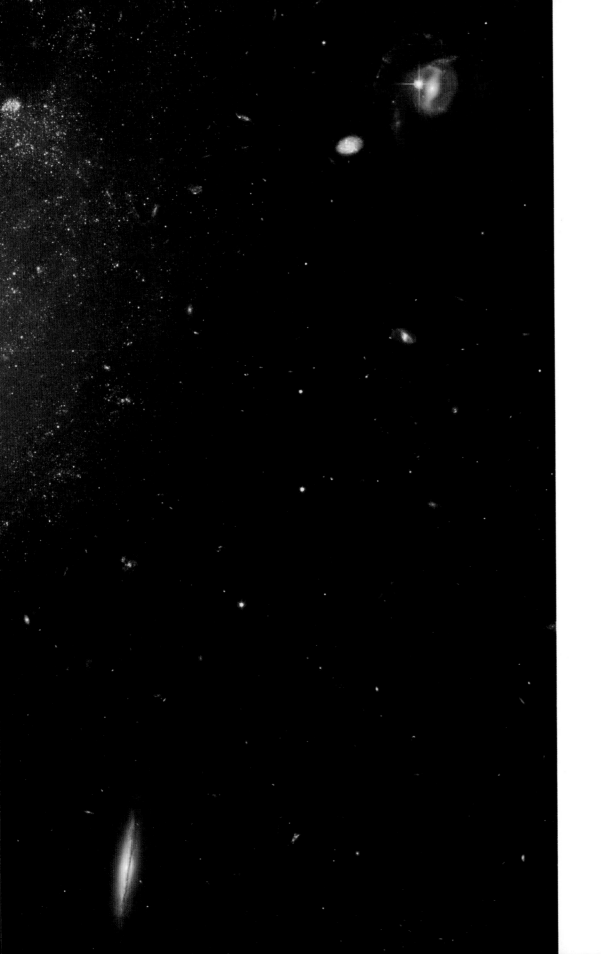

Like no other instrument before it, the Hubble Space Telescope sees remote galaxies with unprecedented clarity and sharpness. Astronomers estimate that the known universe contains at least 100 billion galaxies, each with a cargo of tens to hundreds of billions of stars. NGC3370, the impressive spiral galaxy in the foreground of this Hubble image, is about 98 million light-years distant and 80,000 light-years across (one light-year is 10 trillion kilometers). Many more distant galaxies are visible in the frame; the apparently tiny ones are 10 to 50 times more remote than NGC3370. Most of the celestial pictures throughout this book are similarly flat 2-D portraits of the 3-D reality of the universe.

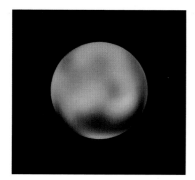

The Hubble Space Telescope's vision is so acute, it can reveal features on our neighbor worlds—in this case, Mars, facing page, and Pluto, above—that are invisible in much larger telescopes located on the Earth's surface. The reason: Hubble orbits completely above the blurring effect caused by the ever present turbulence in the Earth's atmosphere.

30-power spyglass to the heavens and discovered stars in the Milky Way, moons whirling around Jupiter and incontrovertible evidence that Earth orbits the Sun. Hubble research has covered nearly every frontier in deep-space astronomy. Among these are the search for distant supernovas in characterizing dark energy; the apparent link between a galaxy's mass and its central black hole mass; the formation of galaxies just a few hundred million years after the big bang; strange transient events in the atmospheres of planets; and the chemistry for life on planets orbiting other stars.

## RUGGED ROUTE TO THE STARS

Since its launch in 1990, the 12-ton bus-sized Hubble Telescope has made more than one million observations and looked at over 38,000 celestial objects. The Hubble archives are filled with more than half a million images. The amount of astronomical data collected is equivalent to about 50 million books, or five times the printed collection of the U.S. Library of Congress.

During the course of a year, Hubble circles Earth over 5,000 times. It has racked up more than five billion kilometers following its racetrack path around the planet.

Bringing the space telescope to this level of efficiency and performance has been a rugged route to the stars. Time and time again, the orbiting observatory was on the brink of disaster, only to be reinvigorated and made more scientifically powerful than it was before.

This melodrama has played out before the public in numerous articles and television documentaries. And Hubble's enthusiastic supporters have come to the rescue—like the cavalry in an old Western movie—to use their knowledge and influence to save the space telescope.

Hubble got off to a shaky start. In the 1970s, the telescope's construction faced cancellation more than once by a budget-cutting U.S. Congress. Astronomers scrambled to convince legislators that the revolutionary telescope would be worth the cost. Besides, it was a glamorous payload for the future space shuttle. In fact, early plans called for the space telescope to be routinely returned from Earth orbit for servicing and repair, then relaunched aboard the shuttle.

In 1983, NASA named the telescope after the early-20th-century American astronomer Edwin P. Hubble (1889-1953), who is credited with the 1920s discovery that galaxies are huge, remote systems of billions of stars. He then used them to measure the universe's expansion rate. It's hard to believe that in Hubble's time, mule teams struggled up the dirt roads of Mount Wilson, California, hauling parts to build the 100-inch Hooker telescope that Hubble would use to make his discoveries. He could not have imagined that a telescope of comparable size would be hurtling 550 kilometers above Earth three generations later.

This dream of flying a larger observatory-class telescope hit a hiatus in 1986 following the Challenger shuttle disaster. Hubble was the next scheduled payload for Challenger, until the shuttle broke apart shortly after launch in January 1986. The telescope remained in storage on the ground for several years before the shuttle was returned to service.

In April 1990, Hubble thundered into orbit atop the space shuttle's Promethean flame. But just weeks later, engineers were pondering why they couldn' t bring the telescope into sharp focus. What they realized left mission planners incredulous: The telescope's 94-inch-diameter primary mirror had been ground to the wrong prescription, making the images blurry. The starlight could not be brought to a sharp focus.

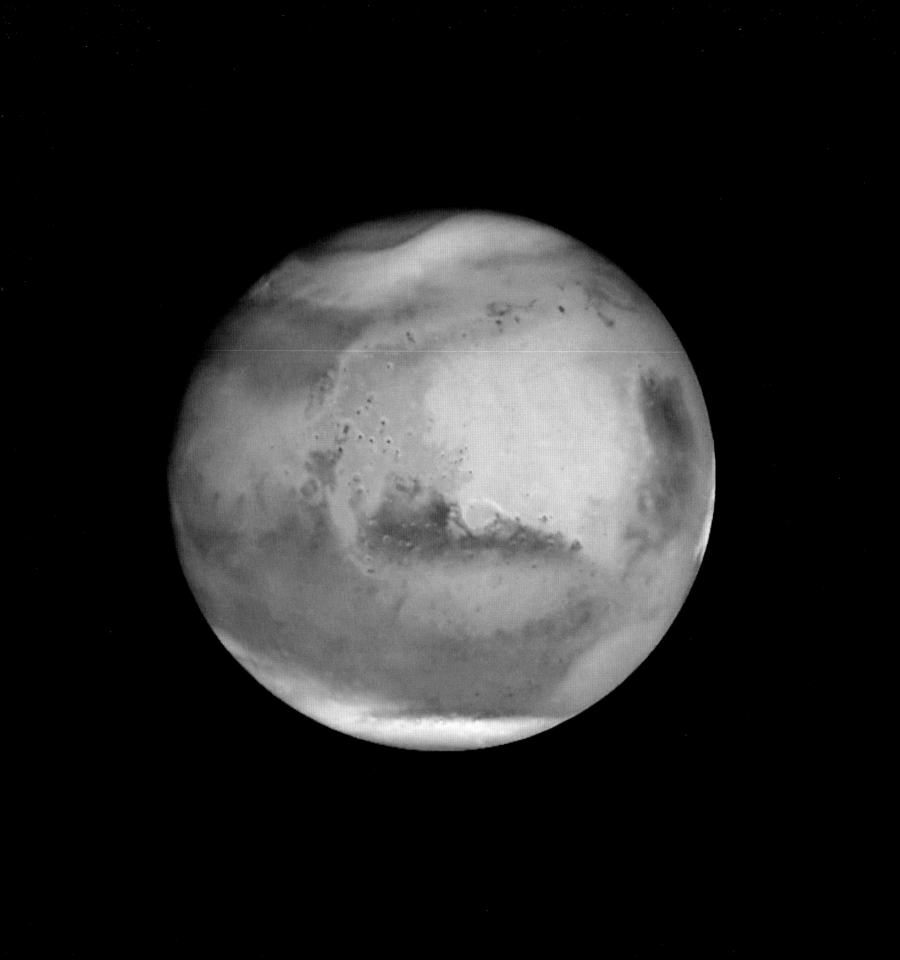

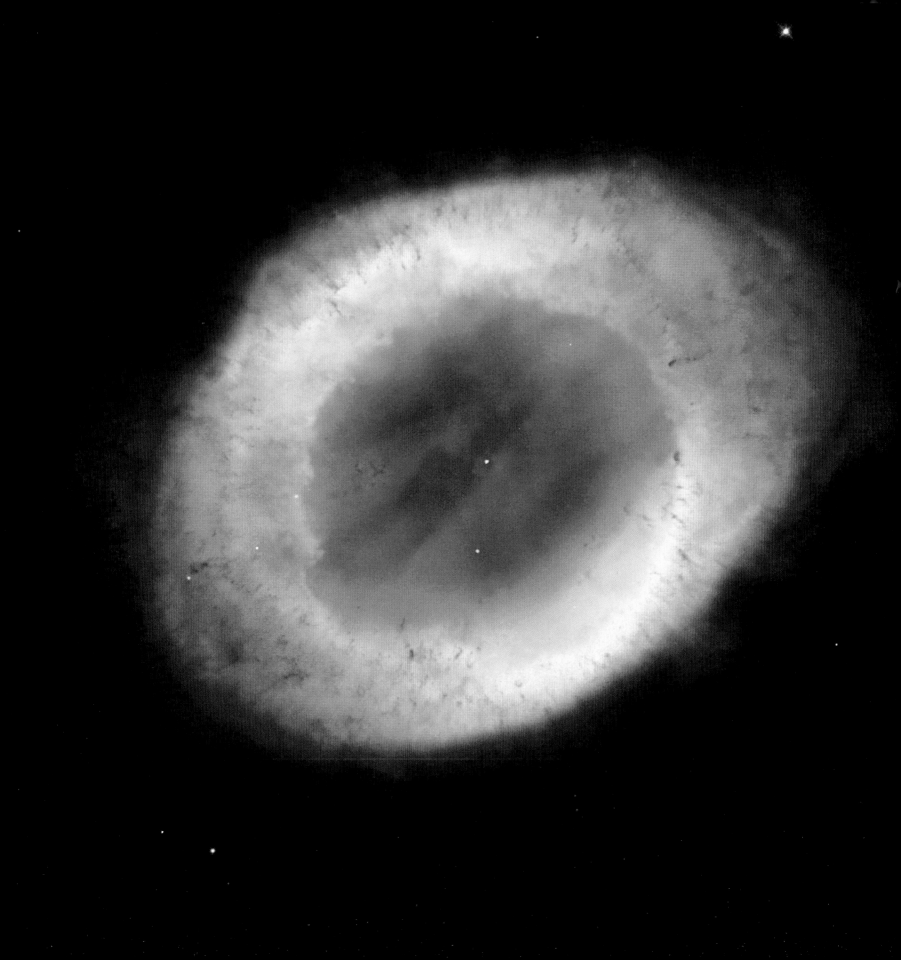

Facing page: Like a ghostly smoke ring frozen in space, the Ring Nebula is actually barrel-shaped. We are looking down one open end of the barrel at gas cast off by an expiring star thousands of years ago. This Hubble photo reveals elongated dark clumps of material embedded in the gas at the edge of the nebula and the dying central star floating in a blue haze of hot gas. The remnant central star has a surface temperature of 120,000 degrees C. About a light-year in diameter, the nebula is located in the constellation Lyra, some 2,000 light-years from Earth. The colors are approximately true colors.

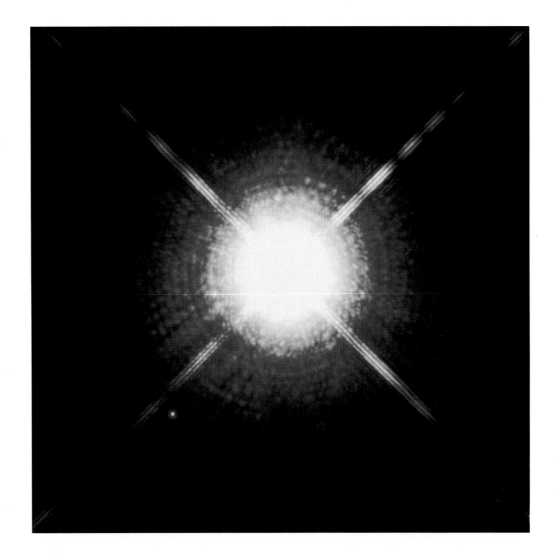

Above: In this Hubble Space Telescope image, Sirius A—the brightest star in our nighttime sky—is shown along with its faint stellar companion Sirius B (the tiny dot at lower left). Astronomers over-exposed the image of Sirius A so that the dim Sirius B could be seen. The cross-shaped diffraction spikes and concentric rings around Sirius A are normal artifacts produced within the telescope's imaging system. The two stars revolve around each other once every 50 years. Only 8.6 light-years from Earth, Sirius A is the fifth closest star system known. Only slightly larger than Earth, Sirius B, a white dwarf, is very faint because of its small size.

It was the ultimate irony for a big-budget space observatory that was supposed to see more than 10 times better than ground-based telescopes. The late-night comics' jokes and political cartoons were withering. Angry astronomers called Hubble the biggest flop in the history of science. A frustrated U.S. Congress put the future of NASA into jeopardy.

Astronomers were somewhat relieved to find that even Hubble's blurry images were sharper than what could be achieved with ground-based telescopes. What's more, computer algorithms could be used to remove much of the blur and make the images crisper, though to do so, much of the smeared light had to be thrown away, crippling Hubble's ability to image the very faint objects it was designed to detect.

More than three years passed before astronauts carrying an ingenious package of corrective optics arrived in orbit in 1993 to fix the problem. This opened the door to follow-on repairs and upgrades during space shuttle missions in 1997, 1999, 2002 and 2009. The ability to service Earth-orbit telescopes remains one of the shuttle program's greatest accomplishments.

During each orbital visit, astronauts carrying extensive to-do lists replaced a host of failed or obsolete hardware, including the original cranky gyroscopes that aim the big telescope. But, most important, Hubble's scientific instruments—cameras and spectrographs—have been repeatedly swapped out with newer state-of-the-art equipment. These high-tech, high-wire performances have kept the observatory viable for over two decades. By the early 2010s, the Hubble Telescope was functioning better than its original designers had ever imagined.

However, it wasn' t easy. Hubble faced a premature end in 2004, when NASA administrator Sean O'Keefe cancelled the last planned Hubble servicing mission. This decision was made following the tragic Columbia space shuttle accident, when the Earth-returning orbiter disintegrated in flames over Texas, killing the crew of seven astronauts. O'Keefe's rationalization was that going to Hubble was too dangerous compared with the relatively safe undertaking of flying to and docking with the International Space Station, which was the space shuttle's only other destination at the time.

But the public outcry at the thought of discarding Hubble was overwhelming. The beloved telescope was more than just a piece of space hardware. It was our emissary to the universe, empowering us with vision to pursue the most fundamental questions ever conceived. Abandoning Hubble was as shocking as was the demotion of Pluto as one of the solar system's nine planets to the lesser category of dwarf planet. Hubble was the "people's telescope."

In 2006, with new NASA administrator Michael Griffin at the helm, the final servicing mission was reinstated. The most powerful space camera ever built was installed on Hubble in 2009, along with many other fixes. Hubble was now performing at a higher level than ever before.

The great cathedrals were monuments to the glory of God; the Great Pyramids were monuments to the fear of death; the Great Wall of China was the ultimate defense fortress; the Panama Canal was a timely revolution in global transportation. The Hubble Space Telescope has earned a place in history texts that will be written a millennium from now. It will be remembered as one of the greatest manifestations of pure human curiosity.

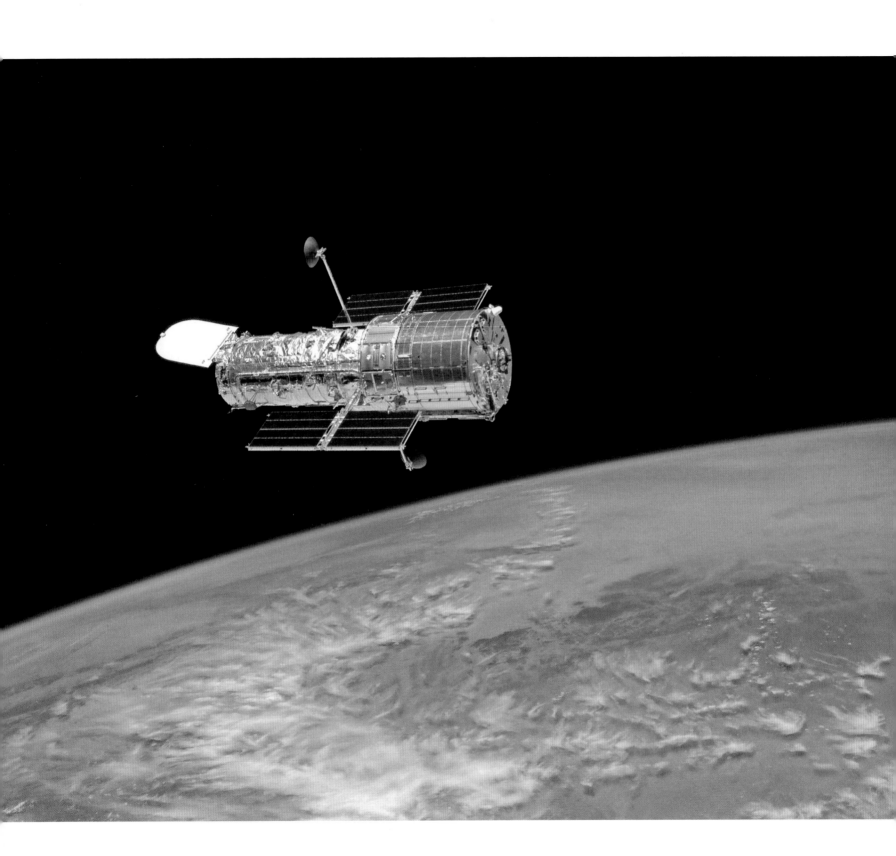

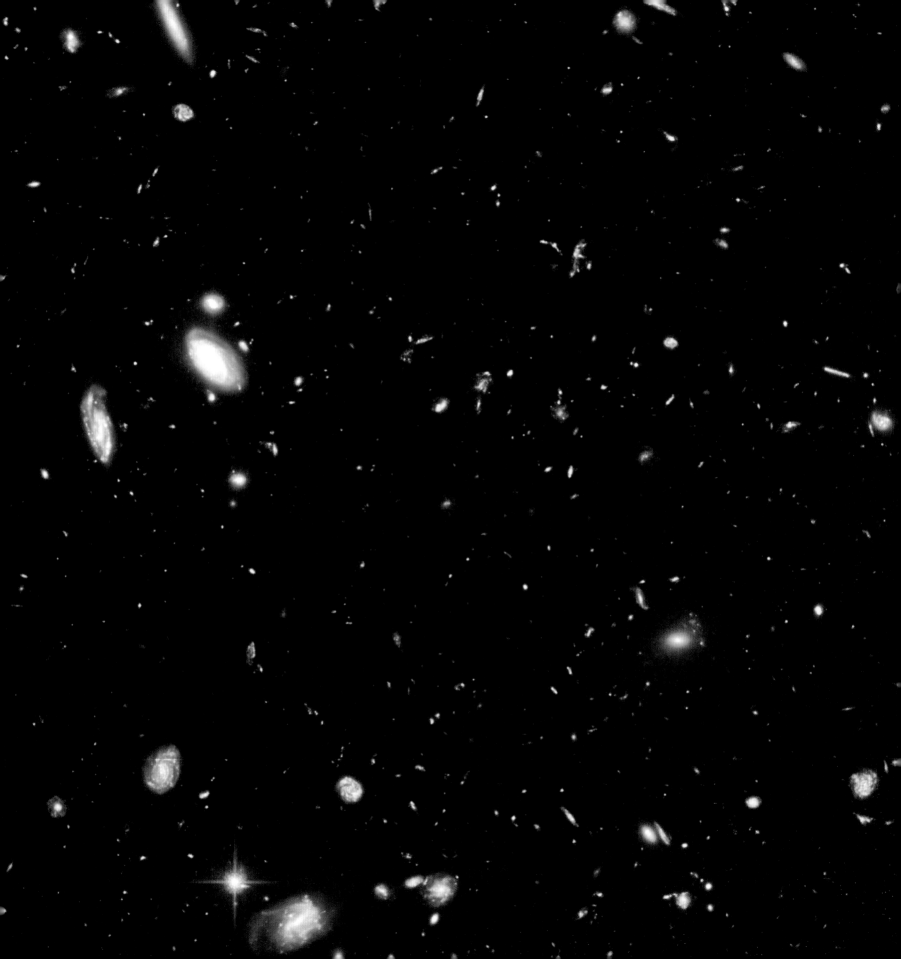

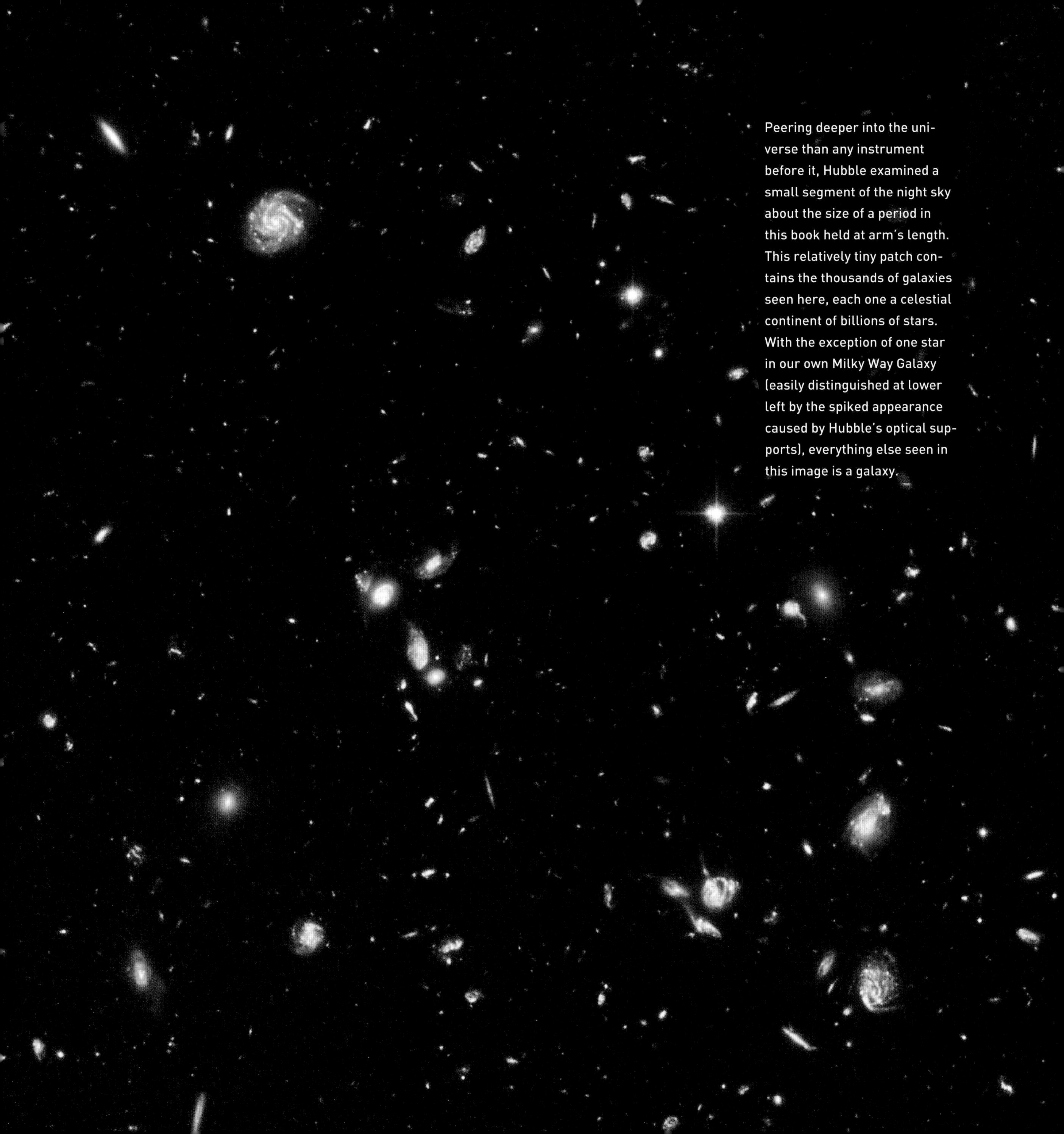

Peering deeper into the universe than any instrument before it, Hubble examined a small segment of the night sky about the size of a period in this book held at arm's length. This relatively tiny patch contains the thousands of galaxies seen here, each one a celestial continent of billions of stars. With the exception of one star in our own Milky Way Galaxy (easily distinguished at lower left by the spiked appearance caused by Hubble's optical supports), everything else seen in this image is a galaxy.

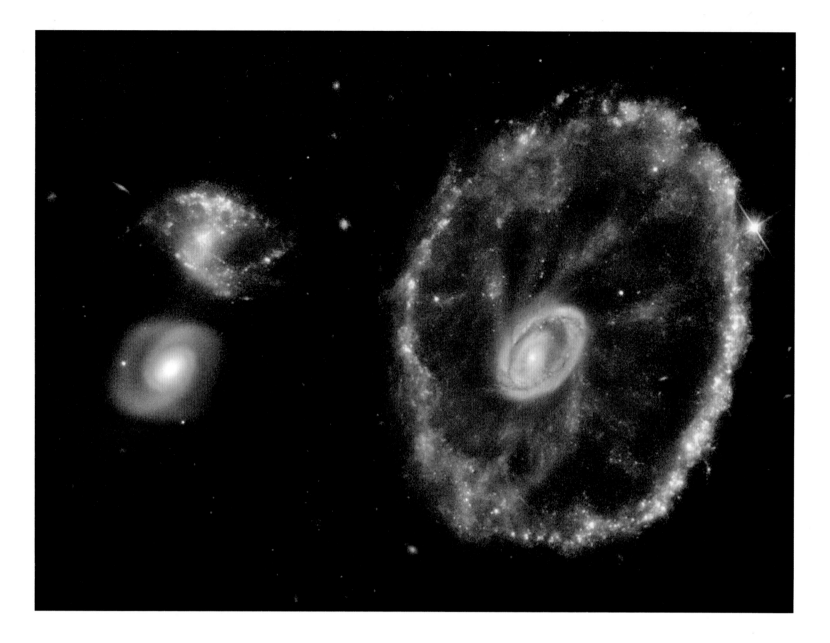

Above: Approximately 500 million light-years distant, the Cartwheel Galaxy is the result of two galaxies colliding. It took 200 million years for the cartwheel shape to develop after the collision. The outer ring shines blue with the light of newborn stars.

Right: A majestic mountain of cosmic gas and dust lies at the center of the Carina Nebula, the largest star-forming region in the Milky Way Galaxy. Stars form from collapsing clouds of cold hydrogen gas. As a star grows, it gravitationally attracts more matter. Eventually, as dust within the disk clumps together, planets may arise around their new sun.

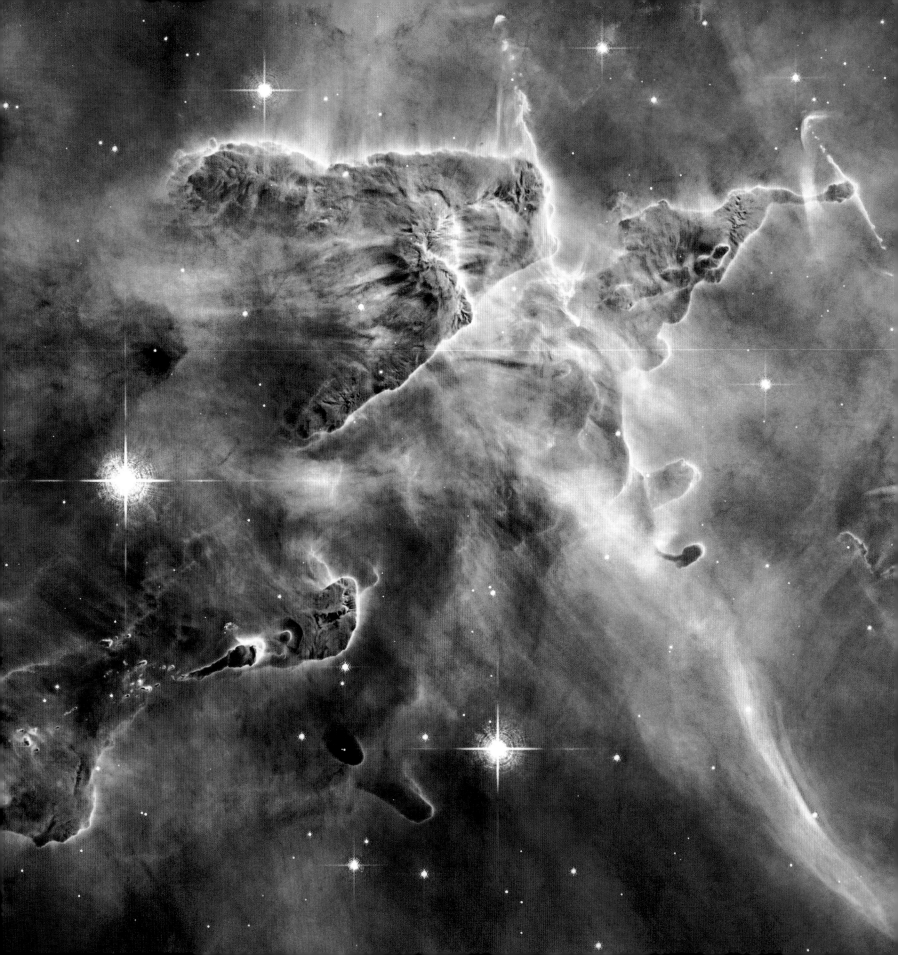

# Hubble Reboot

Servicing the Hubble Space Telescope via the space shuttle led to a quarter-century of cutting-edge astronomical discoveries

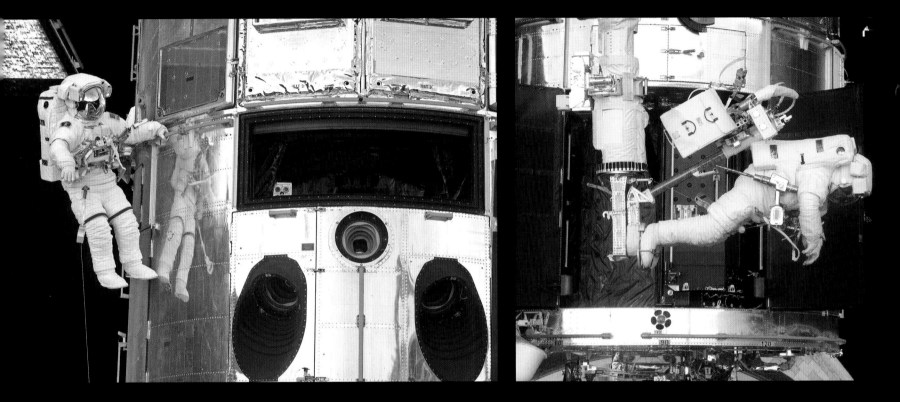

Above left: Astronaut John Grunsfeld works on the Hubble Space Telescope in the first of five spacewalks on May 14, 2009, that kicked off a week of fine-tuning the orbiting observatory. Grunsfeld, a spacewalk veteran who enjoyed a long relationship with the telescope, participated in three spacewalks during Servicing Mission 4. Note the handrails positioned around the telescope specifically for astronaut use and the tension line that secures Grunsfeld when he needs to work with both hands.

Above right: Stabilized on a foot restraint at the end of Atlantis's robotic arm, astronaut Andrew Feustel participates in Servicing Mission 4's third spacewalk, on May 16, 2009. Over a period of six hours and 36 minutes, Feustel and Grunsfeld removed the corrective optics array, installed the new Cosmic Origins Spectrograph and repaired the Advanced Camera for Surveys. This was the final space shuttle visit to Hubble, a highly successful mission that left the big telescope more powerfully equipped than ever.

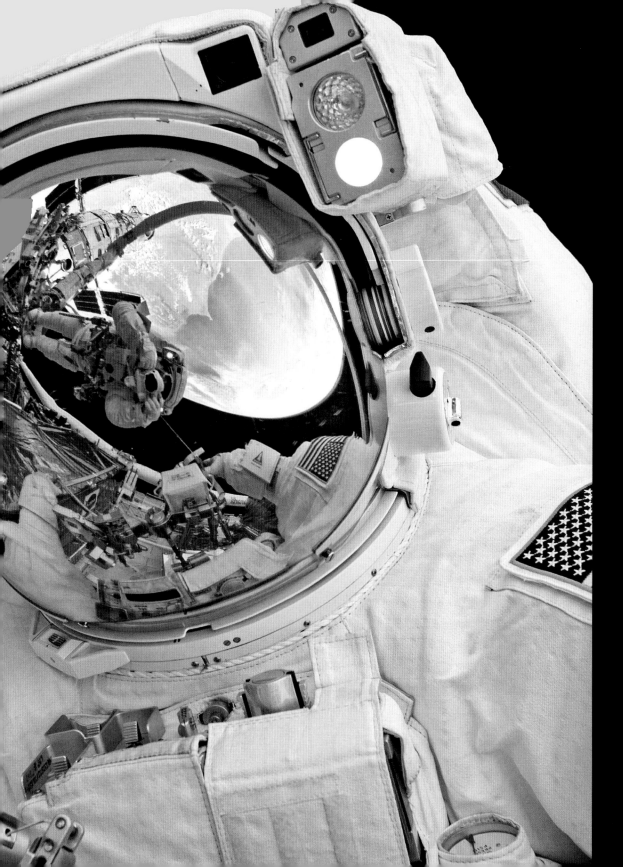
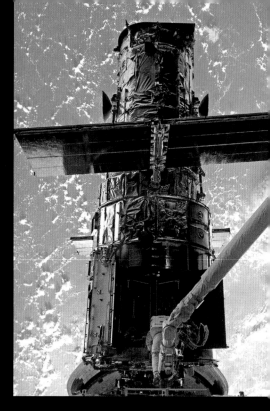

Above: Astronaut Michael Good rides Atlantis's robotic arm to the exact position he needs to be to continue work on the space telescope during Servicing Mission 4, in May 2009. In addition to installing two new instruments and repairing two others, the mission involved replacing Hubble's batteries, fine guidance sensors and protective blankets.

Left: This close-up of astronaut John Grunsfeld shows the reflection of astronaut Andrew Feustel, taking the picture while perched on the robotic arm.

25

Right: The high-resolution spectrograph is removed in preparation for a new instrument during the second servicing mission, in 1997. Hubble can hold four instruments the size of a telephone booth and four the size of a piano.

Below: During Hubble's first servicing mission, in 1993, an astronaut prepares to be elevated to the top of the space telescope.

Below right: The Wide Field Camera 3 (WFC3) is ready to be installed during a 2009 spacewalk.

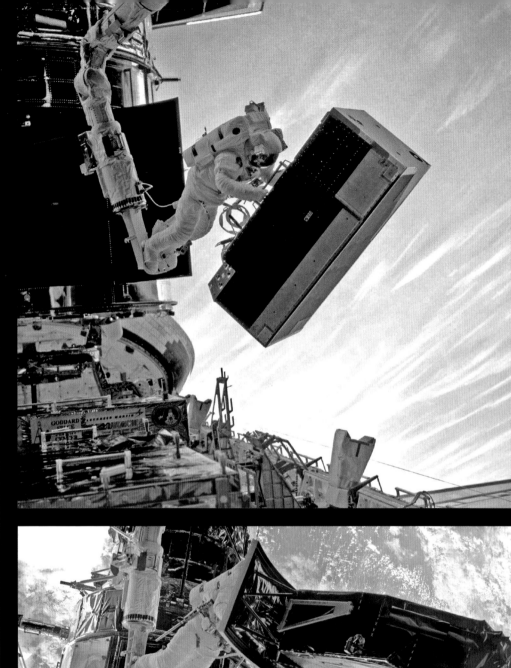

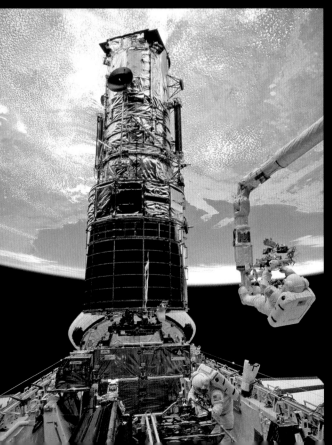

Moments away from being released to resume its travels around Earth, the refurbished Hubble Space Telescope is lifted from Atlantis's cargo bay by the robotic arm on May 19, 2009. This maneuver concluded Servicing Mission 4, the fifth and final astronaut visit to the telescope. Barring any major malfunctions, Hubble should continue doing remarkable science into the 2020s.

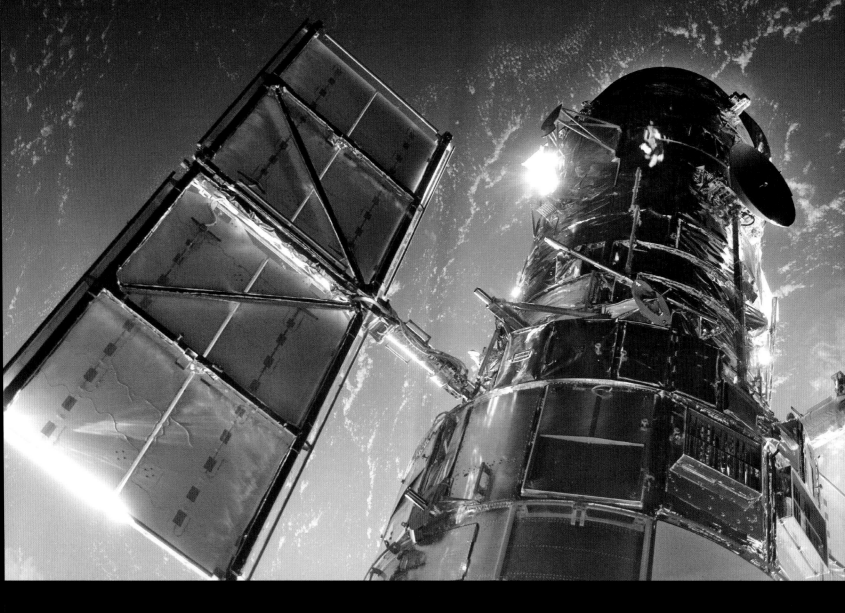

Just after the Hubble Space Telescope was captured by the robotic arm of the space shuttle Atlantis on May 13, 2009, a Servicing Mission 4 crew member snapped this picture. Hubble was about to undergo an upgrade and repairs. The underside of one of the two solar panels can be seen at left.

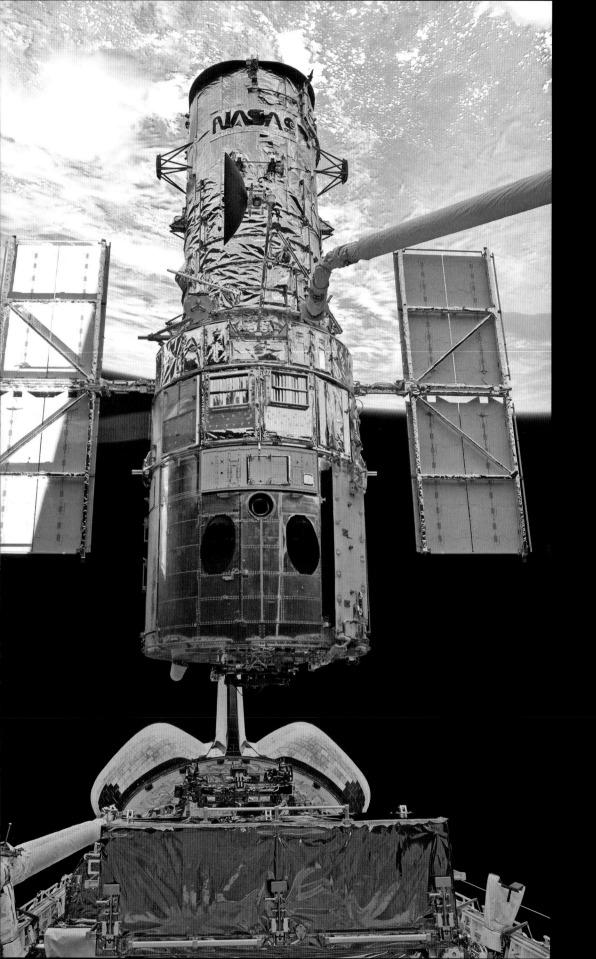

Launched by the space shuttle in 1990, the Hubble Space Telescope is now widely regarded as one of the most successful and productive scientific instruments of all time.

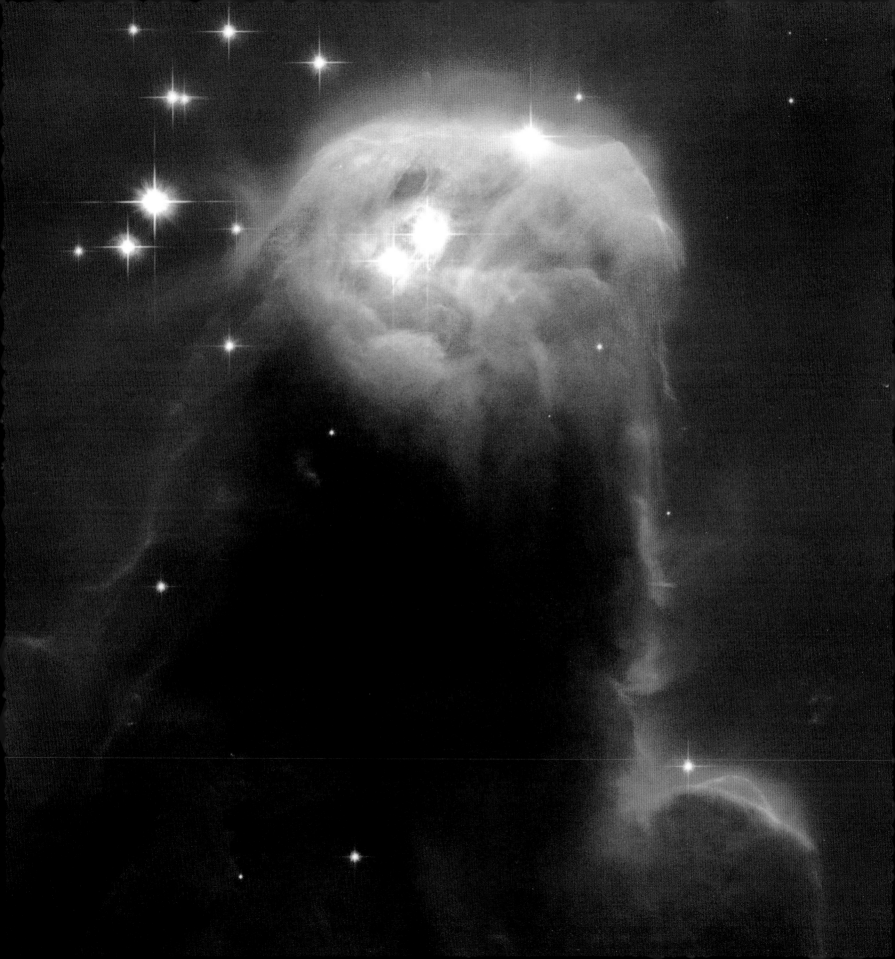

## CHAPTER 2

# HUBBLE'S TOP SCIENCE ACCOMPLISHMENTS

The universe was a different-looking place in 1990, the year the Hubble Space Telescope was launched. The most powerful telescopes on Earth could see only halfway across the universe. Astronomers didn't know whether planets orbited other stars. Even the age of the universe was uncertain by a large margin. There was a zoo of powerful energetic phenomena at great distances, but there was no observational evidence for whatever was powering the fireworks.

Following its deployment into low Earth orbit from the space shuttle Discovery in 1990, the Hubble telescope provided a rapid-fire series of discoveries and breakthroughs. Later joined by its sister Great Observatories—the Spitzer Space Telescope and the Chandra X-ray Observatory—Hubble became the trailblazer of a new golden age of space astronomy.

By today's standards, Hubble's 2.4-meter main mirror is small compared with the 8- to 10-meter monolithic mirrors and segmented mirrors found at major mountaintop observatories. But Hubble's pristine view from above the Earth's atmosphere makes all the difference. Hubble consistently gathers images 10 times sharper across a wider field than the largest ground-based telescopes.

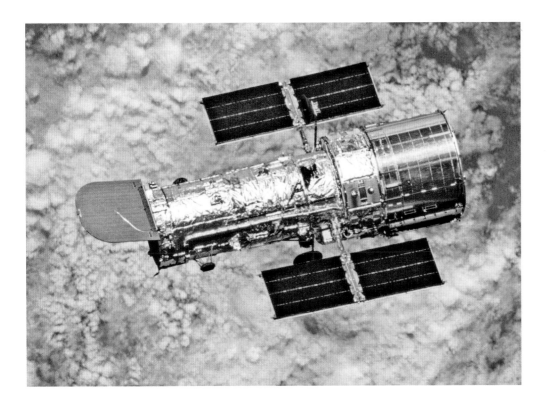

Facing page: Resembling a nightmarish beast rearing its head from a crimson sea, this monstrous object is actually an innocuous pillar of gas and dust called the Cone Nebula. This Hubble Space Telescope view shows the upper 2.5 light-years of the nebula, a distance that equals 23 million round trips to the Moon. The Cone Nebula resides 2,500 light-years away, in the constellation Monoceros. Over millions of years, radiation from hot, young stars [located beyond the top of the image] has slowly eroded the nebula.

Left: The crown jewel in NASA's fleet of orbiting observatories, the Hubble Space Telescope conducts observations of the cosmos regardless of the weather conditions on the Earth's surface.

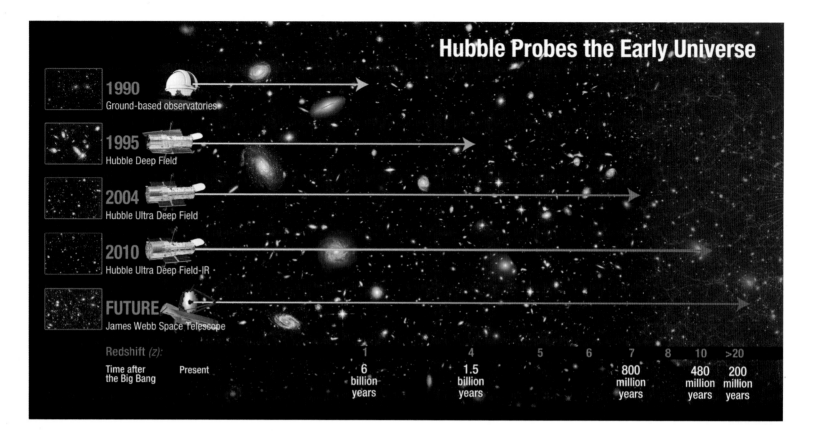

**Hubble Probes the Early Universe**

**1990**
Ground-based observatories

**1995**
Hubble Deep Field

**2004**
Hubble Ultra Deep Field

**2010**
Hubble Ultra Deep Field-IR

**FUTURE**
James Webb Space Telescope

| Redshift (z): | | 1 | 4 | 5 | 6 | 7 | 8 | 10 | >20 |
|---|---|---|---|---|---|---|---|---|---|
| Time after the Big Bang | Present | 6 billion years | 1.5 billion years | | | 800 million years | | 480 million years | 200 million years |

As our telescopes penetrate ever deeper into the universe, we are seeing further back in time. The light from those remote regions is a record of the conditions that prevailed when that light left its source. In effect, we are seeing a snapshot of the early universe, within a few hundred million years of the big bang. The laws of nature have, in effect, created a time machine that allows us to peer back to our universe billions of years before the Earth's formation.

Through the use of adaptive optics, which measure turbulence in the Earth's atmosphere along the line of sight of large telescopes on the Earth's surface, much of the atmospheric blur can be removed. This has allowed ground-based telescopes to partly catch up with the Hubble Telescope. But Hubble has a unique ability to see very faint objects, because there is no sky background glow to mask their feeble light. Even on the darkest night from the Earth's surface, the atmosphere has an inherent faint background light called airglow. Further, as the Earth's population continues to grow, light pollution from cities is becoming an increasing problem, even for telescopes tucked away on mountaintops.

Hubble's view is optically stable: The quality of the observing conditions never changes from day to day or even from orbit to orbit. We can revisit celestial targets knowing that they can be studied with the same acuity and image quality every time. This is crucial for precision observations, when astronomers are trying to detect minute changes in the light, motion or other behavior of a celestial object. They need to ensure that the changes are due to the object itself and not to something that was altered by the seeing conditions in the Earth's atmosphere.

Hubble is able to see across a wide swath of the spectrum and can collect ultraviolet light from very hot stars. (This radiation, potentially dangerous to life as we know it, is prevented from penetrating our atmosphere by the upper ozone layer.) Hubble can also see at near-infrared wavelengths, allowing it to peer through dust in space to see otherwise hidden stars. The big payoff in Hubble's infrared vision to date has been to allow us to see the farthest object in the universe.

# HUBBLE'S TOP DISCOVERIES

## 1. GALAXIES EVOLVED FROM SMALLER STRUCTURES

Before Hubble was launched, there was plenty of room for conjecture and theoretical modeling about how galaxies must have evolved if the universe was born in the big bang. According to the big bang theory, the early universe was a chaotic sea of photons and subatomic particles. Eventually, the subsequent expansion of the universe would have allowed it to cool to the point where structure arose out of that chaos.

In the quest to probe this early history, astronomers prior to Hubble could detect normal galaxies out to, in astronomical parlance, a redshift of 0.7 (stated as $z = 0.7$), which corresponds to a distance of seven billion light-years. But this wasn't deep enough to establish which of several competing theories best described how galaxies formed and evolved in the early universe. Some astronomers were concerned that the light from objects more than seven billion light-years away might be too feeble ever to be picked up by telescopes. In 1985, a committee of top astronomers planning to use Hubble worried that devoting a lot of precious orbits to taking a "deep exposure" of the universe might be fruitless. They assumed that the light from galaxies at great distances would be spread out, making them too diffuse to be seen by Hubble.

Fortunately, nature cooperated. Hubble observations taken even before its optical repair in 1993 showed that galaxies at a then record-breaking redshift of $z = 1.5$ (corresponding to a distance of nine billion light-years) are more compact than the nearby galaxies we see today. Therefore, their light is concentrated into a smaller area. This made distant galaxies detectable to Hubble.

Bottom left: Since its origin in the big bang almost 14 billion years ago, the universe has evolved over time. The launch of the Hubble Space Telescope has greatly extended our reach into the universe's past.

Below: Before the Hubble Space Telescope, astronomers had to rely on ground-based telescopes for views of the universe at large. This research telescope's comparatively fuzzy galaxy images is typical of pre-Hubble portraits of distant galaxies from the 1980s. Compare with the images on pages 36-39.

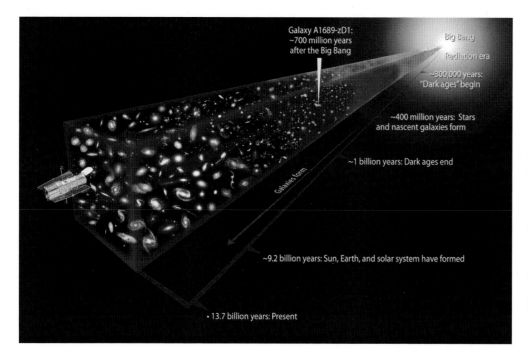

Galaxy A1689-zD1: ~700 million years after the Big Bang

Big Bang

Radiation era

~300,000 years: "Dark ages" begin

~400 million years: Stars and nascent galaxies form

~1 billion years: Dark ages end

Galaxies form

~9.2 billion years: Sun, Earth, and solar system have formed

· 13.7 billion years: Present

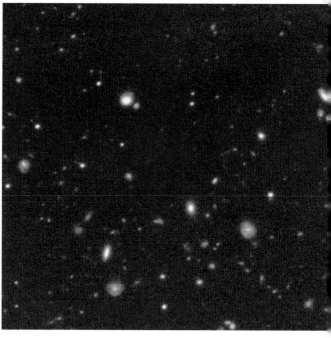

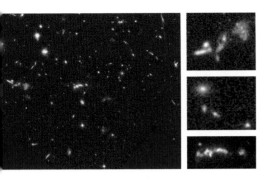

Above: Thousands of galaxies are revealed in one of Hubble's first deep views. Insets show galaxies colliding, sideswiping and interacting with one another in an earlier era of the universe.

Right: This deep Hubble picture taken in 1994 demonstrated that Hubble could see galaxies 12 billion light-years away. Hubble uncovered a zoo of odd-shaped galaxies that are evidence for galaxy mergers and interactions.

Facing page: Our deepest ever view of the universe required a time exposure of more than 270 hours. This Hubble image contains 10,000 galaxies at various stages of evolution and covers a patch of sky smaller than the size of a pinhead held at arm's length. Examine the four pages showing the full detail of this image to see our best view of what the universe at large really looks like. To photograph the entire sky in this detail would require 13 million Hubble images with comparable exposure.

Hubble had entered a cosmic Serengeti of bizarre compact fragmentary objects considered the ancestors of our Milky Way Galaxy. Hubble images show the early universe awash in strange-shaped "pathological" galaxies dubbed "tadpoles" and "train wrecks." This is the land before time—an era undreamt of before Hubble's cosmic unveiling. The Hubble Telescope revealed the shapes of these remote objects by resolving structures a fraction of the size of our Milky Way Galaxy. This has allowed astronomers, for the first time, to discriminate among various types of distant galaxies and trace their evolution.

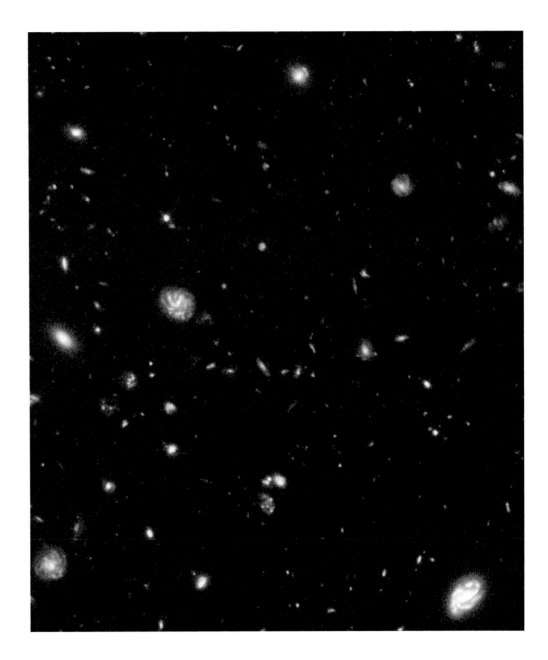

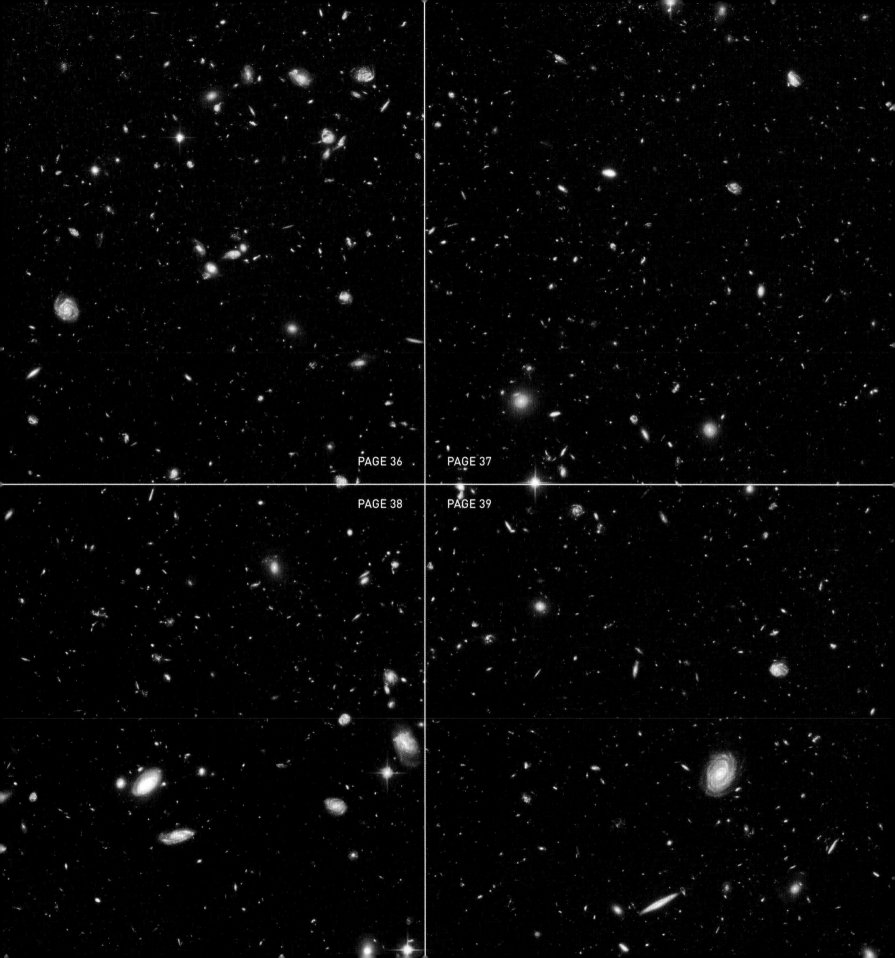

PAGE 36

PAGE 37

PAGE 38

PAGE 39

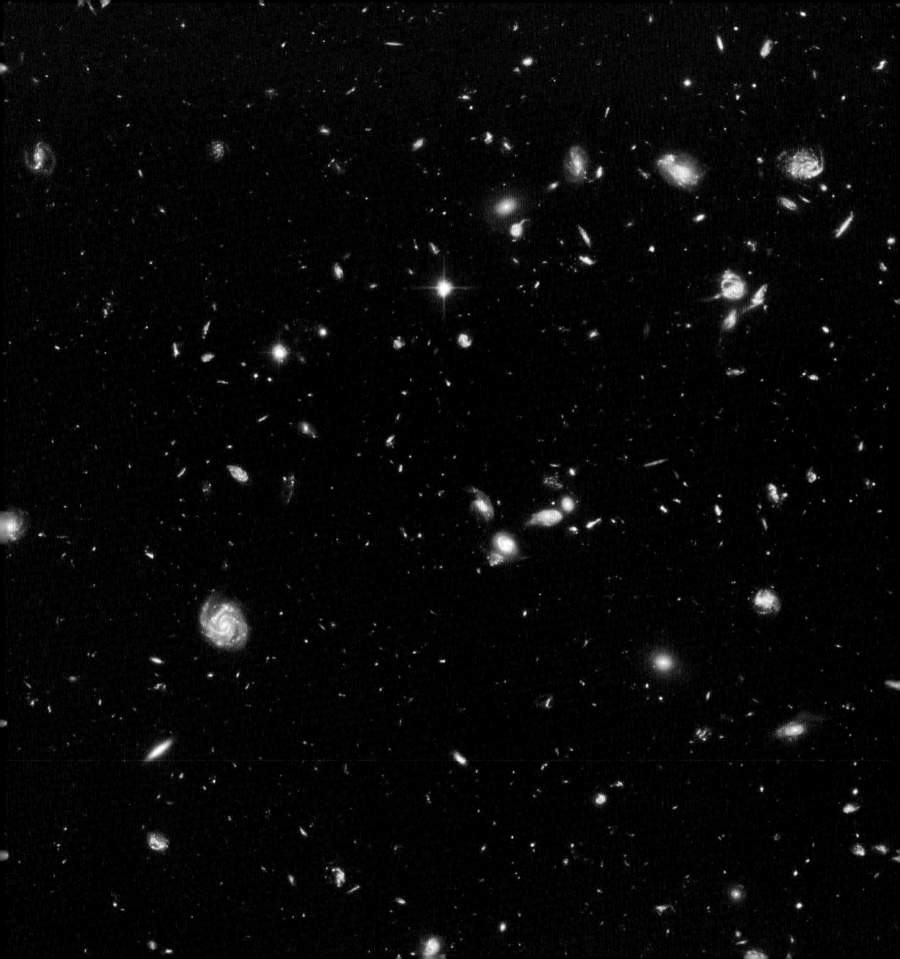

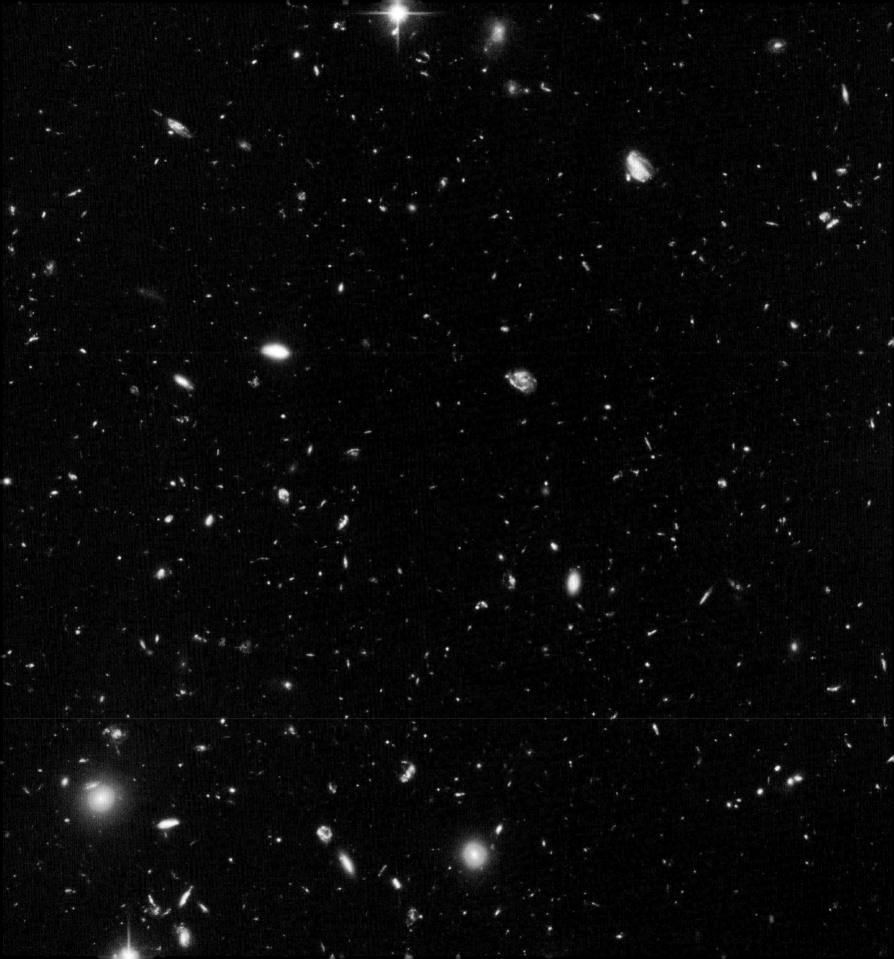

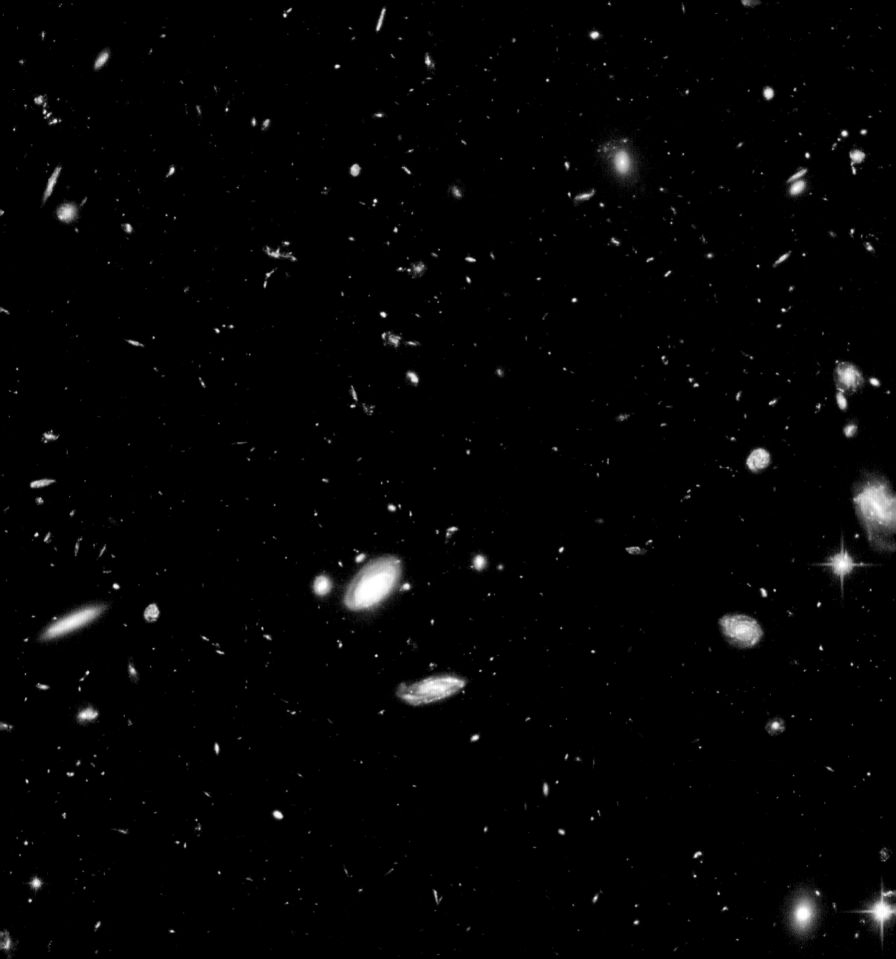

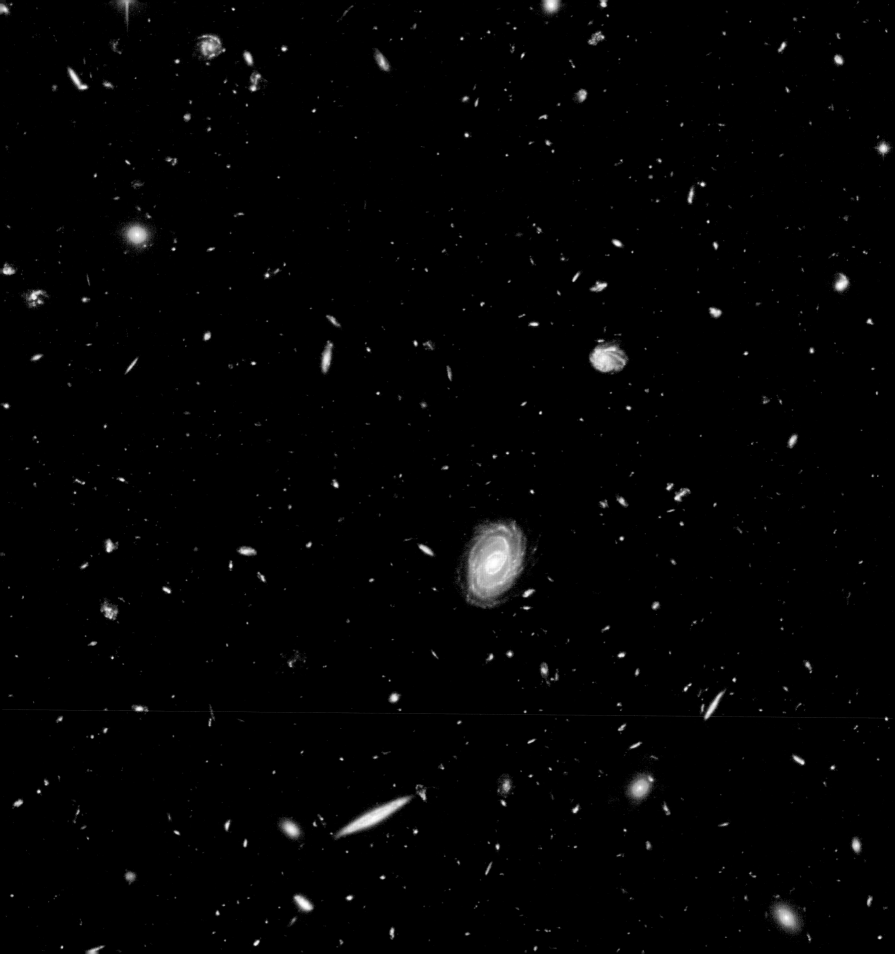

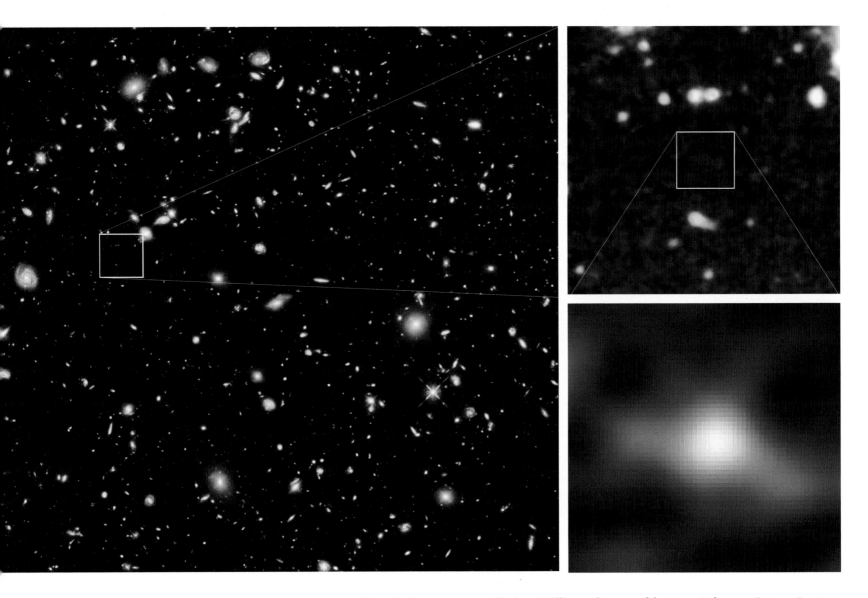

One of the most distant objects ever identified in the universe, UDFj-39546284 is so remote that its light traveled 13.2 billion years to reach Hubble. The dim object is a compact galaxy of blue stars that existed 480 million years after the big bang, only four percent of the universe's current age.

These findings encouraged Robert Williams, director of the Space Telescope Science Institute (STScI) from 1993 to 1998, to devote a large chunk of his director's observing time to take the deepest ever view of the universe, akin to obtaining the deepest ever core sample through layers of sediment in the Earth's crust. Hubble reached galaxies that were many times fainter than what could be seen by the ground-based telescopes of the mid-1990s.

When Hubble's Advanced Camera for Surveys was installed in 2002, the next STScI director, Steve Beckwith, pushed further to achieve the Hubble Ultra Deep Field (HUDF) to ensure that astronomers weren't being fooled by seeing only compact objects and missing larger galaxies at great distances. The HUDF reached deeper still and found only fragmentary developing galaxies.

Observations from Hubble's Wide Field Camera 3, installed in 2009, have pushed into near infrared, identifying objects that existed when the universe was only 450 million years old.

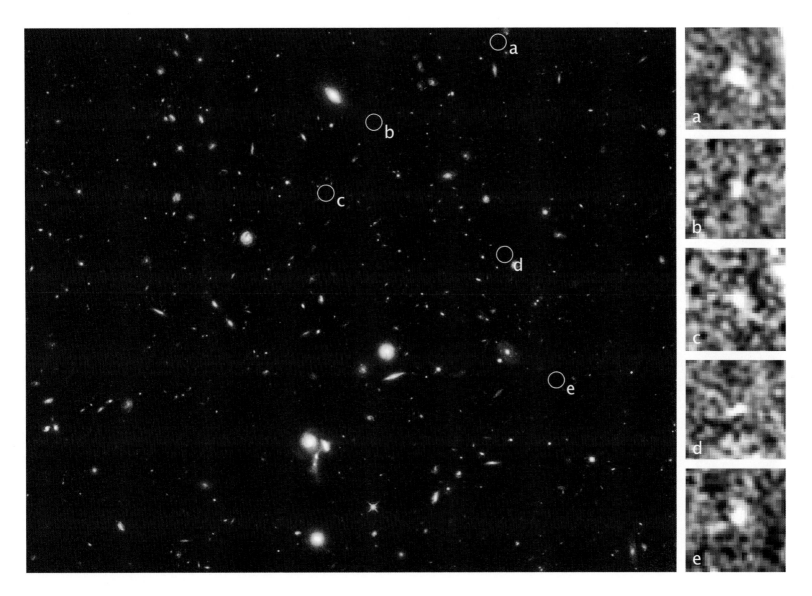

When astronomers compare galaxies at different distances, it's like watching individual frames of a motion picture. The Hubble deep surveys reveal the emergence of structure in the infant universe and the subsequent dynamic stages of galaxy evolution, largely through mergers. Before Hubble, nearby colliding galaxies were simply a curiosity. But deep views show that in the early days, galaxy smashups were more the rule than the exception. This provides compelling, direct visual evidence that the universe is truly changing as it ages.

In an image that pushes Hubble to its limits, the circles identify members of the farthest known cluster of galaxies. We see them forming when the universe was just 600 million years old.

## 2. SUPERMASSIVE BLACK HOLES ARE COMMON IN GALAXIES

One of the eeriest and most attention-getting concepts in modern astronomy is the black hole. In fact, the idea goes back to 1786, when scientists applying Newton's laws of gravity hypothesized

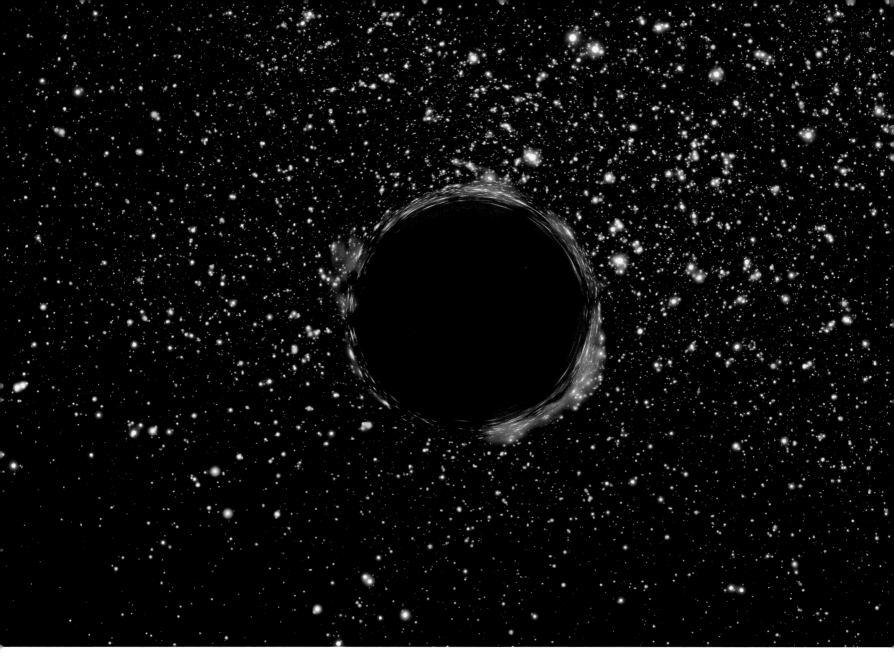

Artist concept of a supermassive black hole. The dark space is the event horizon around the black hole, where light cannot escape.

that there could be "dark stars" so massive, even light could not reach escape velocity. In the 1960s, astrophysicist John Wheeler coined the term "black hole" to describe a gravitationally collapsed star that is so dense, light would be unable to escape. Several years later, X-ray astronomers confirmed the existence of a stellar-mass black hole orbiting a normal star. Other examples have since been identified. A black hole is created when one star in a double-star system explodes and collapses. The black hole can feed on its companion star, producing a firestorm of X-ray emissions.

Astronomers suspected that far more massive black holes must be the "gravitational engines" powering a wide range of extraordinary energetic phenomena seen near and far: Seyfert galaxies, BL Lac objects, blazars and—most important—quasars (quasi-stellar objects).

The brilliant beacons of the quasars can be seen out to 10 billion light-years, even with ground-based telescopes. Although discovered in the early 1960s, quasars remained a mystery for decades, puzzling astronomers as to their exact nature. But they were clearly more abundant long ago. Hubble observations in 1996 showed that quasars live at the cores of a remarkable variety of galaxies, many of which are colliding. The collisions provide galactic gas and dust to fuel the black holes; thus engorged, quasars erupt as blazing blowtorches shooting out of the galaxy's core.

But precision spectroscopy was needed to "weight" a black hole to see whether the amount of hidden, or "dark," mass far exceeded what mass could be attributed to stars alone. In 1994, astronomers aimed Hubble at the nearest "mini-quasar"—the brilliant core of the giant elliptical galaxy M87, in Virgo. As with far more distant quasars, a telltale jet of material is being ejected from the core of the galaxy at nearly the speed of light. The behavior of an accretion disk swirling around an ultradense central object is the best model for the jet "engine." The disk provides the fuel for the jet. The black hole is the engine. And powerful entangled magnetic fields shape a nozzle for the jet, confining the ejected material like the nozzle on a garden hose.

Hubble measured a three-billion-solar-mass core in M87. This was possible because Hubble's Faint Object Spectrograph made velocity measurements of a never-before-seen spiral-shaped

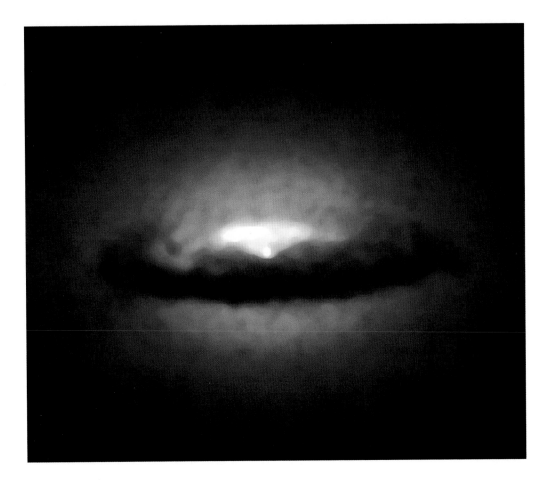

This giant "hubcap" in space is a 3,700-light-year-diameter accretion disk surrounding a black hole at the core of the galaxy NGC7052. As it swirls around and into the hole, matter is compressed, causing a fierce storm of radiation to emerge, which blocks the 300-million-solar-mass black hole from view. NGC7052 is 200 million light-years distant.

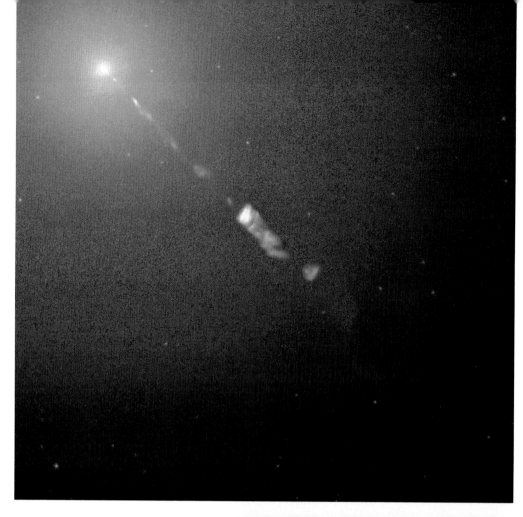

This is the signature of a black hole, as measured by clocking the velocity of gas trapped near it. The spectral zigzag is evidence of the rapid rotation of an entrapped gas disk.

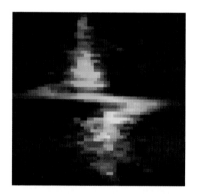

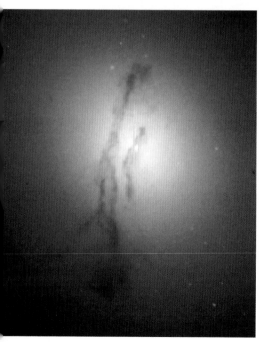

Above: Fired from a 6.6-billion-solar-mass black hole, a jet of gas streaks across space at 200 million miles per hour.

Right: Hubble discovered that quasars live at the cores of active galaxies.

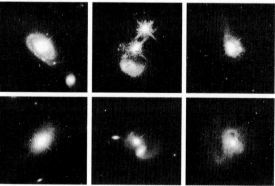

whirlpool of hot gas orbiting around the black hole. The disk's velocity indicates a high concentration of mass. But the mass has no luminous source, as stars do.

A 1997 census of 27 nearby galaxies found that they all had central black holes. This led astronomers to conclude that supermassive black holes are so common, every major large galaxy has one. Even more profound, Hubble discovered that the mass of a black hole is directly related to the mass of a galaxy's central bulge of stars. The bigger the bulge, the more massive the black hole. This links galaxy evolution to the growth of the black hole at its center through some un-

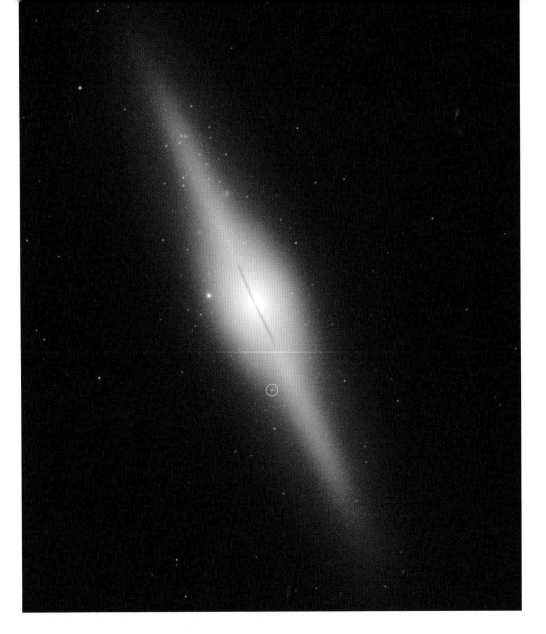

The small circle on this image indicates an intermediate-sized black hole, about 500 times the mass of our Sun. It might once have been the core of a dwarf galaxy that was eaten by the edge-on galaxy in which the black hole now resides. Some of the blue light comes from a cluster of stars encircling the black hole.

known feedback mechanism. While there are six viable theories at the present time, no one is absolutely sure what the black hole–galaxy connection is.

## 3. DARK ENERGY EXISTS

A key project for Hubble astronomers was to measure the amount of deceleration the universe is undergoing. In the late 1920s, Edwin Hubble first discovered that the universe is expanding in all directions, like an inflating balloon. Cosmologists used this evidence to build the big bang theory, which predicts that the universe was once dense and hot and has been expanding ever since. The confirming evidence came in 1990, when NASA's Cosmic Background Explorer precisely measured the cooling afterglow of the big bang and found that it fit the predictions perfectly. The conventional wisdom was that after the big bang, gravity was exerting some sort of drag on the

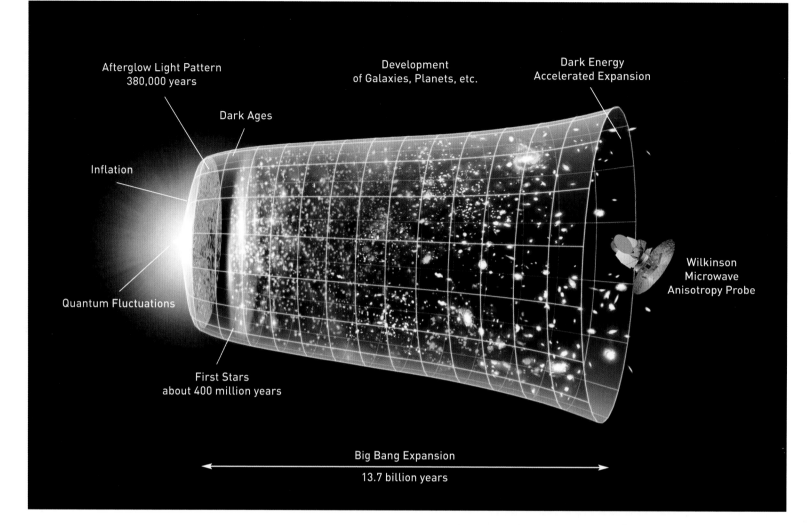

Afterglow Light Pattern
380,000 years

Dark Ages

Development
of Galaxies, Planets, etc.

Dark Energy
Accelerated Expansion

Inflation

Quantum Fluctuations

First Stars
about 400 million years

Wilkinson
Microwave
Anisotropy Probe

Big Bang Expansion
13.7 billion years

Above: The essential stages in the universe's evolution since the big bang, 13.7 billion years ago.

Right: Temperature ripples in the big bang's afterglow (see above) were observed by the Wilkinson Microwave Anisotropy Probe in 2003. The ripples are thought to be the seeds from which galaxy clusters arose in the early universe.

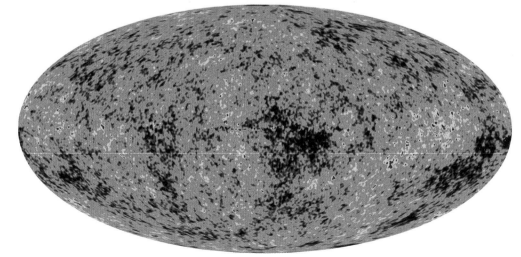

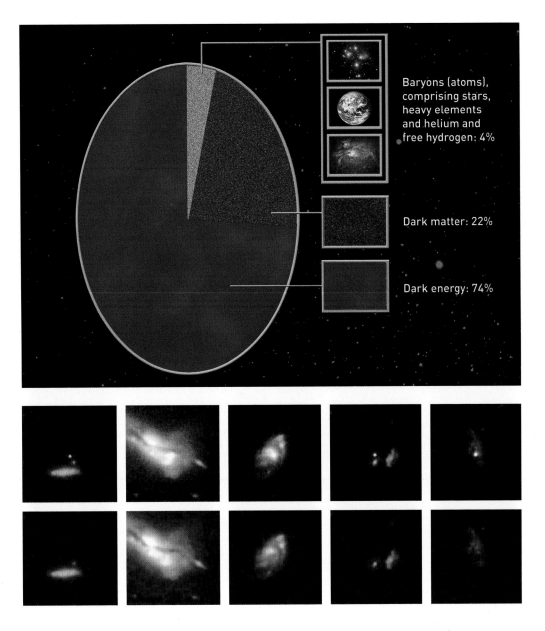

Baryons (atoms), comprising stars, heavy elements and helium and free hydrogen: 4%

Dark matter: 22%

Dark energy: 74%

The big picture of the composition of the cosmos finally came into focus in the late 1990s and early into the first decade of the 21st century. The part of the universe with which we are most familiar—planets, moons, stars, nebulas and galaxies—accounts for less than 1 percent of the universe's total contents. Nonluminous matter (mostly hydrogen dispersed throughout space) is about 3 percent, while the rest is the largely mysterious dark matter and the completely mysterious dark energy.

Looking deep into the universe, Hubble captured supernovas exploding in remote galaxies. The bottom row of images at left shows each galaxy before its supernova went off, while the top row, taken months later, reveals the starlike beacon of the supernova.

expansion of space, like a ball rolling up an incline and gently slowing down. The question as to whether there was enough gravity to halt the universe's expansion persisted for decades. The Hubble Space Telescope's ability to see distant supernovas and precisely measure their distance allowed astronomers to look further back into time to measure the expansion rate of the universe in its early years.

In 1998, when Johns Hopkins University astronomer Adam Riess wrote a computer program to calculate the universe's deceleration rate using supernova survey data collected by Hubble, the computer kept coming up with a "negative mass" for the universe. Riess at first thought this was simply a programming error. But then he realized that the computer program was trying to make

One of the most distant supernova explosions in the universe offers evidence that the universe's expansion was once slowing down, until dark energy overcame gravity. Magnified views of this Hubble image show no supernova (left) and the supernova as seen by Hubble on October 10, 2010 (right).

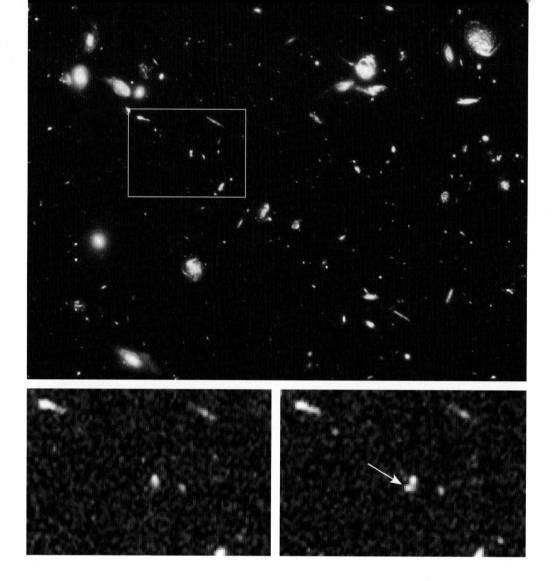

sense out of the nonsensical: There was a repulsive force in space. It's ironic that the computer had reached this conclusion in a methodical way before Riess realized the awesome ramifications.

In California, a team led by Saul Perlmutter of Lawrence Berkeley National Laboratory independently discovered a similar acceleration to the expansion rate of the universe. His team likewise found that distant supernovas are dimmer than predicted, which means there is more space between them and us than if the universe were slowing down or even "just coasting." Therefore, the universe must now be expanding at a faster rate than it was earlier in time.

Both groups had stumbled on Albert Einstein's cosmological constant, a 75-year-old theory of a ghostly counterbalancing force to the universe that would prevent it from imploding. The phenomenon is now simply called dark energy.

The reality of dark energy was bolstered by later Hubble observations of a supernova 10 billion years into the past that was anomalously brighter than expected. This provided evidence that the universe had been decelerating a very long time ago but that between now and then, it has sped up. This transition from "push me, pull you" happened about seven billion years ago. Astronomers are now pursuing observations to better characterize dark energy and to see whether it really does behave like Einstein's cosmological constant. Hubble Space Telescope observations so far show that dark energy has been stable for the lifetime of the universe. If it were unstable, the universe might literally blow apart or go into reverse gear and implode.

Several approaches have been proposed for next-generation telescopes, including surveys to find more supernovas and measuring acoustic oscillations imprinted on the sky that were triggered by gravitational attraction and gas pressure in the primordial plasma of the big bang.

## 4. THE UNIVERSE'S EXPANSION RATE NAILED DOWN

Scientists of the late 1800s suspected that Earth must be much older than previously thought. Geologic evidence and Darwin's emerging concept of biological evolution required over one billion years for the slow changes in the Earth's geologic activity as well as the emergence of different animal species to take place—a far cry from the biblical 10,000 years or Lord Kelvin's simple thermal calculations that predicted an age of 400 million years. Einstein concluded that the universe must be static and perhaps, therefore, eternal. Otherwise, according to his theory of gravity, it would have blown apart or collapsed.

In 1929, Edwin Hubble provided the first observational evidence for the universe having a finite age. His Hubble Constant showed that the farther a galaxy is, the faster it appears to be racing away from us, which indicates that space is expanding uniformly in all directions. By precisely

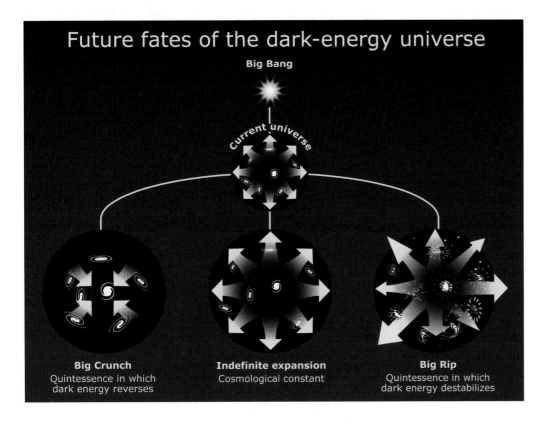

**Future fates of the dark-energy universe**

Big Bang

Current universe

**Big Crunch**
Quintessence in which dark energy reverses

**Indefinite expansion**
Cosmological constant

**Big Rip**
Quintessence in which dark energy destabilizes

When astronomers first realized the universe is accelerating, the conventional wisdom was that it would expand forever. But until we better understand the nature and properties of dark energy, there are other possible scenarios for the fate of the universe. If the repulsion from dark energy is or becomes stronger than Einstein's prediction, the universe may be torn apart by a future "big rip," during which the universe would expand so violently that first galaxies, then stars, then planets and, finally, atoms would come unglued in a catastrophic end of time. Currently, while this idea is very speculative, it is being pursued by theorists. At the other extreme, a variable dark energy might fade away and then reverse in force so that it pulls the universe together, rather than pushing it apart. This would lead to a "big crunch," in which the universe would ultimately implode. According to experts, this is currently considered the least likely scenario. (See illustration, next page.)

The more scientists study dark energy, the more it looks like the repulsive force that Einstein theorized to balance the universe against its own gravity. Even if it turns out that Einstein was wrong, say Hubble researchers, the universe's dark energy probably won't destroy the universe any sooner than about one trillion years. Although dark energy appears to comprise about 70 percent of the universe, cosmologists understand almost nothing about it. They are seeking to uncover its two most fundamental properties: its strength and its permanence. There are two leading interpretations for dark energy as well as many more exotic possibilities. It could be an energy percolating from empty space, as Einstein theorized in his "cosmological constant," an interpretation which predicts that dark energy is unchanging and of a prescribed strength. An alternative possibility is that dark energy is associated with a changing energy field dubbed quintessence. This field would be causing the current acceleration—a milder version of the inflationary episode from which the early universe emerged.

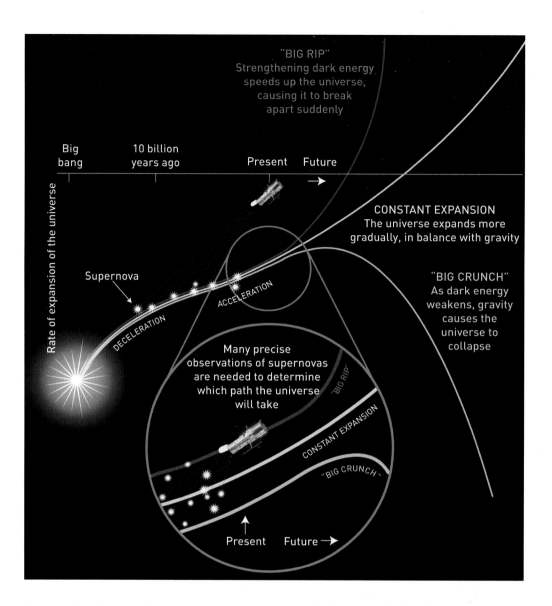

determining the expansion rate, astronomers can rewind the cosmic clock and calculate the age of the universe. But the age estimate is only as reliable as the accuracy of the distance measurements. A precise value for the Hubble Constant is a critical anchor point for calibrating other cosmological parameters for the universe (and, in hindsight, for characterizing dark energy, which was unknown in 1990).

This was identified early on as a Hubble Space Telescope Key Project, because Hubble could resolve Cepheid variable stars—important cosmic milepost markers—out to much greater distances from Earth than ground-based telescopes could.

When Hubble was launched, the uncertainty over the expansion rate of the universe was off by a factor of two. Estimates ranged from 50 kilometers per second per megaparsec (1 megaparsec is 3.26 million light-years) to double that, which meant the universe could be as young as 8 billion years or as old as 16 billion years. The smaller value presented a huge problem: The universe would be younger than its oldest known stars.

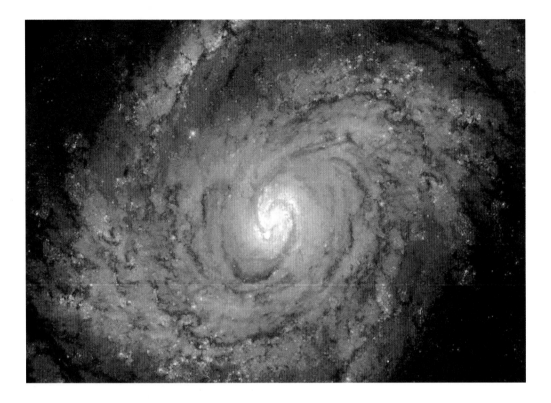

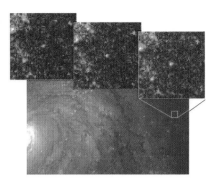

An early target for the Hubble Telescope was galaxy M100 (left). Hubble pinpointed a class of pulsating star in M100 called a Cepheid variable (above). The top three close-up frames, taken roughly a week apart, reveal that the star (in center of each box) changes brightness slightly. Cepheids go through these pulsations rhythmically. The interval it takes for the Cepheid to complete one pulsation is a direct indication of the star's intrinsic brightness. This value can be used to make a precise measurement of the galaxy's distance.

In 1994, Wendy L. Freedman of the Hubble Space Telescope Key Project on the Extragalactic Distance Scale announced a value of 80 kilometers per second per megaparsec, suggesting a universe about 10 billion years old. The results were perplexing, because they, too, indicated a universe that is younger than the oldest stars. It looked as if stellar-evolution models were inaccurate.

By the late 1990s, the refined value of the Hubble Constant was reduced to an error of only about 10 percent. Adam Riess and collaborators continued to streamline and strengthen the construction of a cosmic "distance ladder" by calibrating ever more Cepheids in far-flung space. This allowed the expansion rate to be precisely measured as 74.3 kilometers per second per megaparsec, narrowing it to a value with an uncertainty of no more than 3 percent. Factoring in dark energy, this yields an age of 13.7 billion years for the universe—old enough for the measured ages of the oldest stars. In hindsight, it was almost predictable that astronomers might end up splitting the difference between the 50 and 100 kilometers per second per megaparsec values.

## 5. SAMPLING THE ATMOSPHERES OF EXTRA-SOLAR PLANETS

Although long a staple of science fiction stories, planets around other stars were not discovered until five years after Hubble's launch. This was accomplished by ground-based telescopic observations measuring stellar wobbles (toward and away from Earth) caused by the gravitational pull of one or more unseen planets. Later, the dimming of a star due to a planet crossing directly in front of it (transiting) became a prolific method.

But direct imaging of exoplanets was elusive—even for Hubble. Planets are lost in a star's blind-

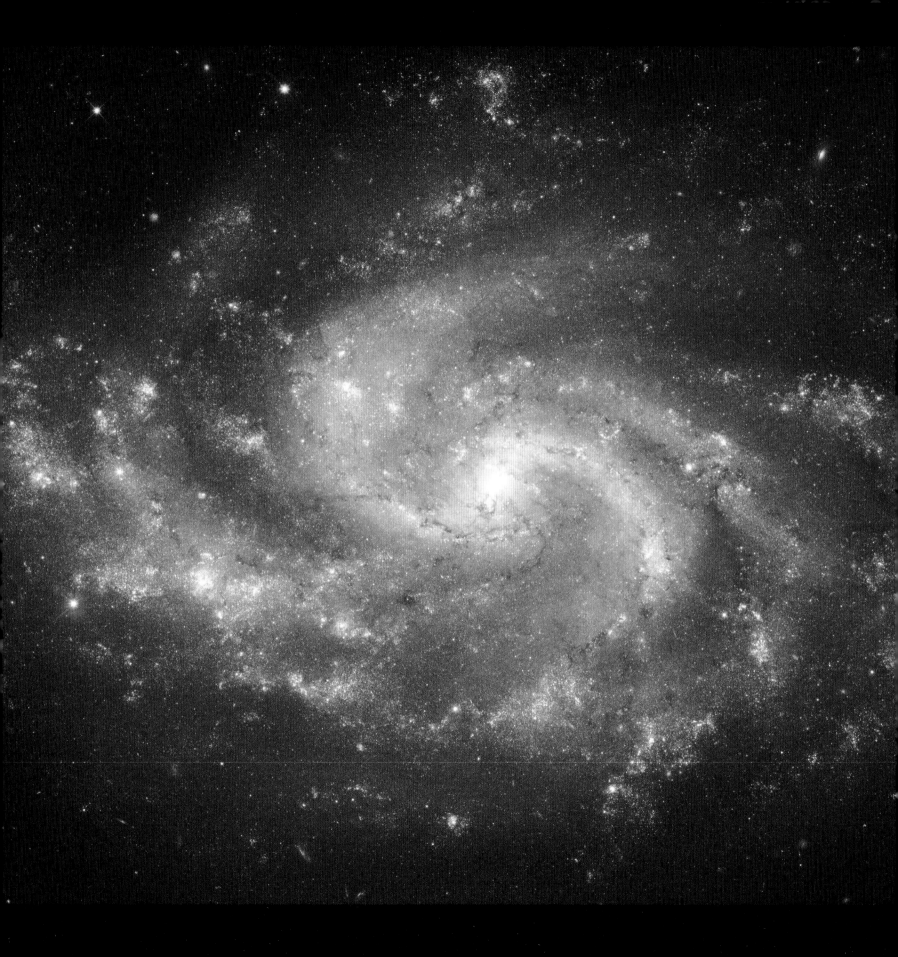

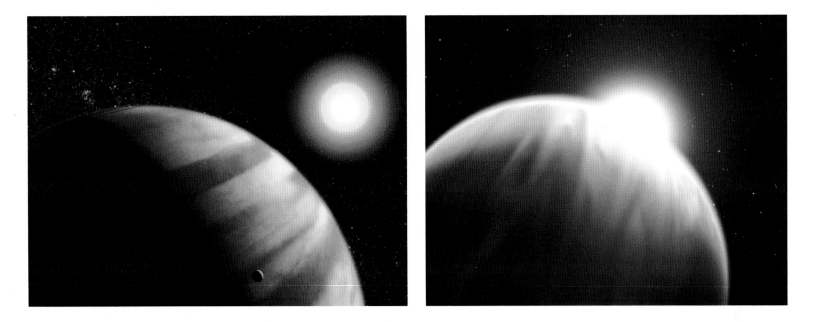

ing glare. Seeing one directly wasn't accomplished until 2008, when Hubble made the first ever visible-light image of a young gas-giant planet orbiting the star Fomalhaut.

The only information provided from indirect detection techniques is the planet's orbital period, diameter and mass. But by the late 1990s, astronomers recognized that detecting extra-solar planets transiting in front of their stars opened the possibility for characterizing conditions on these worlds, and they quickly applied Hubble's unique capabilities to the exoplanet sweepstakes.

Hubble made the first measurements of the atmosphere of an exoplanet. In a landmark observation, the Hubble Telescope was used to spectroscopically measure the parent star's light as it filtered through an exoplanet's atmosphere. Sodium was detected in the atmosphere of HD 209458a.

In subsequent observations, Hubble found carbon dioxide, methane, oxygen and water vapor. The hot Jupiter-class planets (Jupiter-like planets in tight orbits close to their suns) studied by Hubble are no doubt lifeless, but Hubble's ability to look for atmospheric biotracers on apparently Earth-like planets plays an important role in the search for extraterrestrial life.

Hubble also discovered that the hottest-known planet in the Milky Way Galaxy might also be its shortest-lived world. The doomed planet, 40 percent more massive than Jupiter, is being cannibalized by its parent star and may have only another 10 million years left before it is completely devoured. The planet, called WASP-12b, is so close to its Sun-like star that it is superheated to nearly 1,800 degrees C and stretched into a football shape by enormous tidal forces. The atmosphere has ballooned to nearly three times Jupiter's radius and is spilling material onto the star.

Another planet in trouble is HD 209458b. It is orbiting so close to its star that its heated atmosphere is escaping into space. Hubble observations suggest that powerful stellar winds are sweeping the cast-off material from the scorched planet and shaping it into a cometlike tail.

Besides characterizing exoplanets, Hubble has conducted the farthest ever surveys for them. In 2004, Hubble peered at the globular star cluster 47 Tucanae and failed to find any evidence of planets orbiting close to its member stars. The absence of any sign of planets remains a total

Above, left: Artist's concept of a hot Jupiter-class planet.

Above, right: Artist's concept of a hot Jupiter-class planet on which Hubble detected the organic molecule methane.

Facing page: NGC5584, one of the most beautiful spiral galaxies in our sector of the universe, is rich in pulsating stars called Cepheid variables. Astronomers use Cepheid variables as reliable distance markers. NGC5584 lies 72 million light-years away and is one of eight galaxies Hubble astronomers studied to measure the universe's expansion rate.

mystery. Because the stars in globular clusters are so crammed together, their gravitational tugs on one another might disrupt the planet-formation process. Or perhaps there are too few heavy elements to construct planets.

In 2006, a Hubble survey called the Sagittarius Window Eclipsing Extrasolar Planet Search looked across a staggering 26,000 light-years, deep into the crowded central bulge of our galaxy. Hubble found 16 extra-solar planets transiting their stars; some had orbital periods of less than a day! That's not a large number considering that as of 2011, after just two years of observations, NASA'S Kepler space telescope had found more than 1,230 planets. But the brief Hubble survey, when extrapolated to the entire galaxy, offers strong evidence for the existence of approximately six billion Jupiter-sized planets in the Milky Way.

## 6. DARK MATTER MATTERS

In 1933, Swiss-American astronomer Fritz Zwicky encountered a mystery while studying the motions of distant galaxies. Zwicky estimated the total mass of a group of galaxies by measuring their brightness. But when he measured the effects of gravity on the galaxies' velocities of motion, he came up with an estimate that was several hundred times greater than his calculation based on brightness. Zwicky had stumbled on what has become known as the "missing mass" problem.

The mystery lingered until scientists began to realize that only large amounts of hidden mass could support theories attempting to explain the structure of the universe. Galaxy clusters might be tied to an invisible scaffolding of so-called dark matter. In fact, we now know that this invisible

Right: Artist's concept of the gaseous planet WASP-12b being stripped down by the gravitational pull of its companion star.

Below: Artist's concept of the gas-giant planet dubbed HD 209458b, which is orbiting so close to its star that its heated atmosphere is escaping into space, creating a cometlike tail.

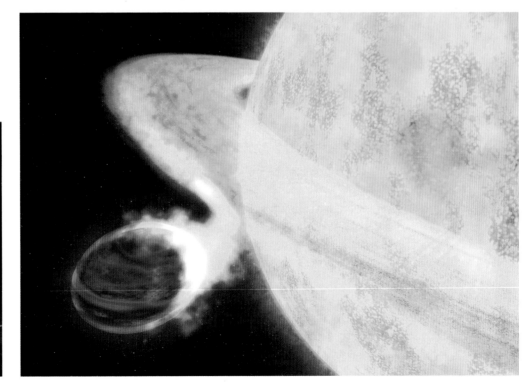

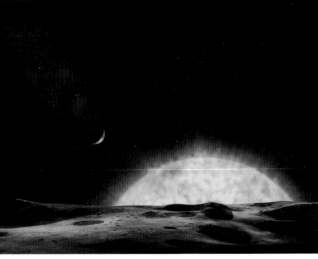

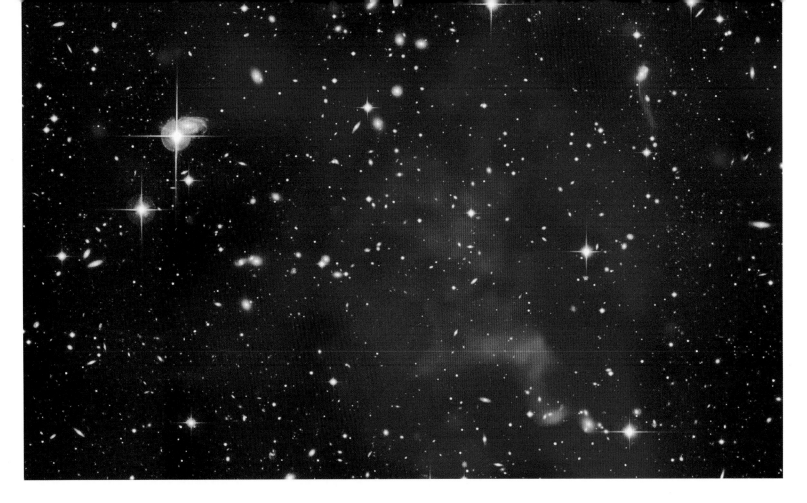

form of matter comprises most of the universe's mass and forms its basic underlying structure.

In 1994, Hubble astronomers quickly ruled out the possibility that faint red dwarf stars constitute dark matter. By simply counting the number of faint red stars in our galaxy's outer halo, scientists determined that the Milky Way has too few of these stars to account for dark matter. They found that despite the fact that dim red stars are by far the most abundant type of star in the universe, they make up no more than 6 percent of the mass in the halo of the galaxy and no more than 15 percent of the mass of the Milky Way's disk.

This left dark matter as a problem for particle physicists. The best dark matter candidates are exotic elementary particles which have yet to be discovered with accelerator experiments and which interact very weakly with familiar particles and with themselves.

Nevertheless, astronomers can probe the distribution of dark matter in space through gravitational lensing. According to general relativity, mass "warps" space-time. Dark matter's gravity distorts space and thereby alters the images of background galaxies. Just as pebbles at the bottom of a pool appear distorted due to the refraction of light through water, the light of distant galaxies passes through the gravitational influence of intervening clusters, resulting in images that are stretched or misshapen.

Aided by Hubble's sharp view, astronomers use this gravitational-lensing technique to reconstruct the large-scale, three-dimensional distribution of the dark matter responsible for the distortions. In one study, astronomers constructed the distribution of dark matter (in one

The galaxy cluster Abell 520 is a cosmic wreck, the result of three galaxy clusters smashing together. The artificial pink color traces the glow of hot gas as seen by an X-ray telescope, while the artificial blue cloud is the dark matter. The purplish center of the cluster identifies a pocket where, for some unknown reason, the mysterious dark matter has pooled.

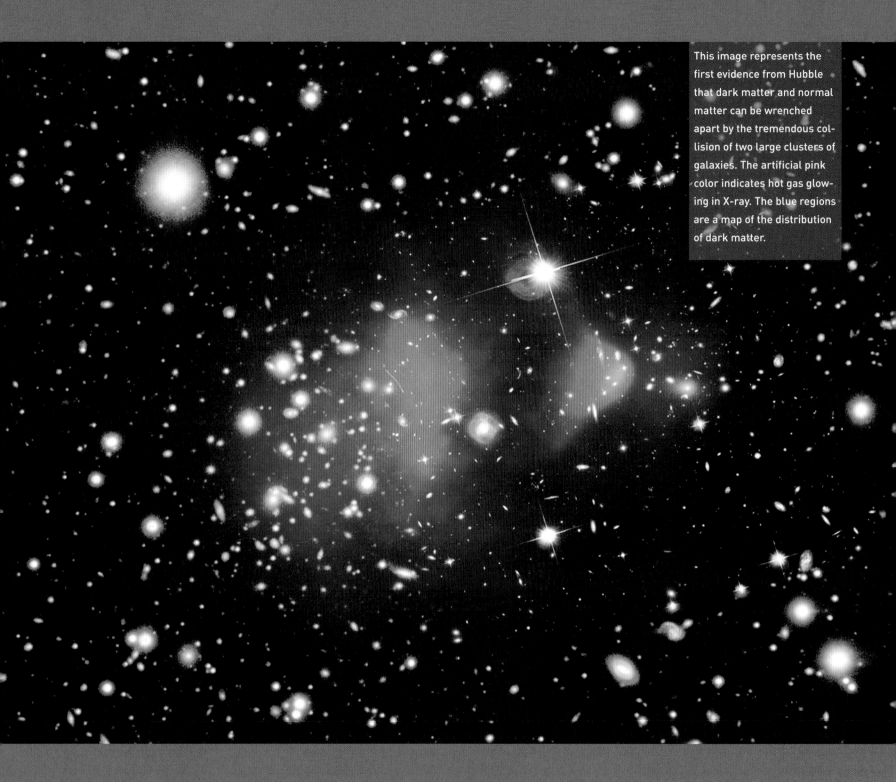

This image represents the first evidence from Hubble that dark matter and normal matter can be wrenched apart by the tremendous collision of two large clusters of galaxies. The artificial pink color indicates hot gas glowing in X-ray. The blue regions are a map of the distribution of dark matter.

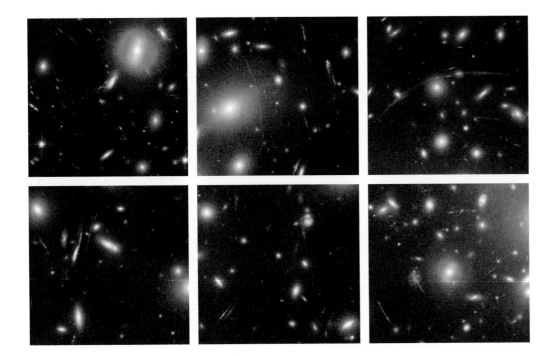

direction) 3.5, 5.0 and 6.5 billion years ago and found that dark matter clumping becomes more pronounced over time.

By studying the warped images of half a million faraway galaxies using this gravitational-lensing technique, astronomers were able to construct a three-dimensional map. The map provides the best evidence to date that normal matter, largely in the form of galaxies, accumulates along the densest concentrations of dark matter. Stretching halfway back to the beginning of the universe, the map reveals a loose network of dark matter filaments.

Astronomers also used Hubble to observe dark matter's distribution in the titanic collisions of clusters of galaxies. In one case, a combination of Hubble and Chandra X-ray observations suggests that dark matter and normal matter, in the form of hot gas, were pulled apart by the smashup between two groupings of galaxies known as the Bullet Cluster.

In a Hubble gravitational-lensing study of a different galaxy cluster, astronomers discovered a ghostly ring of dark matter that formed long ago during a clash between two groups of massive galaxies. Astronomers have long suspected that such clusters would fly apart if they relied only on the gravity of their visible stars. The ring is a ripple of dark matter that was pulled away from normal matter during the collision.

Computer simulations of galaxy-cluster collisions show that when two clusters smash together, dark matter flows through the center of the combined cluster. As it moves outward from the core, it begins to slow down and pile up under the pull of gravity.

Over the past two decades, Hubble results have rewritten astronomy textbooks. The six landmark areas of research mentioned in this chapter have greatly accelerated the rate of discovery and pure revelation in tackling the fundamental mysteries of the universe. More Hubble images illustrating these six areas of research are displayed in the following chapters.

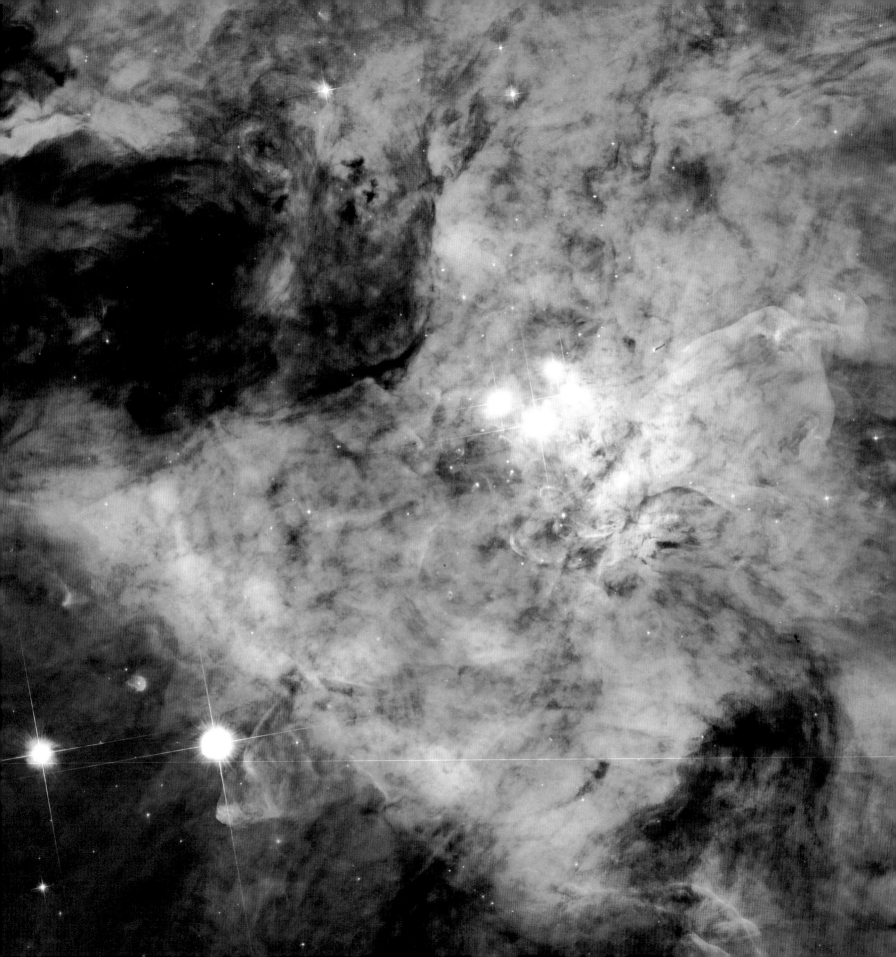

# THE MESSAGE OF STARLIGHT

## CHAPTER 3

Astronomy is a preeminently visual science. Astronomers cannot collect rocks for analysis, dissect organisms or test chemical reactions in a lab. Everything must be deduced from the light that is emitted or reflected from far away in space.

In medieval times, light was considered divine. Biblical scripture declares, "God is Light." Therefore, when Galileo reported his first celestial observations with the newly invented telescope in the early 1600s, there was some skepticism as to whether the telescope was depicting reality or an illusion.

But scientists soon teased out the physical nature of light. Isaac Newton said that luminous bodies radiate energy in particles and that these particles then act on the retina of the eye to produce the sensation of vision in the brain. With the use of a prism that refracts white light into different wavelengths of energy, Newton divided the visual spectrum into the seven colors learned by schoolchildren: ROYGBIV (red, orange, yellow, green, blue, indigo, violet).

By 1800, astronomer William Herschel realized that there were invisible forms of light. Using a prism to split sunlight into a spectrum, he measured the temperature of each color. He found that the highest temperature reading came from the region beyond the red, where no color could be seen. An invisible form of radiation called infrared light was identified. Just a year later, German physicist Johann Wilhelm Ritter discovered ultraviolet radiation. There was more to light than met the eye. Light is just a small part of the electromagnetic spectrum that spans all the radiation emitted by the universe, from X-rays and gamma rays to radio waves.

As the concept of electromagnetic energy was unfolding in the mid-1800s, Gustav Kirchhoff and Robert Bunsen invented the first spectroscope, a device that could reveal the signature of different types of elements in sunlight and starlight. The whole apparatus was assembled from an empty cigar box, a prism and parts of a telescope. Using the spectroscope, astronomers could deduce the composition of stars and glowing interstellar gas.

More than simply looking at stars, scientists were given a license to go anywhere in the universe and identify the same elements found here on Earth by taking spectrograms of celestial bodies. This was a striking demonstration of the Copernican principle that we do not occupy a special location in the universe. Physicists now had evidence that the entire universe is made of the same material.

## LIGHT BUCKETS

But collecting faint starlight, especially to do spectroscopy, required ever larger "light buckets." Telescopes of the 1700s and 1800s embraced major strides in mirror diameter. A whopping six-foot-wide telescope mirror in the mid-1800s, appropriately called the Leviathan of Parsonstown,

Below: White light contains a full rainbow of color, as seen in this photo of a diffraction grating, which disperses different wavelengths of light.

Facing page: The Trapezium cluster is a four-star powerhouse that illuminates and stimulates to luminescence the vast Orion Nebula.

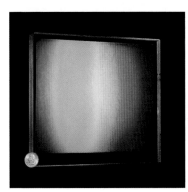

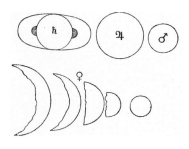

Above: Long before cameras were invented, astronomers had to make pencil sketches of what they saw through a telescope. The most historic are those made by Galileo, recording what he saw (or thought he saw) through the newly invented telescope. Planets that appeared starlike to the unaided eye had visible disks in the telescope. Venus went through phases like our Moon, and Saturn's rings looked like appendages to Galileo.

Right, top: The electromagnetic spectrum is a symphony of radiant energy flooding the universe. Now, for the first time, the full range of this energy can be collected by ground and space telescopes. Hubble covers wavelengths from near-infrared light to ultraviolet light.

Right: Our Sun is seen in living color in this high-resolution spectrum taken with a specialized solar telescope in Arizona. The dark lines etched on the rainbow are the fingerprints of various elements in the Sun's outer atmosphere. Hubble's instruments can make similar measurements of distant stars and galaxies.

revealed a plethora of nebulas that came in all manner of shapes, from spirals to rings to amorphous.

But astronomers could only look at objects. Recording what was seen through the eyepiece was another matter. Galileo's sketches of the wonders he saw in his telescope showed a view of the heavens 10 times sharper than that provided by the eyes alone. He discovered, among other things, the moons of Jupiter, mountains and valleys on the Moon and the rings of Saturn, which he described as "cup handles." But there was no way to store this information short of making pencil sketches, as Galileo did. And the eye-brain connection is subjective.

This was dramatically illustrated at the end of the 19th century when American diplomat-

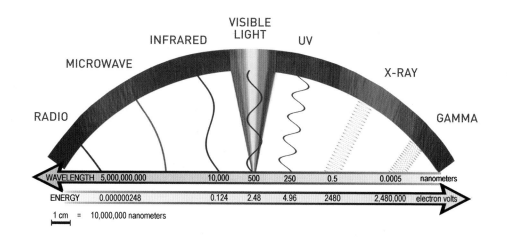

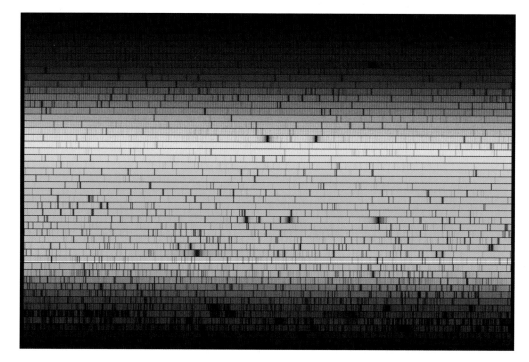

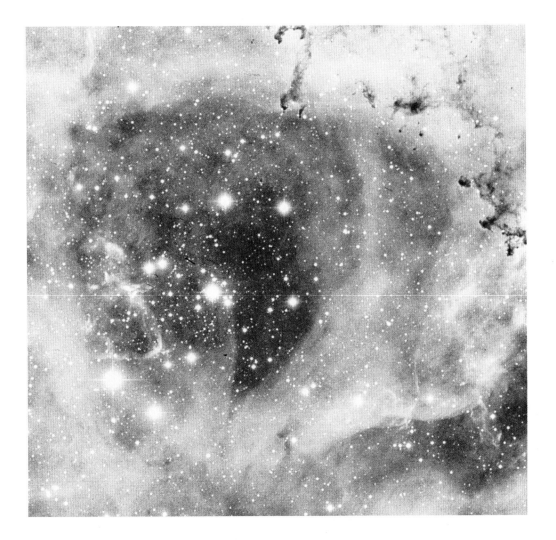

turned-astronomer Percival Lowell sketched what he perceived as a cobweb of lines on the planet Mars. Lowell was convinced that they were evidence of an irrigation canal network built by an advanced civilization on the red planet. But no one else reported seeing quite the same detailed phenomenon, and the "canals" remained highly controversial until the planetary robotic probes of the space age proved that they were nonexistent.

The newly invented process of photography in the mid-1800s allowed astronomers to record what they saw and save it for future generations. It had been known for centuries that salts containing silver would darken when exposed to light. By the early 1800s, crude image recordings were made on materials that had been treated with silver chloride or silver nitrate. Shadow images of overlaid objects were "burned" directly onto a sensitized plate by intense light. But the images faded away over time. In 1819, scientists found that images could be permanently "fixed" if treated with certain chemicals. And in 1839, astronomer John Herschel coined the word "photography."

The first astronomical photographic target was the Moon, which was photographed in 1839 by Louis Daguerre, the inventor of one of the first imaging processes. In 1859, the star Vega was

This photo map, compiled from individual Hubble images of Mars, spans most of the planet's surface. Fine orange dust covers much of the planet. Coarser dust looks darker. This interplay of contrast and fine detail led the late-19th-century astronomer Percival Lowell to perceive linear features through his telescope, which he thought were canals constructed by a Martian civilization. Observations of such detail were historically hampered by the blurring effects of the Earth's atmosphere, but by the 1960s, close-up spacecraft images of the red planet revealed the real Mars: a desert world of sand dunes, volcanoes, craters and canyons—but no canals.

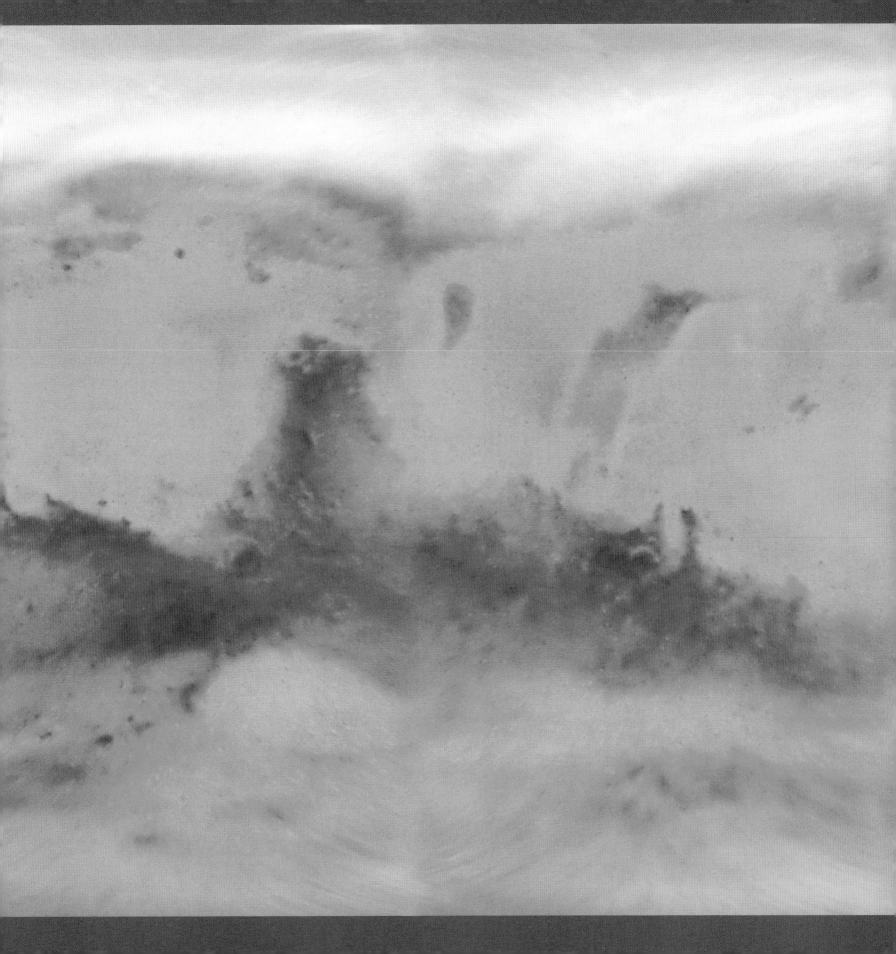

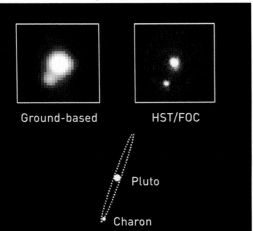

Ground-based    HST/FOC

Pluto

Charon

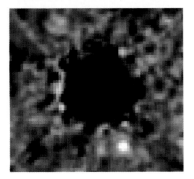

Top: After years of deliberate photographic searching, the icy dwarf planet Pluto was discovered in 1930. Pluto's moon Charon was not identified until 1978 (ground-based photo at left). At right, this Hubble picture, taken shortly after the orbiting telescope's launch, clearly resolves Pluto and its largest moon.

Above: The planet of a distant star appears as a cluster of a few pixels at lower right. The dark patch at center is from a Hubble corona-graph that blocks the star's glare.

photographed at the Harvard College Observatory. The targets had to be bright to provide enough light to record on film emulsion. But they were, nevertheless, blurred because of the long exposure times required and the difficulty involved in tracking objects moving along the celestial sphere.

By the early 20th century, the use of glass photographic plates in astronomy became common, and this made a big difference. Long exposures could record many more stars as well as fine details in wispy nebulas that were too faint ever to be seen by the human eye. This led to many discoveries, two with particularly dramatic payoffs.

Pluto was the first planet discovered by photography. In a hunt for planets beyond Neptune, astronomer Clyde Tombaugh took successive images of small sections of sky. He captured distant Pluto in a pair of exposures taken days apart. Because it was orbiting the Sun, Pluto had drifted against the background stars. Less than a decade earlier, Edwin Hubble had greatly expanded the size of the known universe by discovering, through photography, that galaxies are extragalactic—located far beyond our home galaxy, the Milky Way.

But despite its wonderful properties, photographic film is intrinsically inefficient. Only about two to three percent of the incoming light is actually recorded on the film emulsion. Astronomers boosted this no further than 10 percent by chemically coaxing the emulsion to be more sensitive and lowering its working temperature. This process was reaching its limit around 1975.

At about the same time, the development of solid-state electronics led to the first digital-imaging detectors. Although very expensive, they were vastly more efficient than photographic film, and the data were already in numerical form, making the information easy to store and manipulate by computers. At the heart of this revolution is the charge-coupled device (CCD), which can capture 90 percent of the incoming starlight.

The first CCD was a thumbnail-sized piece of semiconductor material, a silicon chip subdivided into thousands of photosensitive cells arranged in rows and columns like the mesh on a screen door. Each square photo element, or pixel, collects particles of light—photons—like a little rain bucket. The photons build up an electric charge that is stored in the semiconductor pixel "bins" until the exposure is over. The bins of electrons in each pixel are then tallied, and the inventory is fed into a computer memory.

Driven partly by military applications, the size and sensitivity of these devices grew enormously from the 1970s through the end of the century. Computer power grew swiftly too, allowing huge image files to be manipulated and stored.

All this technology converged just in time for the Hubble Space Telescope. Early plans for Hubble had called for astronauts to install photographic film canisters. But by the 1980s, CCD detectors had become more practical and affordable. With each space shuttle servicing mission to repair and upgrade Hubble, instruments were replaced with newer cameras that had increased definition and more sensitive detectors.

## TECHNICOLOR UNIVERSE

People have become accustomed to the rich colors in Hubble's photographic array. One question that is commonly asked is: "Are these objects really so colorful?" If we could fly out to these celestial wonders, would they look this way to our eye? If not, are the Hubble Space Telescope

# The Star That Changed the Universe

Astrophotography was fundamental in the discovery that
there are galaxies far beyond our Milky Way

At the beginning of the 20th century, astronomers believed that spiral nebulas, as galaxies were then known, were part of our Milky Way Galaxy. Two top astronomers, Harlow Shapley and Heber Curtis, held a public debate in 1920 over the nature of the spiral nebulas. Shapley argued that they were much smaller than the Milky Way and, therefore, must be part of our galaxy. But Curtis thought the Milky Way was smaller than Shapley had calculated, leaving room for other island universes beyond our galaxy.

Edwin Hubble was determined to find out just how far away the largest of the spiral nebulas, the Andromeda Nebula (M31), really was. He spent several months scanning Andromeda with the 100-inch telescope on Mount Wilson, California, the most powerful telescope of that era. Andromeda was a monstrous target, about five feet long at the telescope's focal plane. Consequently, it took hundreds of photographic exposures covering dozens of large photographic glass plates to capture the whole nebula.

Hubble made a striking discovery when he compared the plates. One of the stars rhythmically dimmed and brightened over a few days. It was a Cepheid variable star that can be used to calibrate large astronomical distances. The star turned out to be more than one million light-years from Earth, several times the calculated diameter of the Milky Way.

When Shapley received word of the discovery, he told a colleague, "Here is the letter that destroyed my universe."

Thanks to the technological advances of telescope optics and electronic detectors, an amateur astronomer with a modest-sized telescope can duplicate Hubble's discovery by observing M31 from the backyard. The Hubble Telescope is so sensitive, it can find Cepheid variable stars at distances 35 times farther away than the landmark Cepheid that Edwin Hubble found. In commemoration of Edwin Hubble's discovery, the space telescope located the famous Cepheid variable in M31 (upper right image). Hubble's vision is so good, the winking star can be plucked out from the sea of myriad stars surrounding it.

In the 1920s, Edwin Hubble's studies of the Andromeda Galaxy, the nearest large spiral galaxy similar to our Milky Way Galaxy, led to the discovery of a Cepheid variable star that proved the remoteness of the Andromeda Galaxy (see "var!" marked in Edwin Hubble's photographic plate below).

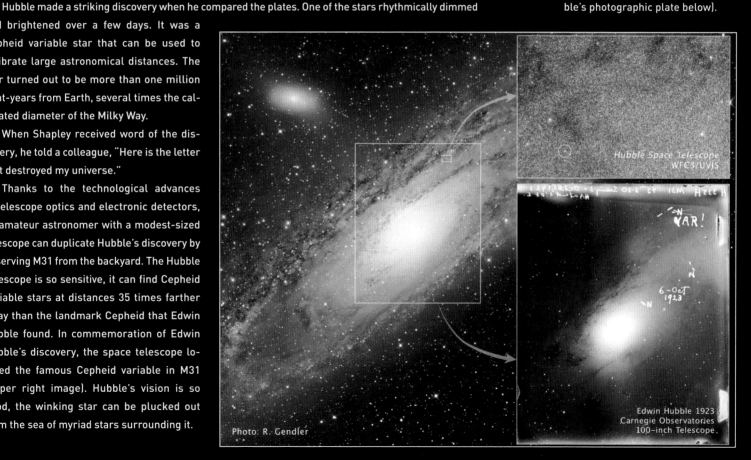

Photo: R. Gendler

Hubble Space Telescope
WFC3/UVIS

VAR!

6-Oct
1923

Edwin Hubble 1923
Carnegie Observatories
100-inch Telescope.

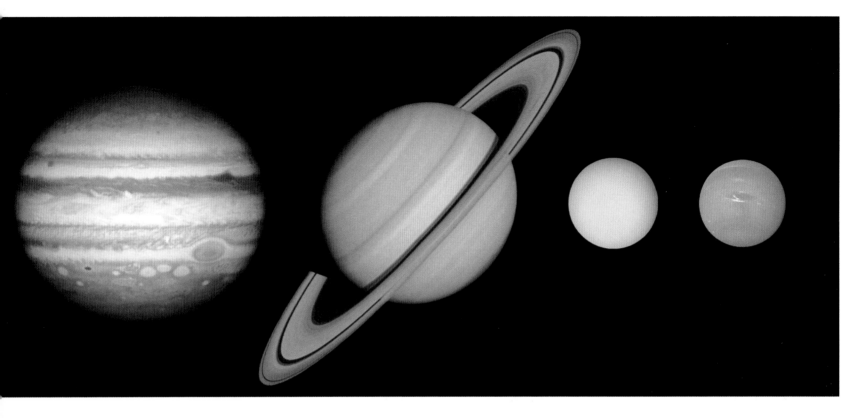

As seen in this family portrait of the four gas giant planets, the solar system is intrinsically very colorful. Jupiter, at left, contains swirls of salmon- and pink-colored clouds. These may come from trace elements, like sulfur, in Jupiter's hydrogen atmosphere. Similarly, elements in Saturn's chilly atmosphere, such as quantities of nitrogen, oxygen and sulfur, produce a pale orange photochemical smog. Methane in the atmospheres of both Uranus and Neptune absorbs red light, giving these worlds a cyan hue.

pictures being overly colorized? What is "truth" when it comes to photographing the universe?

We perceive true astronomical colors three ways. First, the absence of some wavelengths of white light reflecting off an object makes colors. The cyan hues of Neptune and Uranus are due to methane in their atmospheres absorbing red light. Mars' ruddy hue is the result of iron oxides absorbing green wavelengths, and Jupiter's Easter-egg look probably comes from chemical compounds in its turbulent atmosphere.

Second, the emission of light at certain energies or wavelengths makes specific colors. Nebulas glowing with clouds of hydrogen, oxygen and nitrogen are nature's own version of a honky-tonk strip of neon lights. Because the gases glow only in specific colors of light, the colors are very saturated and pure; they are free of being mixed with other colors.

Third, the stars are different colors depending on their temperatures but appear pastel because stars radiate energy across the visible spectrum. Yes, they may look pale reddish like Betelgeuse or blue like Sirius, but they emit light at many other colors, so the resulting hue is subdued. Because stars are thermal objects, their comparative brightness and color behave much like the coils in a toaster. When turned on, the coils heat from dull cherry red to yellow-orange to yellow-white. Similarly, star colors range across a broad spectrum, from hot blue to a comparatively cool ruddy red.

An aesthetic astronomical image that has proper color variation can yield new information and insights. Our eyes are better at distinguishing differing hues, and otherwise hard-to-see features often pop out in a color image versus a black and white image that is just shades of gray.

By definition, telescopes and their CCD arrays are built to extend our vision into areas where

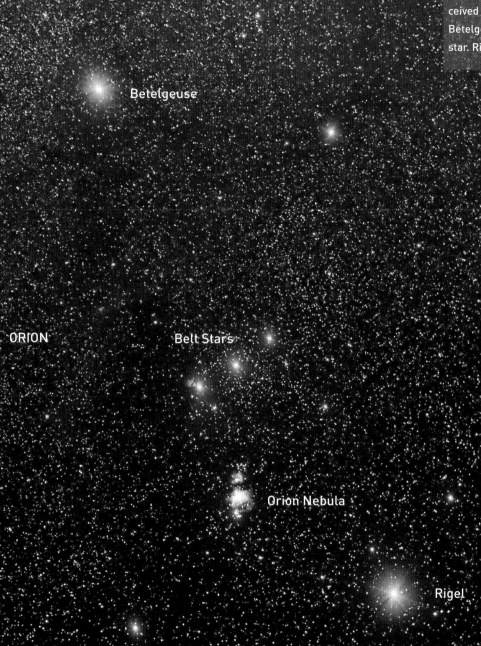

Some stars in the winter constellation Orion the hunter are so bright that colors can be perceived with the unaided eye alone. Betelgeuse is a cool red giant star. Rigel is a blue-white giant.

Betelgeuse

ORION

Belt Stars

Orion Nebula

Rigel

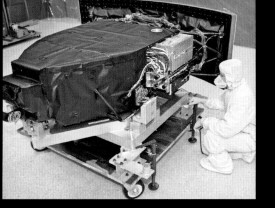

# Hubble's Cameras and Two Companion Orbiting Observatories

## Wide Field and Planetary Camera 2

Above: This camera was installed on Hubble during the first space shuttle servicing mission in 1993 to replace the Wide Field and Planetary Camera 1 (WFPC1). WFPC1 made important observations but was crippled by the fact that Hubble's primary mirror had been optically ground to the wrong curvature and could not bring images into sharp focus. WFPC2 was essentially outfitted with "contact lenses" to correct for the blurriness, or spherical aberration. WFPC2 was the workhorse camera for Hubble until it was replaced in 2009. A large number of Hubble's discoveries and iconic pictures came from this camera. Yet by today's standards, its resolution is modest. Each of the three CCDs was just 800 pixels square. A fourth CCD had double the resolution but covered only one-quarter as much area. This left the camera with odd stair-step-shaped pictures when all four CCDs were used in unison. The camera was sensitive to near-ultraviolet, visible and near-infrared light.

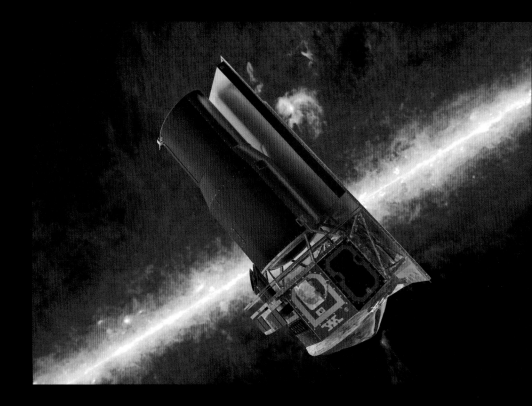

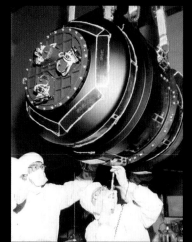

## Spitzer Space Telescope

Above: Launched into orbit around the Sun in 2003, the Spitzer Space Telescope is a cryogenically cooled infrared observatory capable of studying objects ranging from an asteroid in our solar system to distant galaxies. It is the final element in NASA's Great Observatory program.

## Near Infrared Camera and Multi-Object Spectrometer (NICMOS)

Left: Hubble got extended infrared vision in 1997, with the installation of NICMOS. To keep the detector cold enough to record near-infrared wavelengths, the camera had to be chilled with solid nitrogen. NICMOS demonstrated that there was a lot for Hubble to discover at near-infrared wavelengths.

## Advanced Camera for Surveys (ACS)

Right: Installed in 2002, the ACS has a high-definition view of the universe with a pair of CCDs that are each approximately 2000 x 4000 pixels, a huge improvement in sharpness over WFPC2. Because the ACS is more sensitive to light, exposures can be taken in less time. It has a large and complex array of color filters that are fine-tuned for looking at distant galaxies. The camera's power supply failed in 2005. In a bold repair job during Hubble's final servicing mission, the supply was replaced in 2009.

## Space Telescope Imaging Spectrograph (STIS)

Right: Although the STIS is a spectroscopic instrument, it has a small CCD camera with a resolution of 1024 x 1024 pixels that is sensitive to visible and near-infrared radiation.

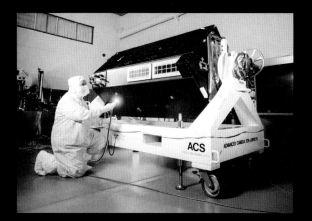

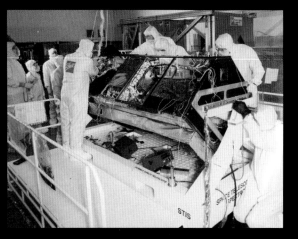

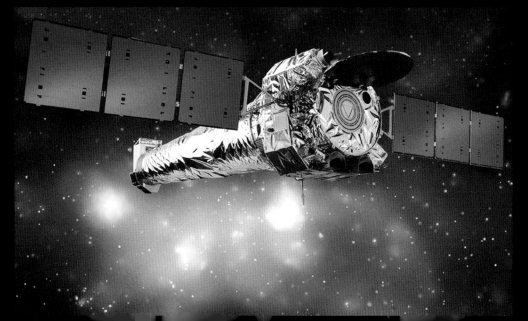

## Faint Object Camera

This first-generation camera (not shown here) was built by the European Space Agency. Its vision improved with the addition of corrective optics in December 2003, during the first Hubble servicing mission. This is the only camera aboard Hubble that utilizes technology predating CCDs, using instead photomultiplier tubes similar to those once used in studio television cameras.

## Wide Field Camera 3 (WFC3)

One of the CCD detectors in this ambitious camera (not shown here) has an impressive sharpness of 16 million pixels and can see in ultraviolet- and visible-light wavelengths. Another detector can see in near-infrared wavelengths with much greater sharpness than the NICMOS. The WFC3 was installed during Hubble's fifth and last servicing mission, in 2009.

## Chandra X-ray Observatory

Left: Operating in high Earth orbit since 1999, Chandra detects and images X-ray sources that lie within our solar system as well as billions of light-years away. The results from Chandra provide insights into the universe's structure and evolution. Along with Hubble and Spitzer, Chandra is part of NASA's Great Observatory program. The picture on page 73 is a combined image from the three observatories.

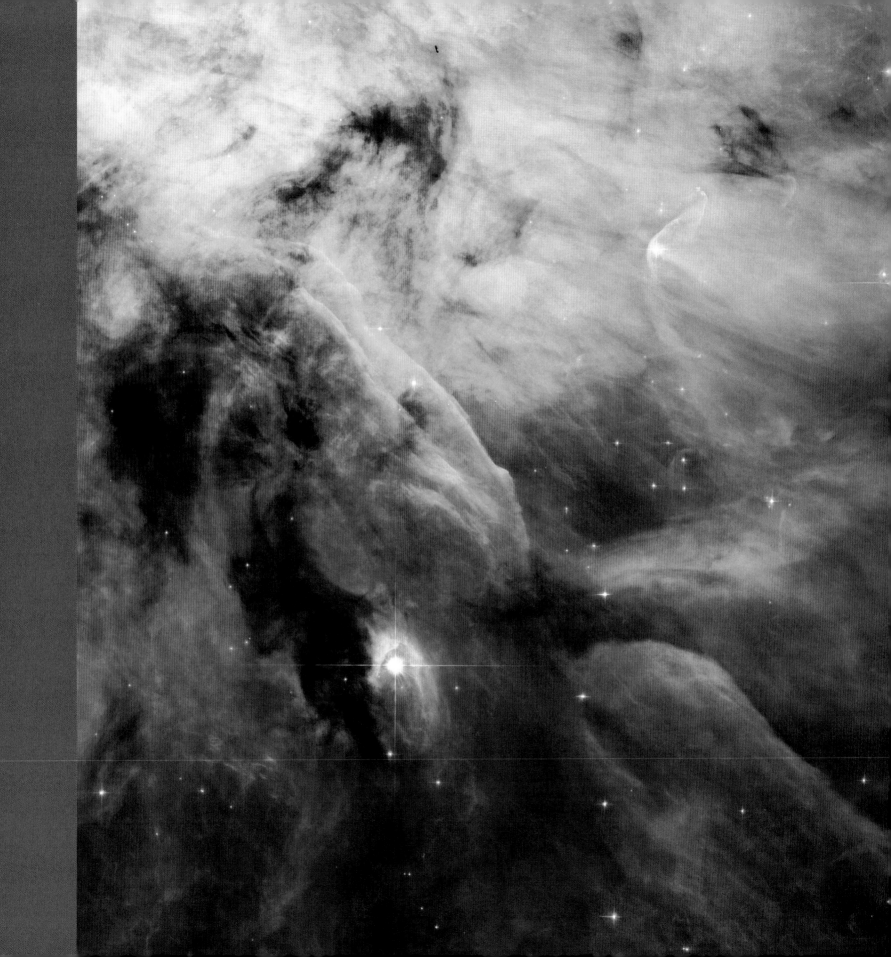

The warm, pastel hues in the Orion Nebula are largely due to glowing hydrogen. There is also a small trace of green from glowing oxygen. A reflection nebula of dust that scatters blue light surrounds LP Orionis, the bright star left of center. Because the light is too faint to stimulate the color receptors in our eyes, the nebula look greenish gray when viewed through a telescope.

we either can't see very well or can't see at all. The colors from these observations are colors you could never hope to see with your own eyes, because the objects are faint to begin with. Further, the light is feeble because it is spread out rather than concentrated into a single starlike point.

Even when we look at these objects through a telescope, the color-sensitive cone cells in our retinas just give up. Only the rods—those receptors sensitive to black and white—still work. Hence a dimly lit room looks monochromatic. Similarly, nebulas look grayish, with just the slightest hint of color. The question of true color becomes largely a moot point if we can't perceive it in the first place, even though it is there.

## ASSEMBLING COLOR PICTURES

Like beauty, color is in the eye of the beholder. The human eye-brain "computer" has its own unique way of collecting and decoding the message of light from color-sensitive cones in our retina.

Over one million Hubble images are archived at the Space Telescope Science Institute, in Baltimore. This database is huge—over 50 terabytes, which is equivalent to about 50 million books, or five times the printed collection of the U.S. Library of Congress.

The archives store "raw" data that represent the exact readouts from the cameras. The images are automatically calibrated when they are requested from the archive. Calibration is needed because no detectors on any of Hubble's cameras respond uniformly to light. The data must be processed to remove the CCD detector's unique "footprint," because some pixels are more responsive to light than others. This is true for home-photography cameras too. But conventional photography is not dependent on the exact value of how much light is hitting each pixel on the detector. This is needed only for precision scientific research.

An early step in Hubble image processing is to remove cosmic rays, which strike the CCD detector and leave hot spots and wormlike trails on the chip. The longer the exposure, the more cosmic-ray hits on the CCD, like snowflakes accumulating on a sidewalk during a storm. Observations are typically composed of multiple exposures of the same object. The pattern of cosmic-ray hits is unique to each exposure, which makes their subtraction easy. The cosmic-ray-removal software searches an identical exposure and removes anything that isn't duplicated in both expo-

This breathtakingly beautiful color composite picture of a pair of colliding galaxies expands our vision into "invisible" colors of the electromagnetic spectrum. The collision, which began more than 100 million years ago and is still occurring, has triggered the formation of millions of stars in clouds of dust and gas in the galaxies. The power of all three of NASA's Great Observatories was needed to produce this image. Chandra's X-ray image reveals huge clouds of hot interstellar gas. The bright, pointlike sources are produced by material falling onto black holes and neutron stars that are remnants of the massive stars. The Spitzer Space Telescope imaging shows infrared light from warm dust clouds that have been heated by newborn stars, with the brightest clouds lying in the overlapping region between the two galaxies. In Hubble's view, older stars and star-forming regions can be seen in yellow and white. Brownish filaments of dust are silhouetted.

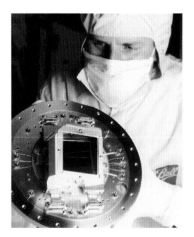

The largest detector on Hubble's Advanced Camera for Surveys (ACS) is a pair of CCDs. Each array measures 4096 x 2048 pixels, allowing the camera to produce extraordinarily sharp 16-megabyte images that are comparable to what can be accomplished with a professional digital camera. The ACS's sharpness is further boosted by the fact that the CCDs are not divided into red, green and blue pixels, which reduces the effective resolution of a consumer camera by one-third. Correct colors based on ACS's filters are added later.

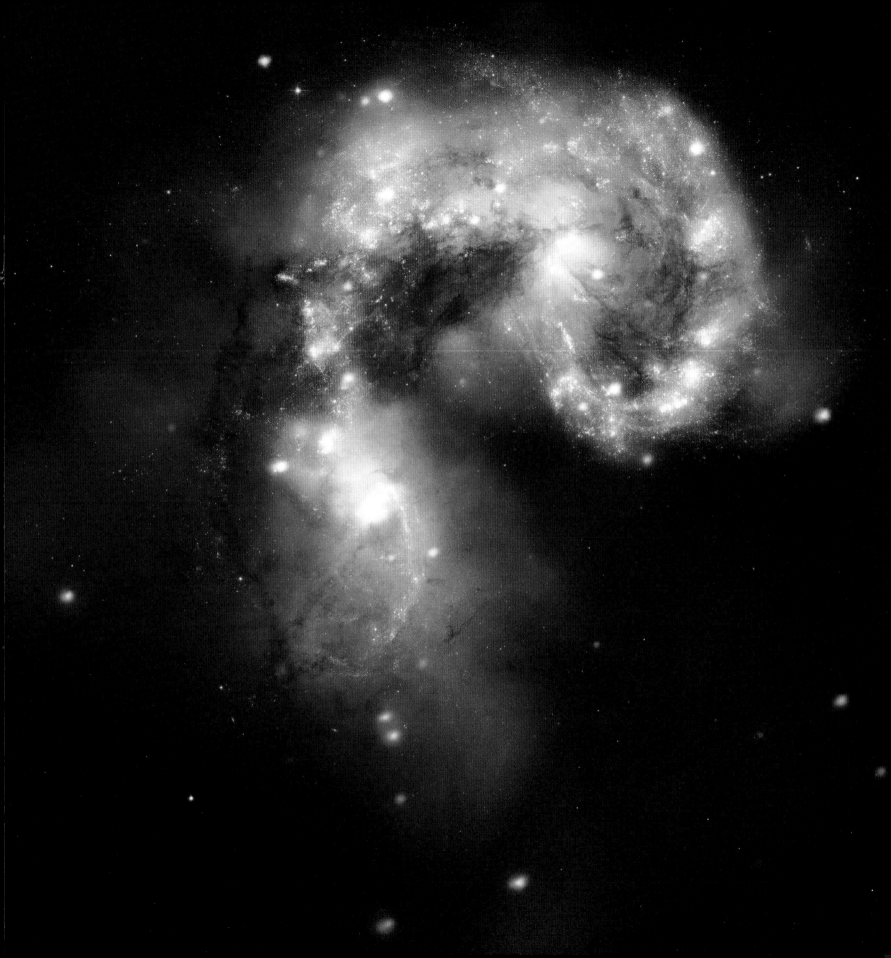

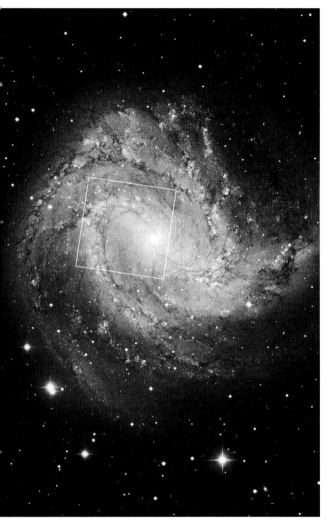

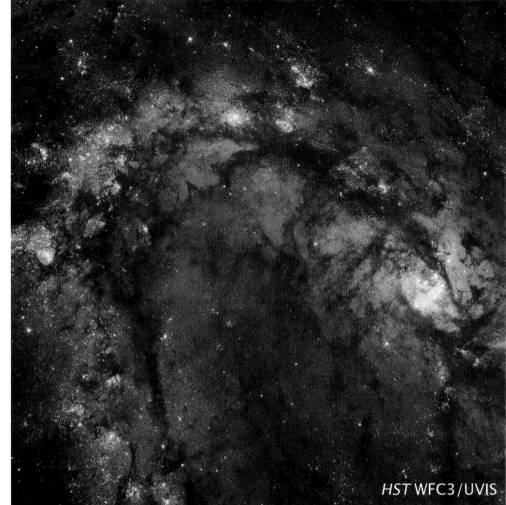

These images demonstrate how colors are stretched for scientific analysis. At left is an image of the galaxy M83 taken at the European Southern Observatory (ESO) with a telescope comparable in size to Hubble. The ESO image presents a natural-color view of the galaxy as seen in visible light. Above is Hubble's close-up view of the myriad stars near the galaxy's core. Hubble's broad wavelength ranges from ultraviolet to near-infrared, showing stars at different stages of evolution and allowing astronomers to dissect the galaxy's star-formation history. The ultraviolet glow of young stars is brilliant blue. The intense glow of hydrogen clouds appears red. Infrared light reveals an underlying population of stars in the galaxy's disk.

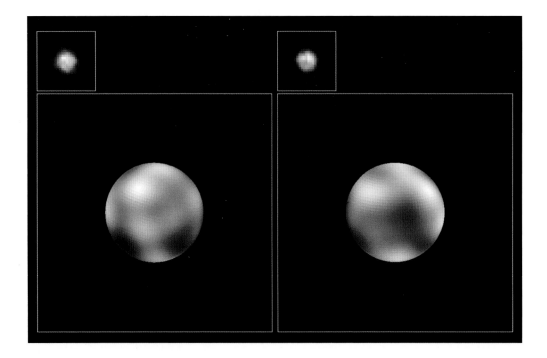

By using computers to parse the distribution of light across single pixels, Hubble images can be further sharpened. This technique was used to create the sharpest-ever images of Pluto. The two smaller inset pictures show the actual images from Hubble's Faint Object Camera. Each square pixel (picture element) represents a span more than 150 kilometers wide on Pluto. At this resolution, Hubble discerns roughly 12 major regions where the surface is either bright or dark. The larger images at left are from a global map constructed by using computer image processing to divide the pixels in the Hubble data into even smaller pixels. The "waffle iron" tile pattern is an artifact of the image-enhancement technique.

sures. Duplicating pictures isn't time wasted on Hubble, because the double exposures can be added to each other to increase the effective exposure time.

Astronomers have learned to sharpen Hubble images through a process called "drizzling." Several images are taken at nearly the same position, but the telescope is nudged slightly by the width of just a few pixels before each of the photos is taken. These images are then aligned. The result is that the image looks as if the effective number of pixels has been increased.

Studio photographers are taught to set up lighting so that the range of brightness between highlights and shadows is never too extreme. Outside, both amateur and professional photographers alike have to contend with a huge range of brightnesses encountered on a sunny day. The most extreme example is trying to take a picture on a bright sunny beach. Shadows are pitch-dark, while the sand is blindingly white.

But, by its very nature, the universe presents an even broader range of brightness and darkness with which astronomers must contend. Brilliant stars appear against inky-black sky, often sharing the view with soft, subtle nebulas. It is challenging to capture the broad range of levels among brightest and faintest features. Much of the image adjustment is done by eye. Image-processing software displays a graph that is helpful in assessing what needs to be done to the gray tones in an image.

At this point, creating a full-color image is as much an art as it is a science. Attractive and evocative science pictures complement the information content of the image. All color-imaging devices mimic the retina's red, blue and green cones by dividing light into its component colors. These are then reassembled into a full-color image. But these devices are not absolutely faithful to how the eye handles color information.

Likewise, Hubble color pictures are assembled from separate exposures taken through red, green and blue filters. Consumer cameras don't need to do this because the CCD simultaneously

Before the colors were composed in the final picture, image processors faced a challenge as they tried to balance the brightness levels in this Hubble image of the massive young star S106. The lower lobe of hot glowing gas ejected from the star is much brighter than the outlying gas and dust in the upper lobe. The contrast in the picture must be compressed to reveal the full tonal range, which could never be perceived by the human eye alone. When the picture is properly balanced, with the correct contrast adjustment, subtle details in the filamentary structure become evident, as do the ripples and ridges that are created as gas slams into the cooler interstellar medium.

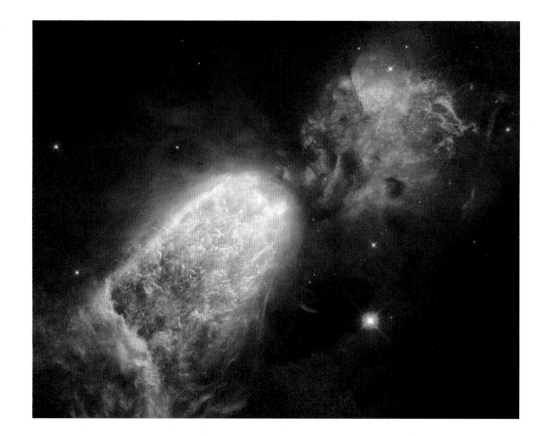

records in red, blue and green light. The trade-off is that the CCD is reduced to one-third of its intrinsic sharpness because information from three pixels is required to construct one full-color pixel. But by taking separate exposures for each needed color, Hubble's cameras do not have to dilute the number of pixels in an exposure. This is critical to achieving the sharpness in the final Hubble images.

This is very similar to the Technicolor film process developed by the motion picture industry in the 1930s. Separate strips of black and white movie film were simultaneously recorded behind red, blue and green filters. This information was then precisely assembled onto color film stock to yield a vibrant and rich image. One of the best-known films prepared like this is *The Wizard of Oz* (1939), with its rich red ruby slippers and the Emerald City.

However, Hubble's cameras also have a huge collection of filters. Some transmit only a narrow range of colors that are fine-tuned to the emission of various glowing gases found in nebulas: hydrogen is pinkish, nitrogen deep red, oxygen green and sulfur blue.

A portrait photographer can balance colors according to flesh tones, because we all know the color of skin. But it's much trickier to arrive at a color to represent galaxies and nebulas that are largely not detectable by human vision. Planets are the easiest to color-match because when seen through an amateur astronomer's backyard telescope, they are bright enough to exhibit color.

Balancing the color in Hubble images of galaxies is fairly straightforward, because the colors follow the predictable colors of stellar temperatures, from hot blue to cool red. Glowing gases in

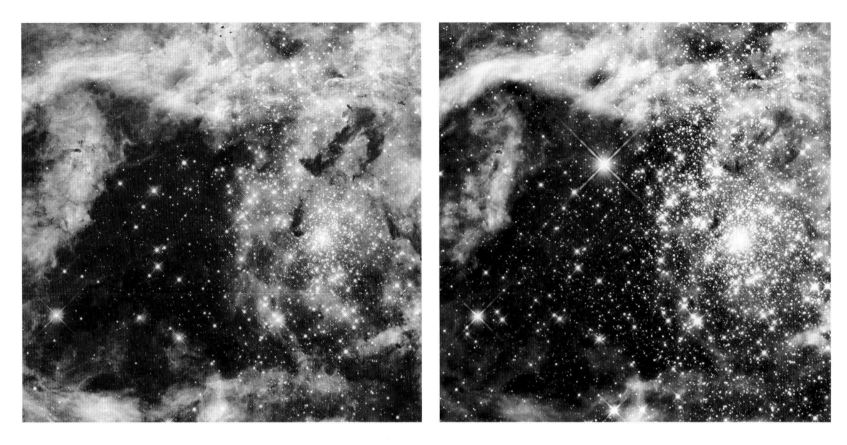

nebulas are well identified by looking at laboratory samples. Therefore, Hubble astronomers instinctively know the proper color for many bright celestial objects.

The most direct Hubble color images are those assembled from three exposures taken through filters that nicely slice the visible spectrum into three equal parts. Some of Hubble's most popular pictures are more problematic when it comes to the meaning of "true" color. Nebulas glow at specific colors that may not faithfully represent the full spectrum.

An image processor can arbitrarily assign red, green and blue to the longest, intermediate and shortest wavelengths of light. The term most commonly used for this approach is "representative color." The three available primary colors are assigned to filters according to their relative placement on the spectrum and may not necessarily reflect actual emission colors.

A much bigger challenge for astronomers is preparing Hubble's infrared snapshots of the universe. There is no such thing as true color in the infrared region of the spectrum, where all wavelengths are invisible to the naked eye. By necessity, an infrared image must be depicted using the color of the visible spectrum. Infrared images are taken through different filters that are sensitive to the short, intermediate and long infrared wavelengths. The most straightforward way to handle the data is to assign blue to the shortest wavelengths, green to intermediate wavelengths and red to the longest wavelengths. This creates a false-color picture that represents different types of emissions at various infrared wavelengths.

Astronomers can also use "paint by number" (or color mapping) to display inherently black

Taken in visible and infrared light, these two images reveal R136, a massive star cluster nestled in the heart of the 30 Doradus Nebula, a turbulent star-birth region in the Large Magellanic Cloud. The stars in the image at left, taken in ultra-violet, visible and red light, are brilliant blue. The green in the nebula is from glowing oxygen, the red from glowing hydrogen. The image at right, taken in infrared wavelengths, penetrates a dusty nebula to reveal many stars that cannot be seen in the visible-light view.

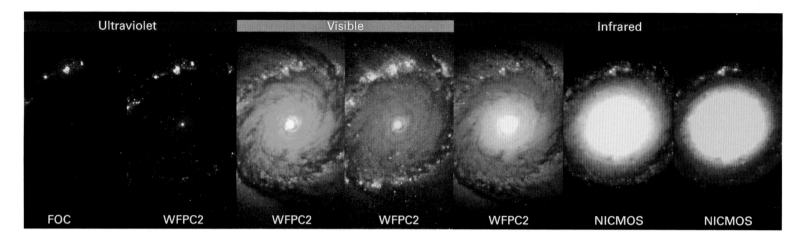

| Ultraviolet | | Visible | | | Infrared | |
|---|---|---|---|---|---|---|
| FOC | WFPC2 | WFPC2 | WFPC2 | WFPC2 | NICMOS | NICMOS |

Three of Hubble's cameras were used to dissect the galaxy NGC1512 into its component colors. The Faint Object Camera picks up the glow of the hottest stars and gas. Color filters used on the Wide Field and Planetary Camera 2 span ultraviolet to near-infrared wavelengths. These two cameras reveal the glow of the full population of stars in the galaxy—from red dwarfs to blue supergiants. The Near Infrared Camera and Multi-Object Spectrometer records the emission of warm dust in the galaxy's disk.

and white images that were not taken through multiple filters. When an arbitrary color is applied to a black and white image, it becomes easier for the eye to distinguish small variations in hue than it is to see these differences in gray values. Here, the colors do not represent intrinsic color differences but changes in the object's brightness. Choosing a color palette that is useful yet visually pleasing is another challenge. Some palettes are terribly garish yellow to green to red; others are heavily blue to magenta. A compelling astronomical picture avoids this "psychedelic" treatment.

After a full-color image is assembled, the Hubble image processor digitally removes seams between the CCD chips. The pixels along the boundaries between each CCD must be carefully "knitted" together. This is especially important in a wide mosaic image of, say, a large galaxy or nebula. The CCDs also have dead pixels that are removed separately. Particularly annoying are linear "bleeds," when a bright star floods a linear row of pixels. The image sometimes has "doughnuts," internal reflections of unfocused stars off mirror elements. Real-world artifacts, such as numerous satellite trails, must be removed as well.

Hubble is so sharp and has been in space long enough that astronomers can assemble movies of objects that are rapidly changing. Among Hubble's movie collections are time-lapse views of glowing gas jets ejected from black holes and young stars; the motion of a star escaping our Milky Way; and the expansion and deformation of material around a supernova that exploded in 1987.

The final Hubble pictures have become cultural icons that have graced not only the covers of popular magazines and nearly every astronomy textbook on the market but also rock album covers, product advertisements, T-shirts, coffee mugs and even a Pepsi-Cola Company corporate report. Hubble pictures have routinely made cameo appearances in numerous science fiction movies and, in fact, have inspired a whole new look to the celestial scenery in Hollywood space operas. A well-known stained glass artist in England is designing a church window based on Hubble imagery. Hubble's images will long be remembered as having ushered in a golden age of popular interest in astronomy and given new views of the universe as culturally powerful and evocative as were the Apollo images of planet Earth taken in the early 1970s.

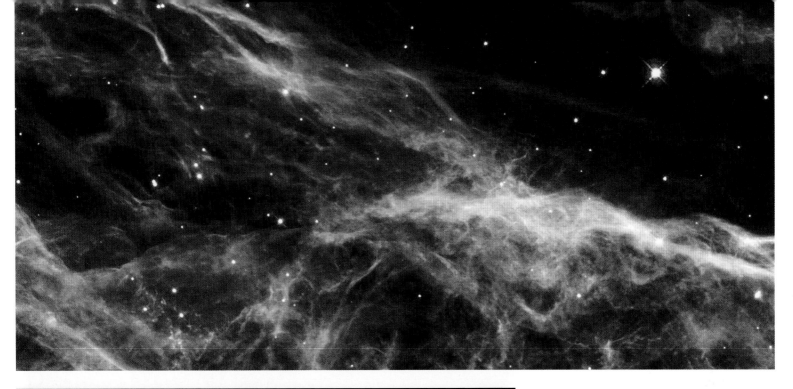

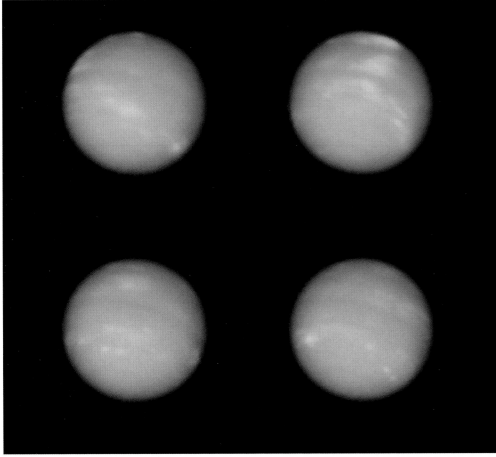

Above: These wispy filaments are a small part of the Veil Nebula, the expanding remnant of a star that exploded thousands of years ago. The colors, obtained by combining images taken through a selection of color filters, correspond to the colorful glow of oxygen (blue), sulfur (green) and hydrogen (red). These elements will be recycled by future generations of stars and planets.

Left: These views of Neptune were taken in ultraviolet to near-infrared wavelengths, which allows astronomers to study the atmospheric structure of the planet. The high-altitude bright cloud features look pink because they are reflecting near-infrared light. The edge of the planet's disk also appears bright in these colors, indicating the presence of a high-altitude haze layer.

Right: These are two dramatically different face-on views of the Whirlpool Galaxy. The image at left, taken in visible light, highlights pink star-forming regions and brilliant blue strands of star clusters. In the image at right, taken in infrared light with Hubble's Near Infrared Camera and Multi-Object Spectrometer, most of the starlight has been removed through image processing. What's left is the Whirlpool's skeletal dust structure that gives the illusion of matter swirling toward the galaxy's core.

Bottom, left: In this Hubble view of the lunar impact crater Aristarchus (lower right) and adjacent Schroter's Valley (top, center), the geologic diversity of the Moon is captured in ultraviolet and visible wavelengths. False colors are then applied to the various wavelengths. The ultraviolet-to visible-color ratio at lower right is false-colored to identify anorthosite, basalt and olivine in the crater Aristarchus. In the image of Schroter's Valley at upper right, the magenta false color indicates titanium-bearing dark mantle material.

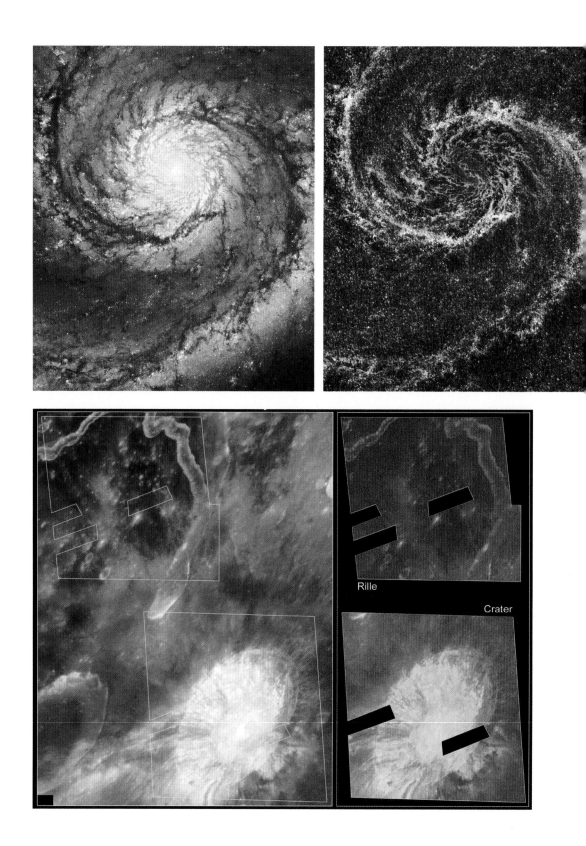

Rille

Crater

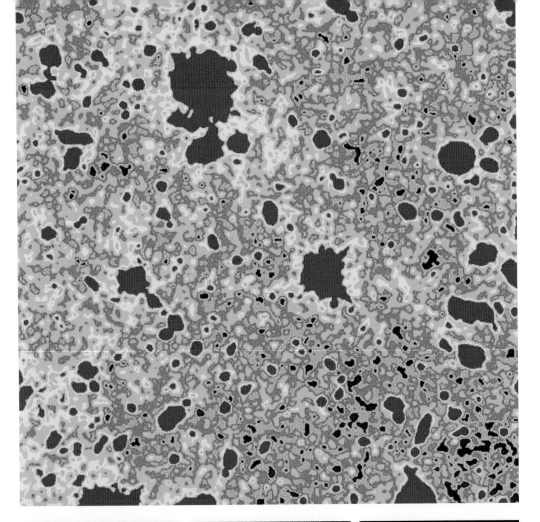

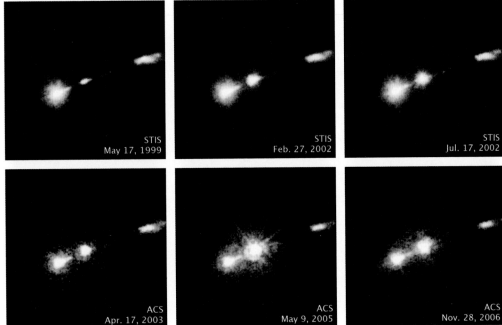

| STIS<br>May 17, 1999 | STIS<br>Feb. 27, 2002 | STIS<br>Jul. 17, 2002 |
| ACS<br>Apr. 17, 2003 | ACS<br>May 9, 2005 | ACS<br>Nov. 28, 2006 |

Left: False color was applied to analyze a patch of the Hubble Deep Field (pages 20-21). The red regions in this figure correspond to galaxies and stars that Hubble has detected. The smallest red patches are galaxies that are four billion times fainter than the human eye can see. Astronomers probed apparently blank patches that lie between the faint galaxies, searching for tiny ripples in the sky brightness that would signal the presence of even more galaxies. They found very little variation in brightness, indicating that most of the visible light filling the universe comes from galaxies like those in the Hubble Deep Field and not from still fainter galaxies.

Bottom: Hubble is sharp enough to measure changes in astronomical objects that occur over just a few years. The jet from the supermassive black hole in the elliptical galaxy M87 is an ideal target, blazing across space at nearly the speed of light. In 2002, Hubble recorded an outburst coming from what might be a blob of material impacted by the jet. It was so bright that it outshone even M87's brilliant core. Astronomers watched the blob for several years and saw it brighten steadily, then fade, then brighten again. The flare-up may provide insights into the variability of black hole jets in distant galaxies, which are difficult to study because they are too far away.

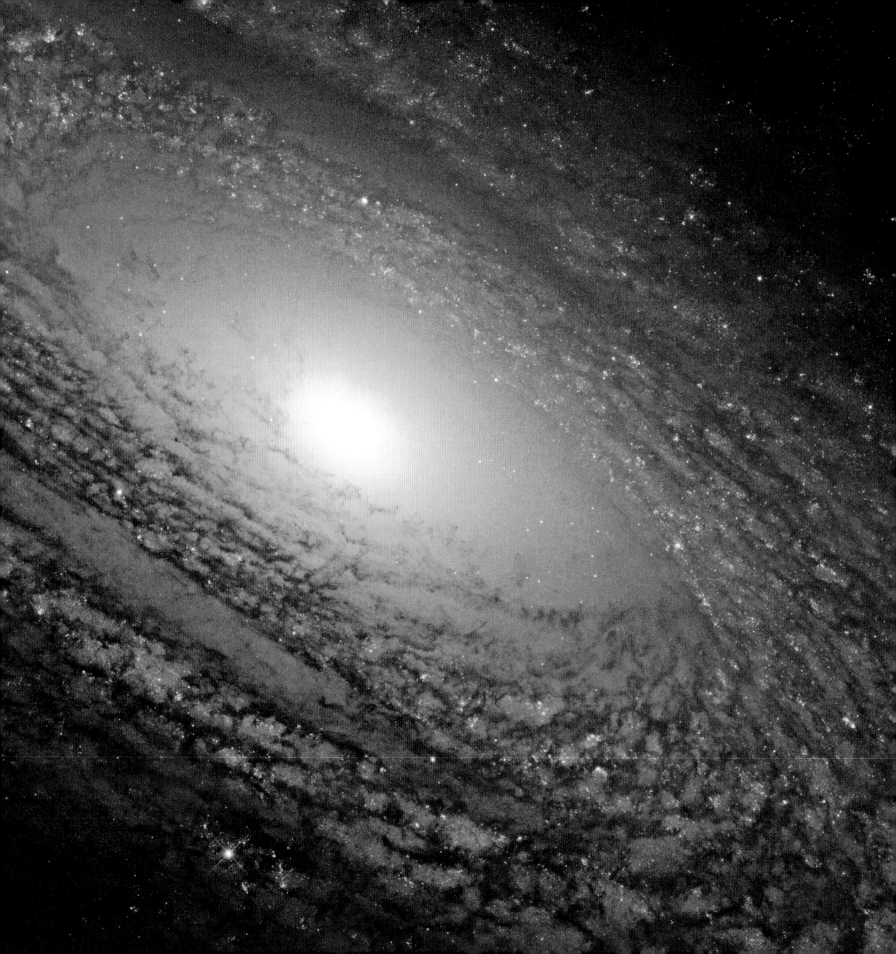

# CRUCIBLES OF CREATION

Our galaxy's industry is making stars. If we could view our galaxy from high above the pancake-shaped stellar disk, it would resemble a sprawling city with a bright downtown hub, burgeoning suburbs of Sun-like stars and avenues of young blue stars and nebulas. Interspersed are the raw materials for making stars: huge clouds of cold hydrogen gas laced with dust.

Bright clouds of gases within galaxies, nebulas are the sites of star formation. Think of them as the lights on a Christmas tree, and the branches of the tree are the skeletal dust structure of a galaxy. Nebulas are firestorms of star birth, an ongoing feature of any spiral galaxy.

But nebulas are just the tip of the iceberg. They are bubbles on the edge of dark, giant molecular clouds that store most of the star-forming mass of our galaxy. The hydrogen in these clouds is so cold that its atoms bond together in pairs (molecules). Hidden from the destructive influence of hot radiation from young stars and chilled to nearly absolute zero, dust and hydrogen can collect. Giant molecular clouds can be hundreds of light-years across and are 100,000 times denser than the normal abyss of space between stars.

Like a summer-afternoon thunderhead, dark molecular clouds are turbulent, clumpy and chaotic. Their movement is sluggish by earthly standards but appears dramatically speeded up in astronomers' supercomputer simulations that mimic the hydrodynamics of the clouds. Light-year-wide pockets buried deep inside these clouds fragment and condense "nuggets" of hydrogen that precipitate a flurry of star formation.

Each construction zone yields a batch of at least 10,000 solar-type stars. The energy of newborn stars eats out gaseous caverns inside molecular clouds. Imagine a block of Swiss cheese in which the holes are the stellar birthing bubbles and the cheese is the cold, comparatively dense interstellar gas. At the edge of a molecular cloud, the bubbles burst out as glowing "blisters," the most famous of which is the Orion Nebula. Despite this prolific star formation, molecular clouds convert only about 10 percent of their gas into stars, leaving vast fuel reserves for future star manufacturing.

Most of the hydrogen gas buried inside a molecular cloud is laced with trace gases, such as oxygen and nitrogen, and fine dust grains of carbon and silicon, which are smaller than smoke particles. This is building material for both stars and the planets that orbit them.

Cycles of star birth and star death seed the galaxy, as one generation of stars leads to another. In this stellar recycling, elements are drawn in to stars, further enriched to make heavier elements, then blown back into space for future generations of stars. Elements such as carbon, nitrogen, sulfur and oxygen come from earlier-generation solar-type stars. Strontium, zirconium and barium are made in aging red giant stars, while iron, gold and uranium are forged in supernovas. The fraction of heavier elements in stars increases the later a star forms in the universe. Our Sun, as measured by its heavy-element content, is a second- or third-generation star.

Above: Masquerading as a flamboyant piece of modern art, a star-forming region within the Carina Nebula is shown in this enhanced-color image, seen in full on pages 104-105.

Facing page: The outer arms of spiral galaxy NGC2841 are especially rich in gas and dust lanes. Young blue stars trace the spiral arms, while older yellow stars crowd the galaxy's hub.

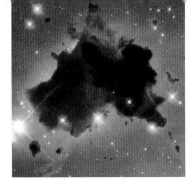

## CHAPTER 4

Above: One of the most stunning collections of Bok globules is called Thackeray's globules. South African astronomer David Thackeray discovered them in 1950, in the southern constellation Centaurus. The largest of the globules, shown here, is actually two separate clouds that gently overlap along our line of sight. Each cloud is nearly 1.4 light-years long and contains enough material to build more than a dozen stars like our Sun. When radio telescopes penetrate these dark islands in space, they reveal that the gas inside the globules is churning. Although a globule is very cold inside, the outside shell is heated by ultraviolet radiation. This temperature difference must be triggering rigorous convection, like oatmeal bubbling on the stove.

Floating within many star-birth nebulas are opaque, dark knots of gas and dust called Bok globules. They were first observed by astronomer Bart Bok in the 1940s. He compared them to insect cocoons. If these dense pockets capture enough gas and dust, they have the potential of creating stars in their cores. However, not all Bok globules will form stars; some will dissipate before they can collapse into stars. They remain some of the coldest objects in the universe.

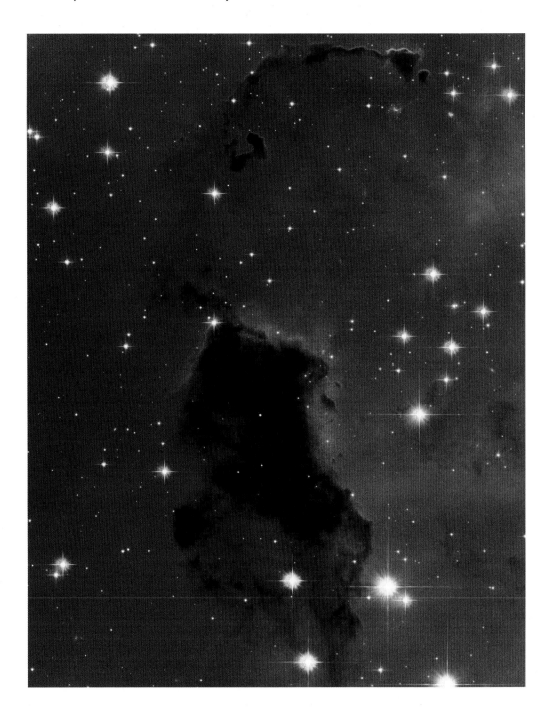

These cosmic columns look strangely organic, which is probably the reason this picture caused such a sensation when it was released in 1995. Nothing like this celestial scene had been seen before. Because it was thought that stars were forming inside the towers, they were dubbed the "Pillars of Creation." These structures were seen with ground-based telescopes long before Hubble's iconic snapshots, but Hubble brought an unprecedented level of detail. It's like knowing the general contours of a continent, then suddenly seeing bays, inlets and coves on closer inspection.

# ORION NEBULA

One of the nearest molecular clouds to the Sun spreads from Orion's belt to his sword. The entire cloud is 1,600 light-years away and hundreds of light-years across. Parts of the nebula can be observed with binoculars and small telescopes. The largest bright patch, the Orion Nebula, is visible to the unaided eye as a small, hazy cloud. One of the first nebulas noticed by skywatchers, it can be found in the winter constellation Orion, just below the belt of the mythological hunter. This interstellar cave is ballooning under the relentless pressure of starlight. A monster star, theta Orionis, is doing all the heavy lifting. Theta is actually a quartet of stars—the four dazzling stars at center that were born together at the heart of the Orion Nebula. Hubble photography and spectroscopy have been used to create a detailed three-dimensional model of the Orion Nebula, which shows that its apparent flatness in images is deceptive. We are really looking into a grotto-shaped cloud of glowing gases. A hint of that can be gleaned from the image of the full Orion Nebula on page 89.

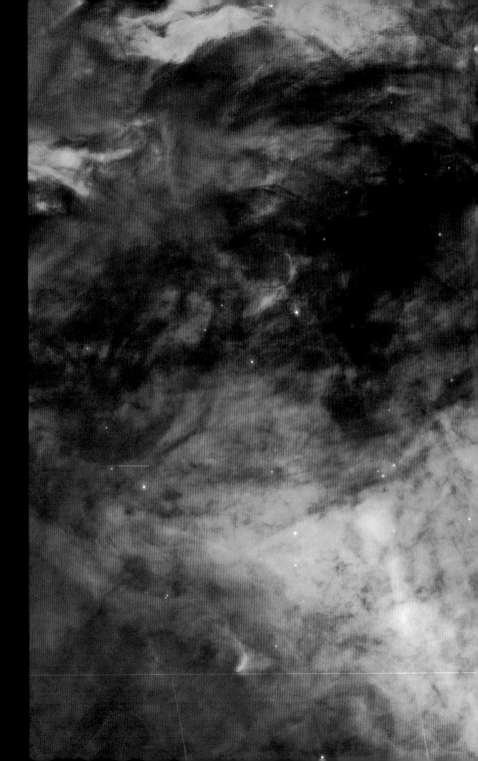

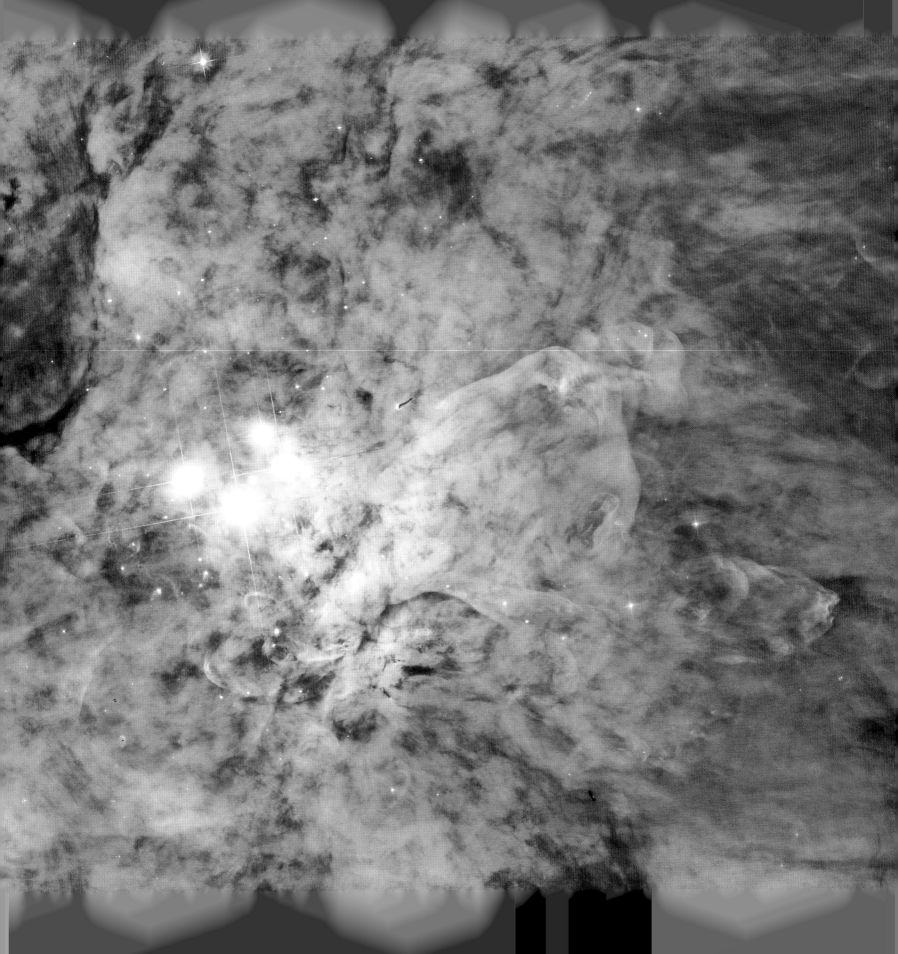

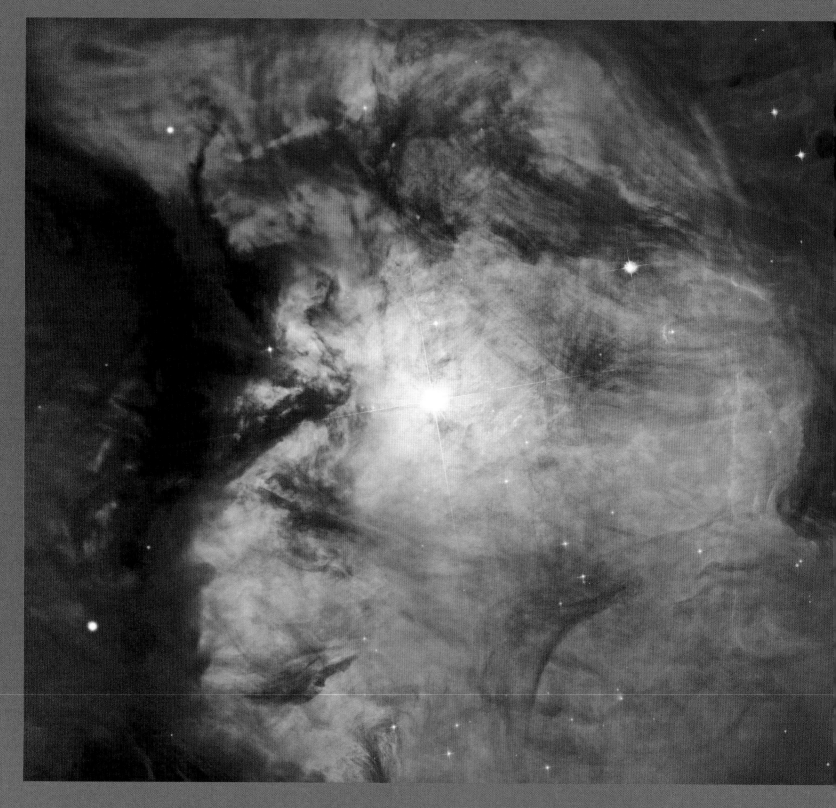

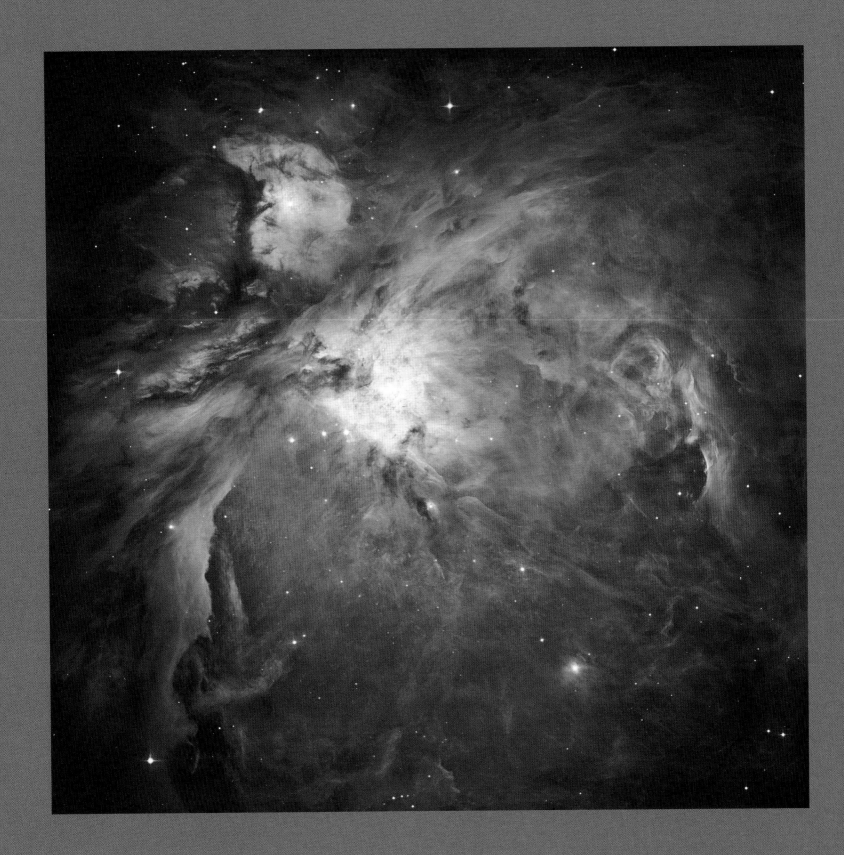

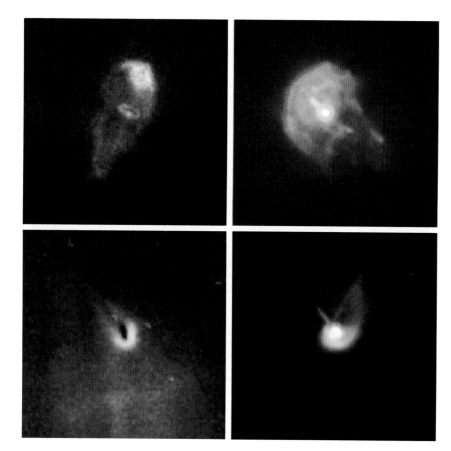

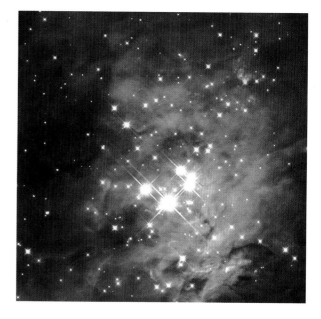

## TRAPEZIUM CLUSTER

The cluster of four stars at the very center of the Orion Nebula is called the Trapezium (it can be seen in a more expansive close-up on pages 86-87). Each of these stars is much hotter and more massive than our Sun, and collectively, they provide enough radiation to make the entire nebula glow. The tadpole-looking objects near the stars are known as proplyds, which are being eroded by powerful stellar winds that are sweeping their protoplanetary cocoon off into space.

## REALM OF THE PROPLYDS

More than 3,000 stars have been detected in the Orion Nebula in a Hubble photographic survey. Hubble's ultrasharp images reveal small, black, circular or elliptical shapes silhouetted against the glowing drapery. At first, astronomers thought these were the tips of towers of cold gas pointing in to the center of the nebula. But some of the objects have the appearance of edge-on Frisbees. They clearly are giant pancakelike disks of dust encircling young stars and represent a new class of astronomical object. Robert O'Dell, their discoverer, dubbed them "proplyds," for protoplanetary disks. As matter at the center of the spinning dust disk heats up, it becomes a new star, while remnants around the outskirts of the disk continue to attract dust. Radiation from the Trapezium cluster impinges on the proplyds, producing a crescent of heated material around the brighter disks. Some disks appear edge-on, while others are tilted face-on in our direction. Some show emerging jets of matter and shock waves that form when stellar winds from the central massive stars collide with the gas. The Hubble images confirmed two centuries of speculation, dating back to Immanuel Kant, that circular disks of dust and gas surround newborn stars.

HH502

This view into a small patch of the Orion Nebula shows the region around a young stellar object known as Herbig-Haro 502 (just left of center). The narrow jet extending to the upper right from the star and the curved bow-shock features to the lower left are caused by the pressure of radiation emerging from a massive star just out of the frame at lower left.

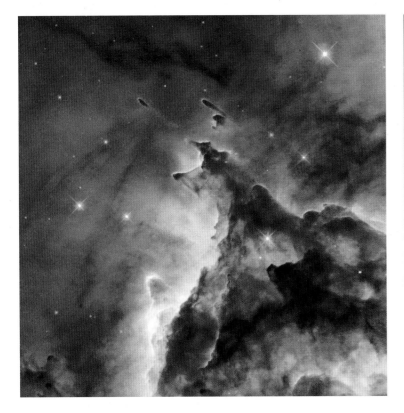

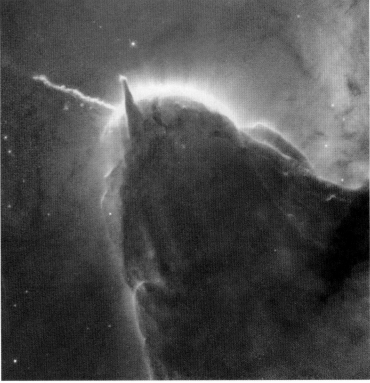

### NGC2174

A violent stellar nursery, NGC2174 is part of the Orion complex. The wall of a molecular cloud is being eaten away by intense radiation, making it glow brightly. The elephant-trunk features all point in the direction of a brilliant hot star, whose energy is sculpting the nebula.

### TRIFID JET

Resembling a creature from a horror movie, this Hubble image shows a small part of a dense cloud of dust and gas in the Trifid Nebula, a stellar nursery located roughly 9,000 light-years from Earth. The stellar jet protruding from the head of the cloud is about 0.75 light-year long, and its source is an embryonic star deep within the cloud. Jets are the exhaust gases of star formation. Radiation from the nebula's central star causes the jet to glow. The vertical fingerlike stalk to the right of the jet points directly toward the star that powers the Trifid Nebula. A prominent example of an evaporating gaseous globule, this stalk has survived because its tip is a knot of gas dense enough to resist being eroded by the star's radiation.

### NGC281

Dubbed the Pac-Man Nebula, this wide-field view of the star-forming region NGC281 was photographed with a 0.9-meter telescope at Kitt Peak National Observatory, in Arizona. As is typical of star-birth regions, NGC281 has large zones of obscuring gas and dust in which stars may still be forming. To investigate further, Hubble zoomed in to take a close-up look at the young open cluster of stars IC1590 deep inside the nebula. See the resulting image on page 99.

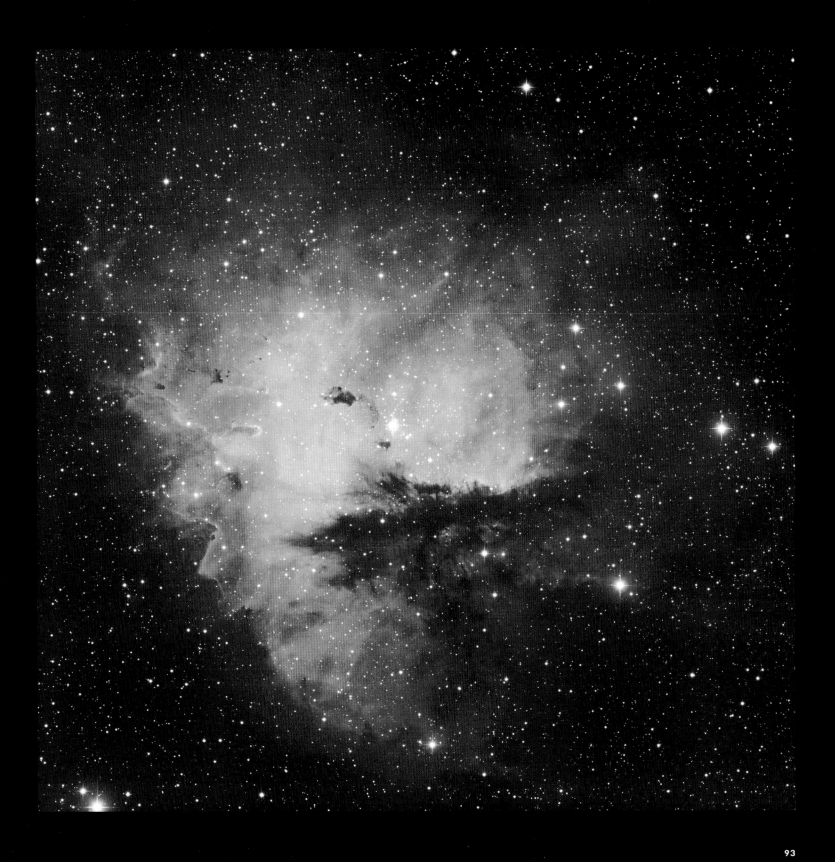

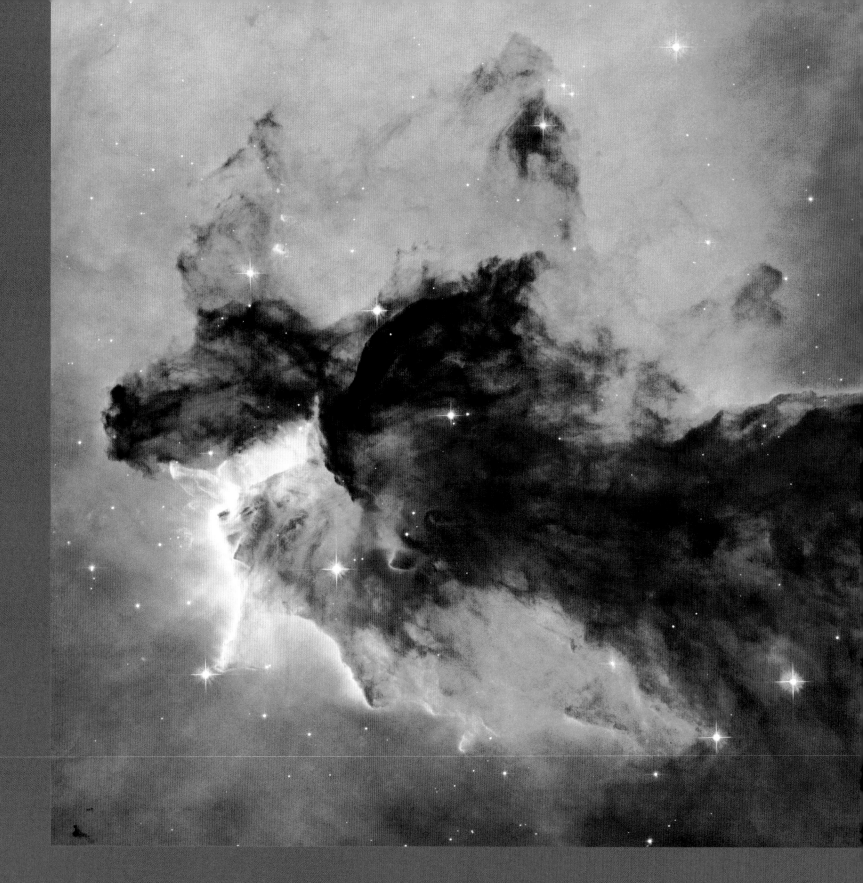

## STELLAR SPIRE

This vast, elongated branch of dust and gas is 9.5 light-years long, about twice the distance from our Sun to the next nearest star. Ghostly streamers of gas boiling off the surface create a haze around the structure, giving it a three-dimensional quality. The column is silhouetted against a background of more distant gas.

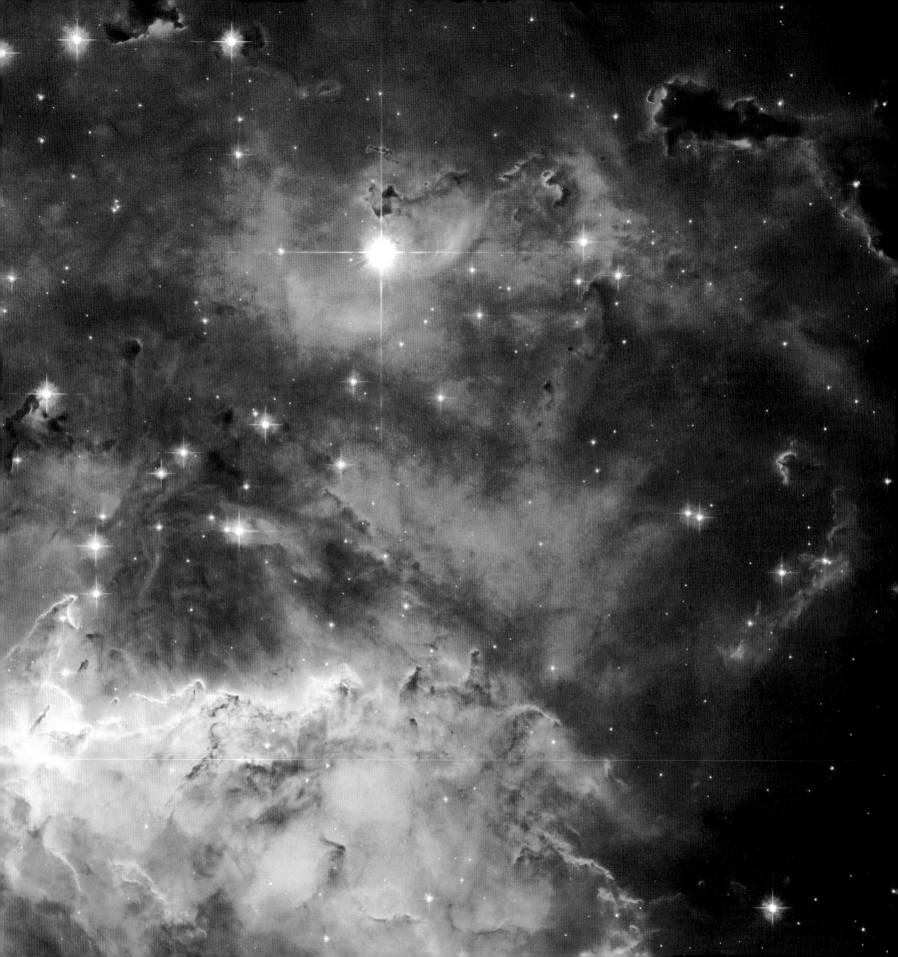

## M17

Hydrogen, oxygen and sulfur gases simmer and glow on the edge of an extremely massive and luminous molecular cloud in this view of a small part of the star-birth nebula M17. Also known as the Omega or Swan Nebula, M17 is located about 5,500 light-years away, in the constellation Sagittarius. The wavelike patterns of gas are sculpted and illuminated by ultraviolet radiation from young massive stars out of the frame at the top of the image. The lighting brings into relief the three-dimensional structure of the gases. The intense heat and pressure create a veil of even hotter green-colored gas as material streams away from the surface.

## NGC2467

Similar to the Orion Nebula, the distant star-forming region NGC2467 (facing page) is 11 times farther away, in the southern constellation Puppis. A churning foam of strangely shaped dust clouds forms the backdrop to the newborn blue stars emerging from the gas and dust. Most of the radiation that is eating away at the cloud is being emitted by the single brilliant massive star near the center of the image. Its fierce radiation has cleared the surrounding area, and some of the next generation of stars are forming in the denser regions around the edge.

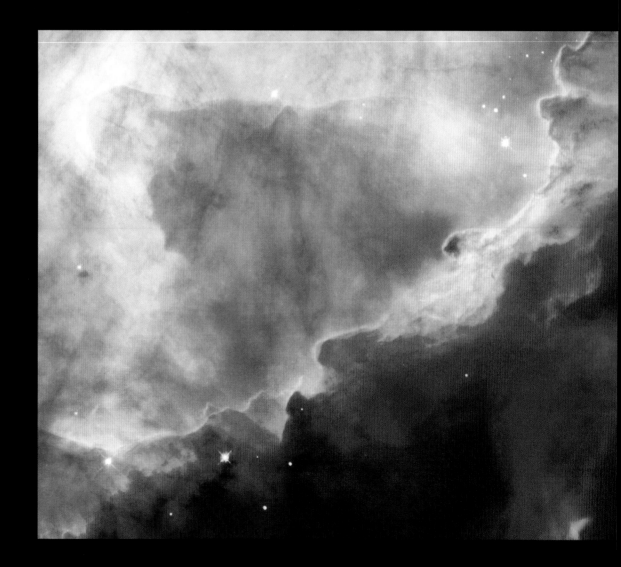

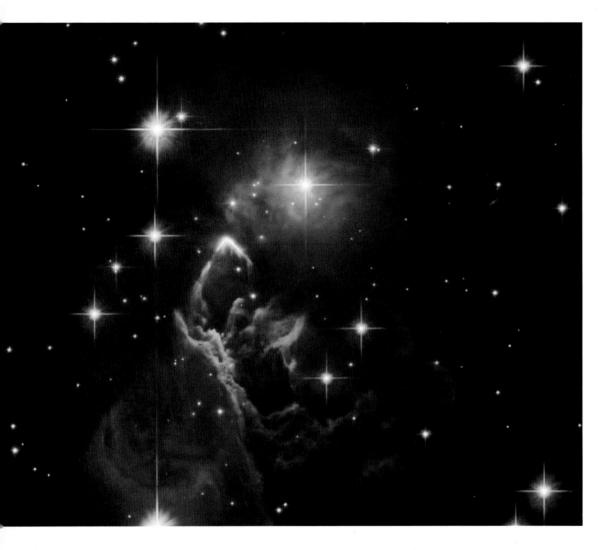

M8
Clouds of hydrogen form an abstract art texture in this view of a small section of Messier 8, the Lagoon Nebula. Radiation from a hot, young star is sculpting the surrounding nebula into pillow-cushion shapes.

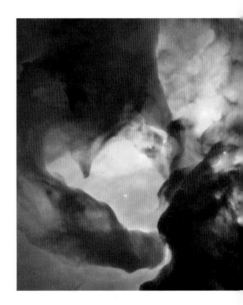

IRAS 05437+2502
This nebula in Taurus is one of the spookiest and most evocative targets ever seen by Hubble. It is not glowing but is reflecting light from neighboring stars clusters, which gives it a remarkably solid appearance, even though it is a near vacuum. It billows out among the bright stars and dark dust clouds that surround it. The faint cloud was originally discovered in 1983 by NASA's Infrared Astronomical Satellite (IRAS), the first space telescope to survey the whole sky in the infrared.

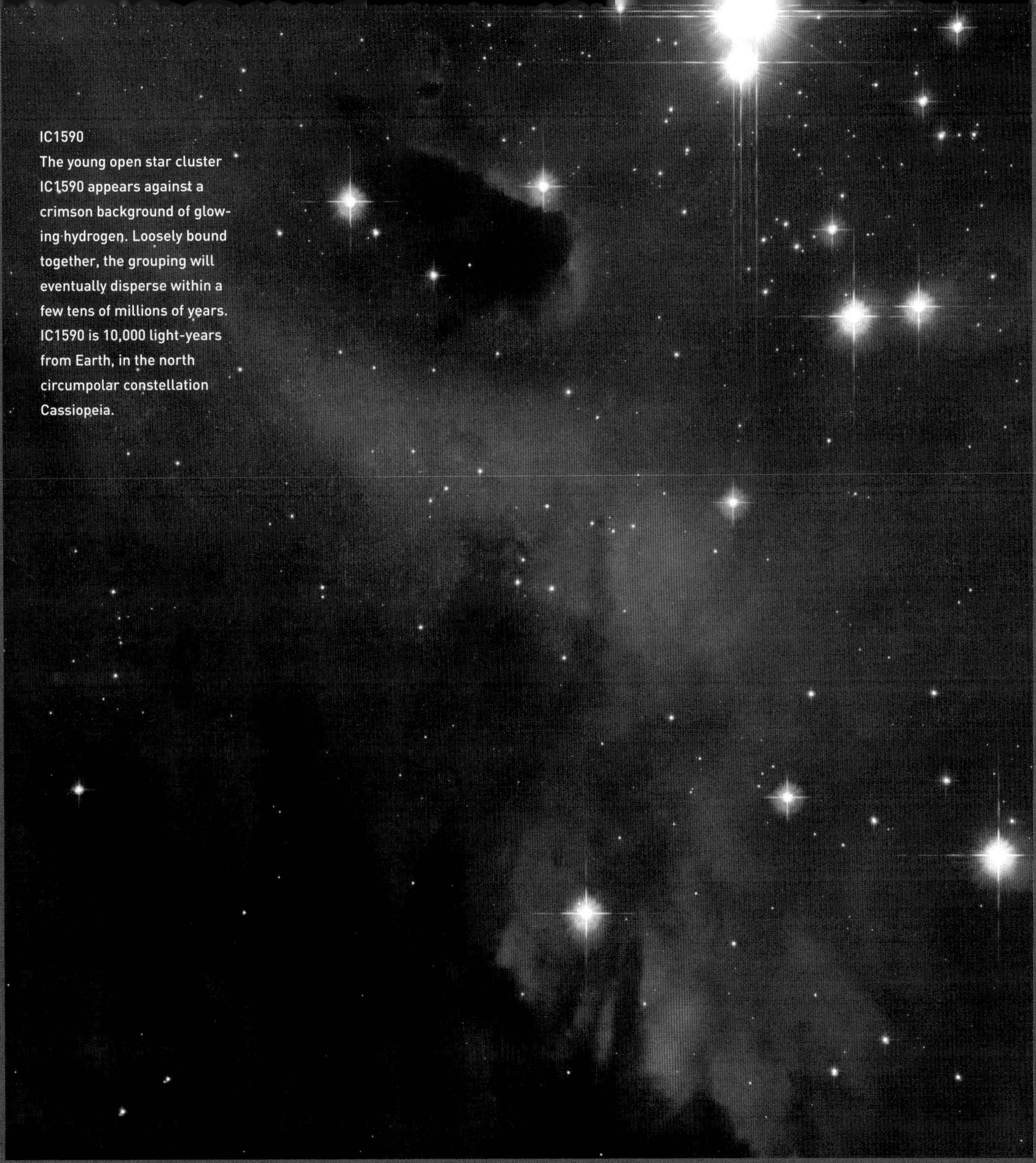

IC1590

The young open star cluster
IC1590 appears against a
crimson background of glowing hydrogen. Loosely bound
together, the grouping will
eventually disperse within a
few tens of millions of years.
IC1590 is 10,000 light-years
from Earth, in the north
circumpolar constellation
Cassiopeia.

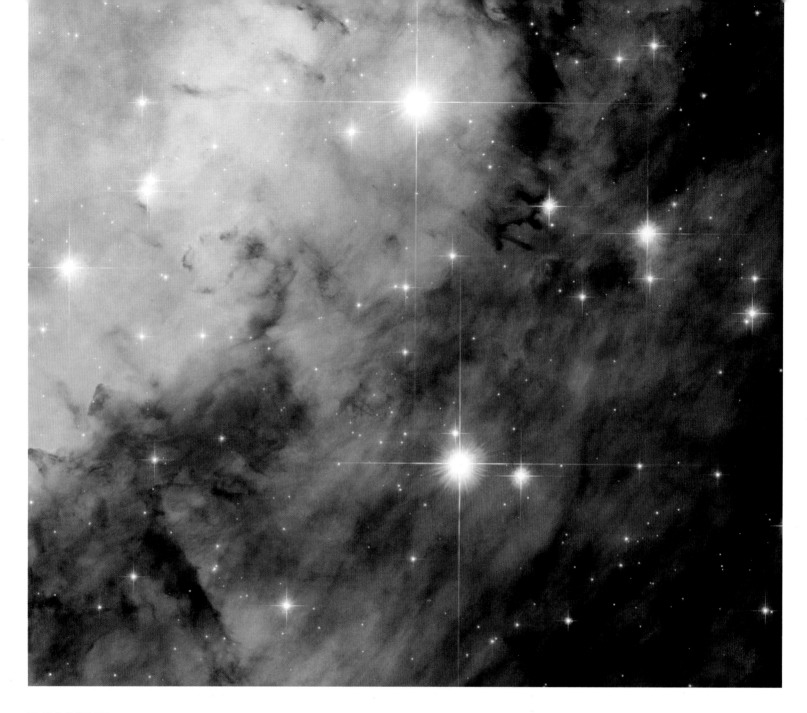

## EAGLE CLUSTER

This spectacular section of the Eagle Nebula is known as NGC6611, an open star cluster that formed five million years ago. It is a very young cluster that contains many hot blue stars, whose fierce ultraviolet radiation makes the surrounding Eagle Nebula glow brightly. The dark patches are comparatively dense regions of gas and dust that block background light like a curtain.

## JEWEL BOX

A torrent of radiation from young stars in NGC3603 sculpts the dark stalks of dense gas embedded in the walls of the nebula. The light-years-tall pillars all point to the central cluster as the source of their towering shapes. This prominent star-forming region is 20,000 light-years away.

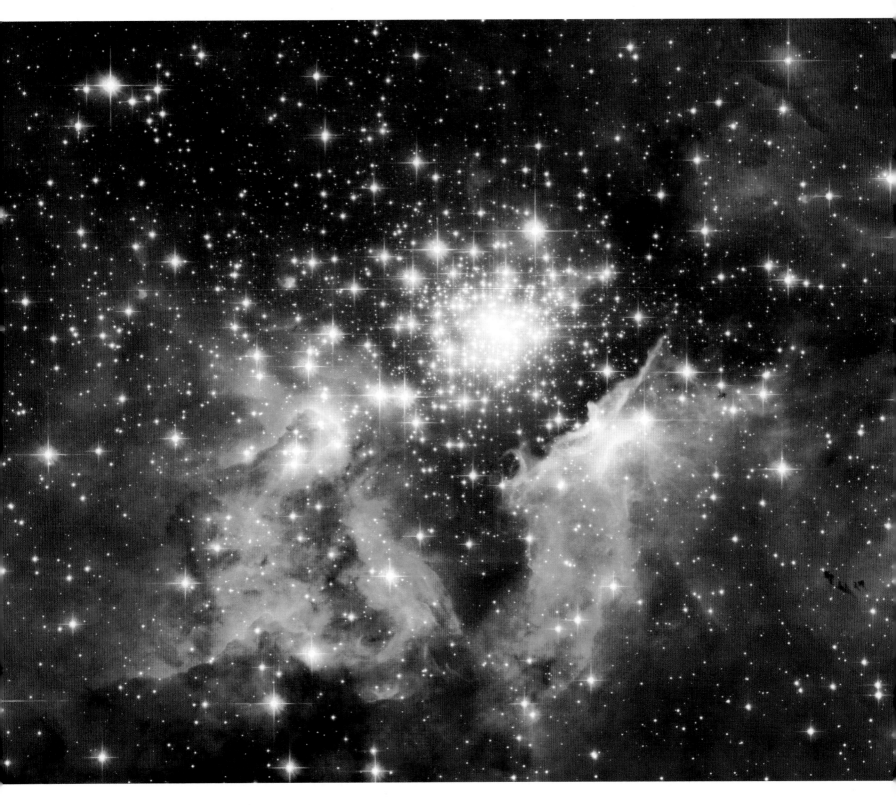

## NGC3324

This Hubble image shows the edge
of the giant gaseous cavity within
the star-forming region NGC3324,
in the constellation Carina. Despite
its substantial appearance, even
the densest part of the nebula is
millions of times less dense than
earthly clouds. NGC3324 has been
carved out by intense ultraviolet
radiation and stellar winds from a
cluster of extremely massive stars,
located outside this image.

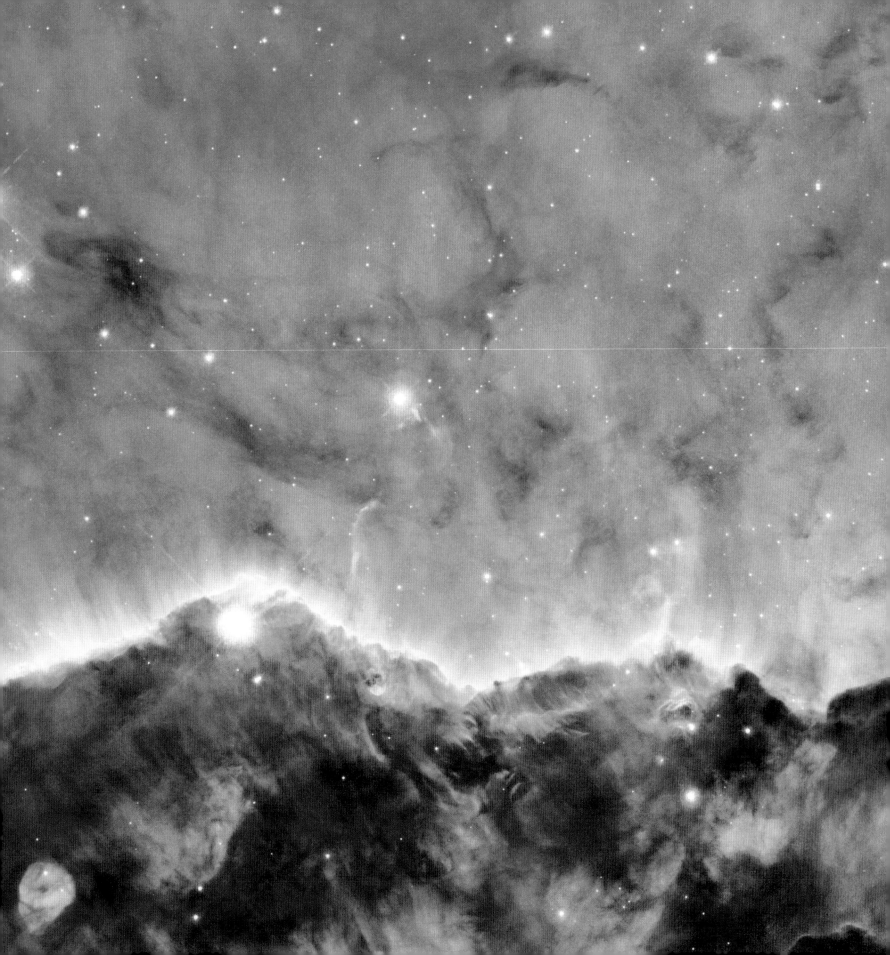

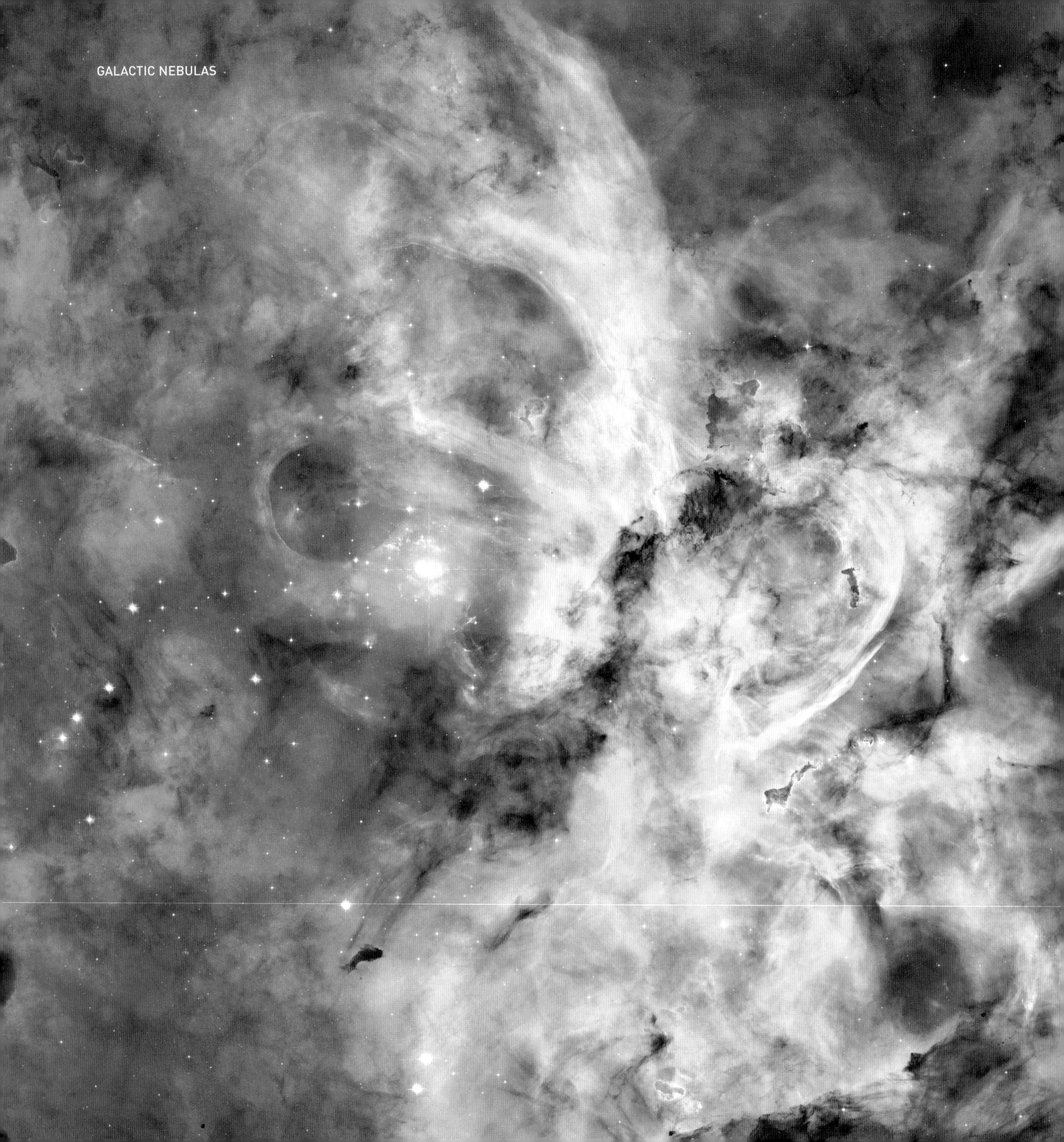

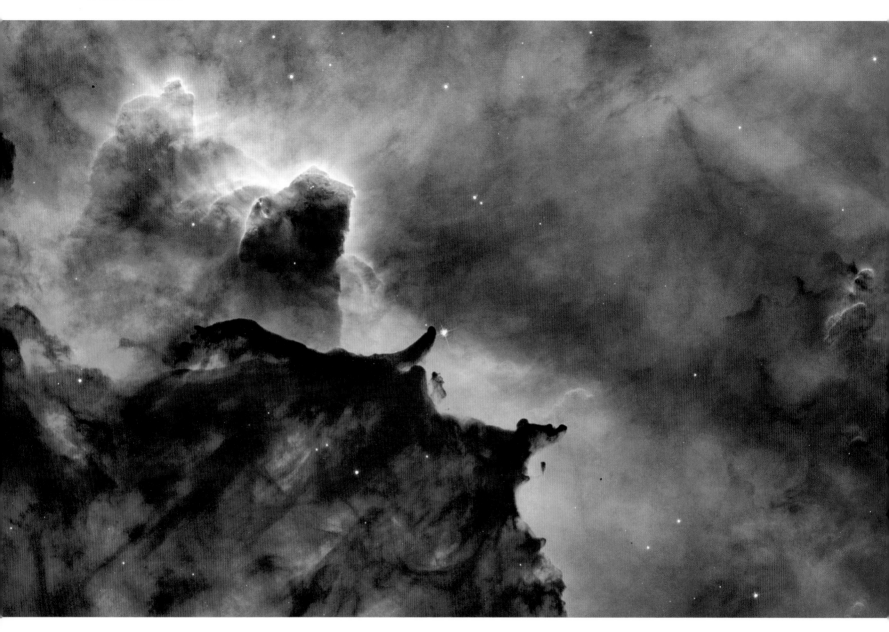

### INSIDE THE CARINA NEBULA

The fireworks in the Carina region started three million years ago, when the nebula's first generation of newborn stars condensed and ignited in the middle of a huge molecular cloud of cold hydrogen gas. The clumps of dark clouds seen close-up here are nodules of dust and gas that are resisting being eaten away by the radiation from bright stars. They look like little islands in the middle of a vast sea of rarefied hot plasma.

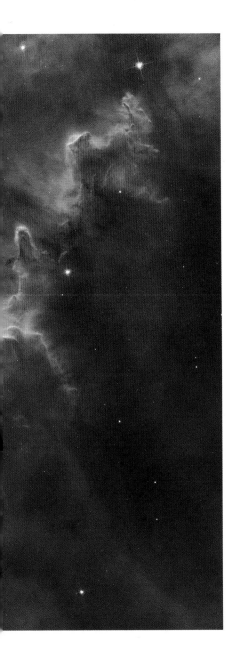

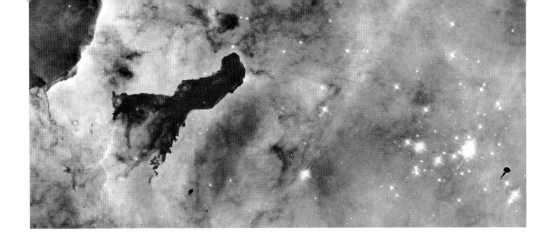

## LOCH NESS

Dubbed Loch Ness, the dark nebula at left looks like the neck and head of the legendary sea monster. The hurricane blast of stellar winds and blistering ultraviolet radiation within the Carina Nebula are compressing the surrounding walls of cold hydrogen gas, seen as the background in this image. The scale here is immense: The length of the "head" of the Loch Ness dark cloud, for example, is about 30,000 times the distance from Earth to the Sun.

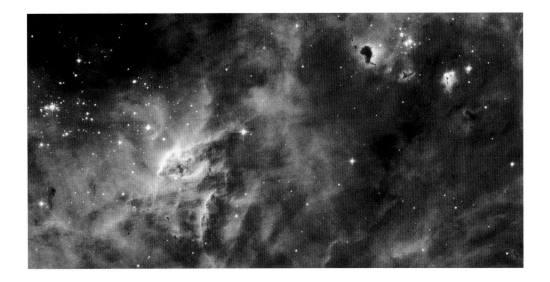

## SEQUENTIAL STAR FORMATION

This Hubble image illustrates sequential star formation, where new star birth is being triggered by previous generations of massive stars. A collection of hot blue-white stars appears near the left side of the image. The region around the cluster of hot stars is relatively clear of gas, because stellar winds and radiation from the stars have pushed back the gas. When the gas collides with surrounding dense clouds, it compresses them, and they collapse under their own gravity and start to form new stars.

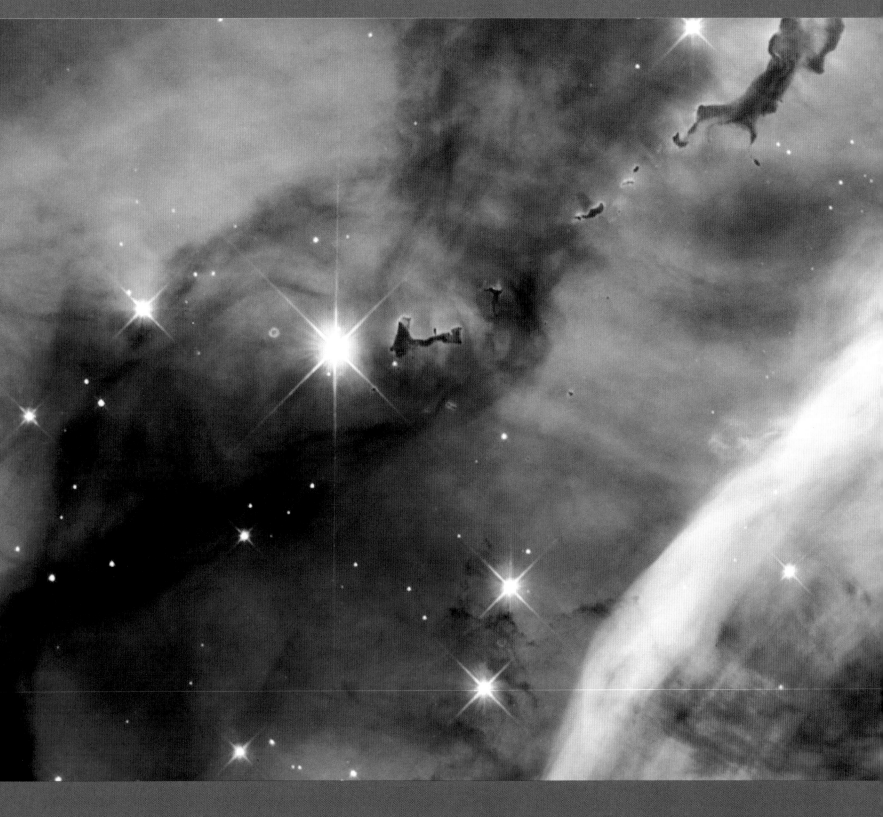

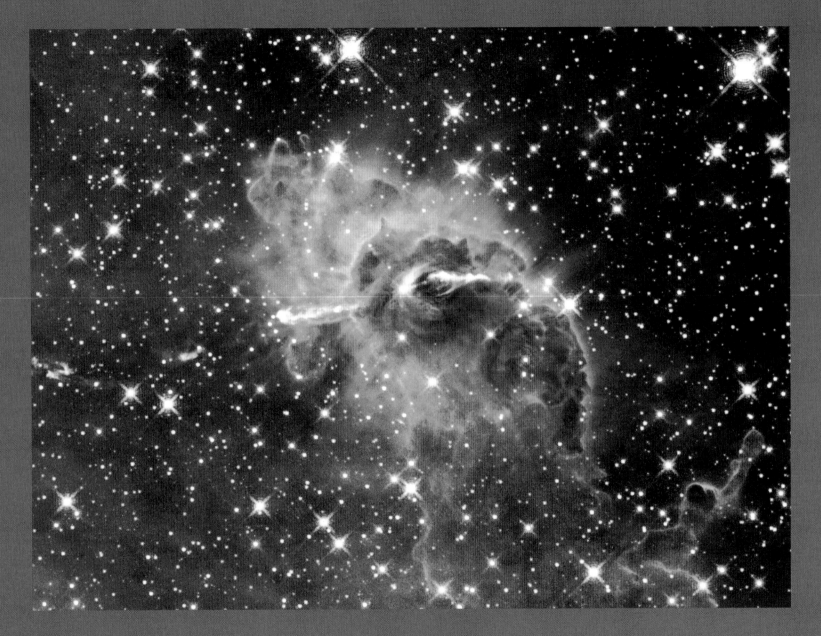

ELEPHANT TRUNK

This giant column of gas and dust in the Carina Nebula is the tip of a three-light-year-long pillar being bathed in the fierce radiation from hot, massive stars. The radiation is sculpting the pillar and causing new stars to form within it. Streamers of gas and dust can be seen flowing from the top of the structure.

CARINA GLOBULES

Dark globules nestled among brilliant young stars in the foreground of the Carina Nebula give this image a stunning three-dimensional look. A twisted lane of dust is backlit by glowing luminescent gas.

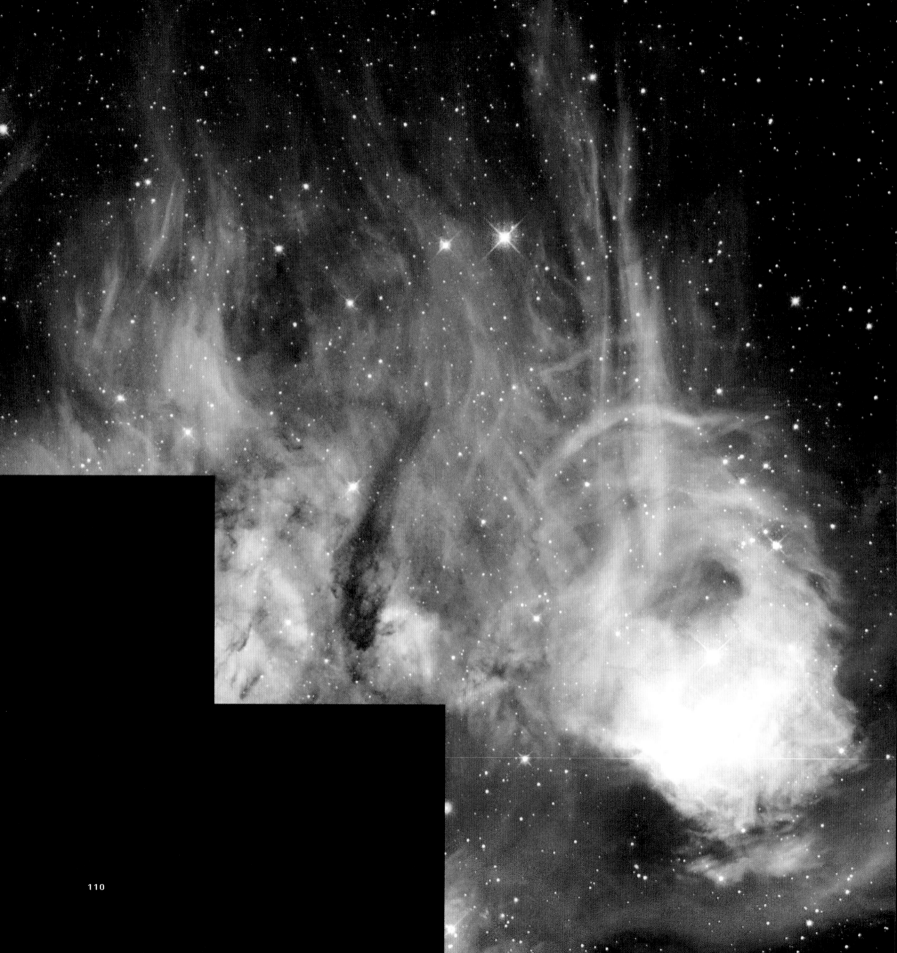

# STAR BIRTH IN SATELLITE GALAXIES

The nearest giant star factories do not lie in the Milky Way itself. Instead, they are 170,000 and 250,000 light-years away, in the Milky Way Galaxy's satellite galaxies: the Large Magellanic Cloud (LMC) and the Small Magellanic Cloud (SMC). First described by Ferdinand Magellan in 1519, during his around-the-world voyage, these star-forming regions in the southern sky are easily visible to the unaided eye, looking like fuzzy patches adrift in the cosmos. The brightest nebula in the LMC is the Tarantula Nebula, so called because of its ghostly spidery appearance in the eyepiece of a telescope (the center of the Tarantula Nebula is shown on pages 116 to 121).

Facing page: These wild-looking streamers of glowing gas are part of N44C, a nebula surrounding a grouping of young stars in the Large Magellanic Cloud. The source of energy for illuminating the nebula is even hotter than a massive star—about 75,000 degrees C. This unusually high temperature may be produced by a neutron star or black hole spewing out energetic X-rays.

Below: Over 10,000 stars can be seen in a region of the Large Magellanic Cloud that is about 130 light-years wide. Curtains of glowing gas and dark patches of interstellar dust are silhouetted against the stars and gas.

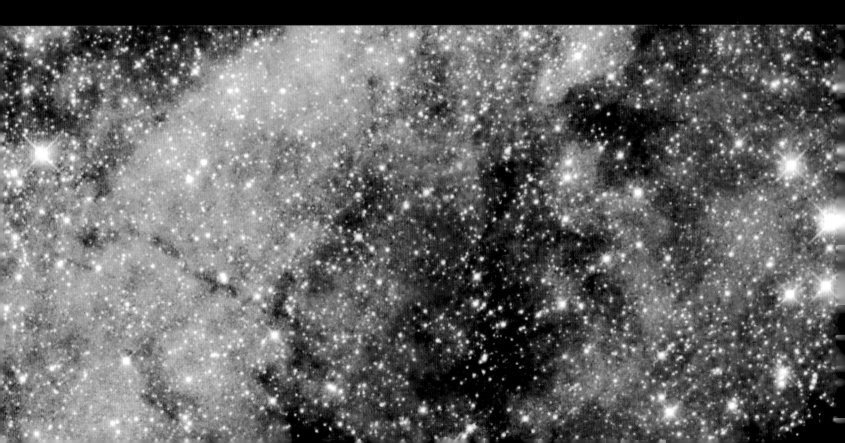

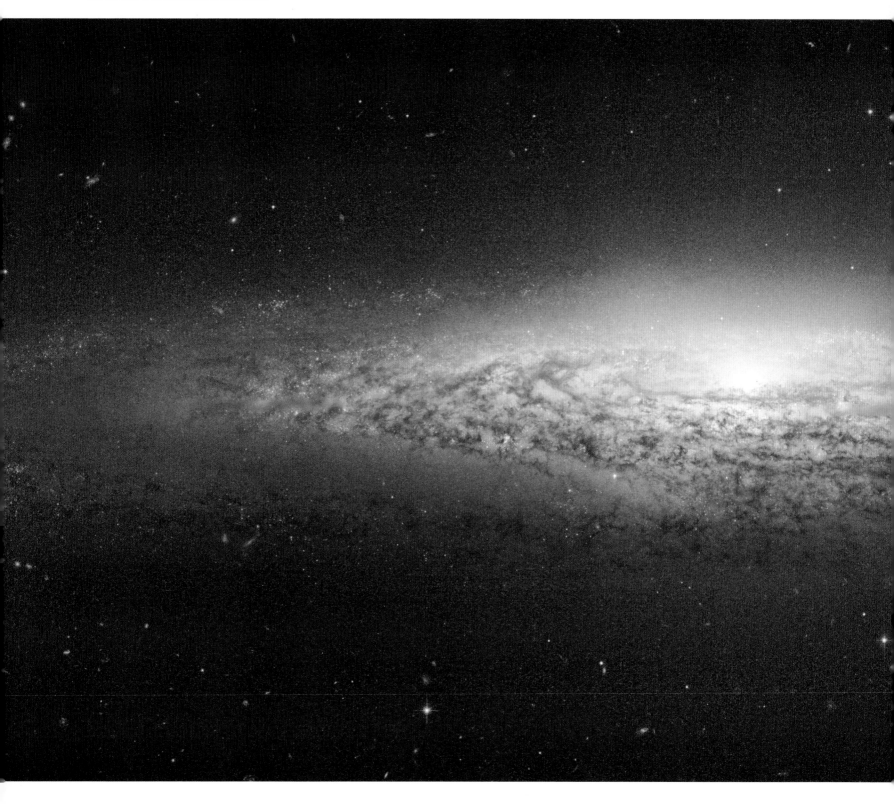

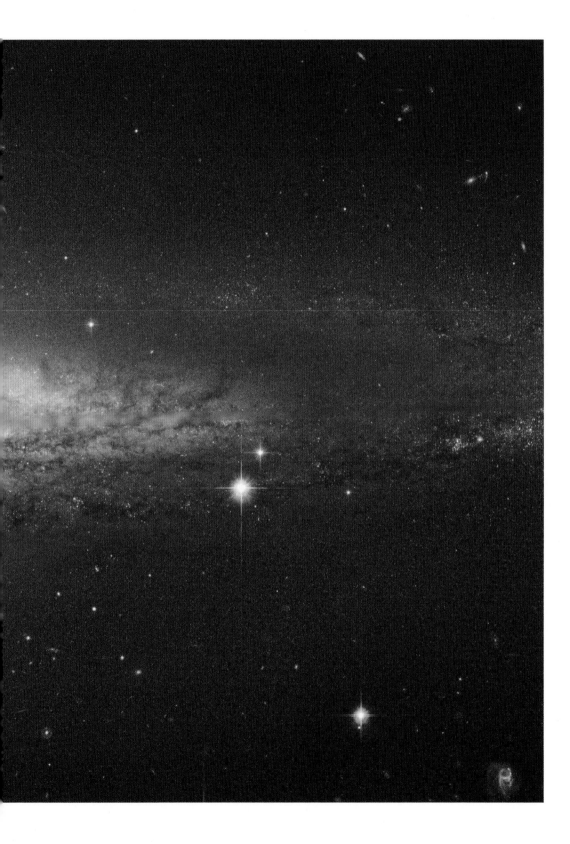

NGC2683

While a face-on view of a spiral galaxy allows us to see detailed structure, a side view, as shown here in NGC2683, reveals the delicate dusty lanes of the spiral arms overlying the golden haze of the galaxy's core. In addition, brilliant clusters of young blue stars scattered throughout the disk map the galaxy's star-forming regions. Two adjacent fields observed in visible and infrared light by Hubble's Advanced Camera for Surveys were used to produce this image. A narrow strip that appears slightly blurred to the right of center has been patched using images from ground-based telescopes.

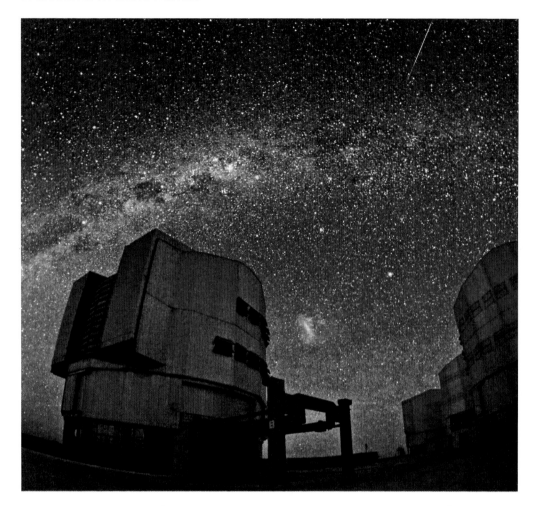

## OUR NEIGHBOR GALAXY

A maelstrom of star birth 10 times the rate of star creation in the Milky Way Galaxy is taking place in the Large Magellanic Cloud (LMC). Although only 5 percent as massive as the Milky Way, the LMC contains the largest and most active star-forming nebula known in our sector of the universe: the mighty Tarantula Nebula. The Tarantula, the brightest and largest pinkish patch on the upper-left side of the LMC in this image, is just visible to the naked eye as a hazy smudge. The LMC is one of many treasures of the southern night sky not visible to observers north of the latitude of Mexico City.

## NIGHT AT THE OBSERVATORY

The middle of Chile's Atacama Desert was selected as the site for the European Southern Observatory's Very Large Telescope because it is astronomy nirvana: the driest, clearest site on Earth for scanning the universe. Four massive telescopes are used nightly for astronomical research, often in collaboration with the Hubble Space Telescope. Above the telescopes in this image, the starry Milky Way shines undiminished by the glow of city lights or the fog of atmospheric dust and pollution. Only the Hubble has a better view. What looks like a detached fragment of the Milky Way just below center is the Large Magellanic Cloud, a satellite galaxy of our home galaxy, shown as a close-up on the facing page.

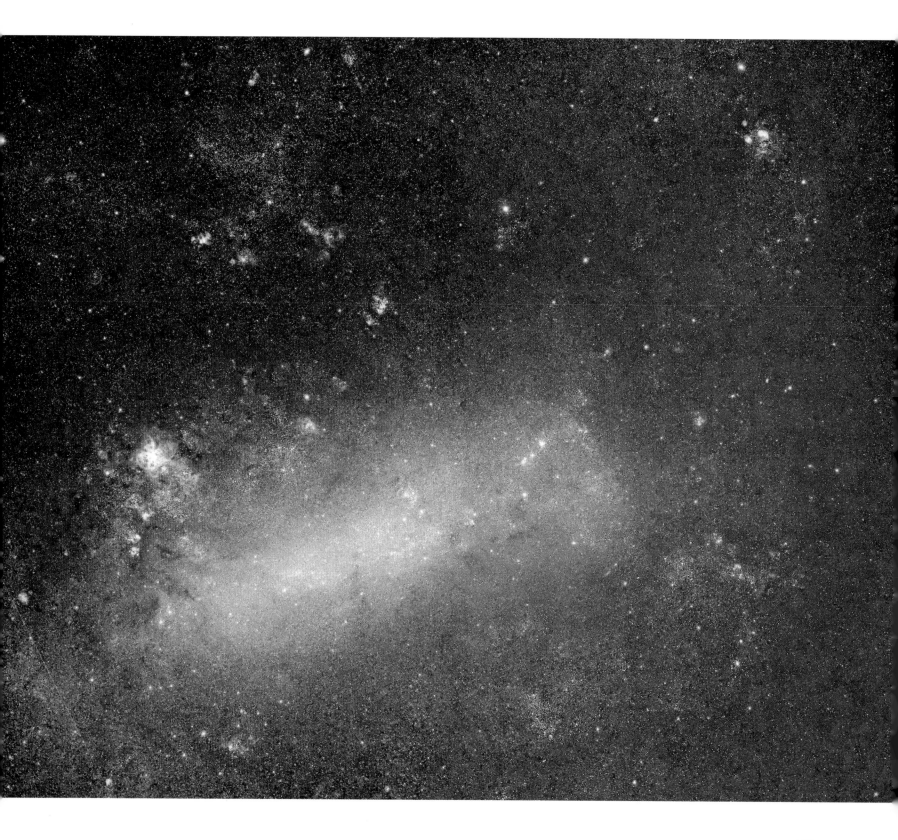

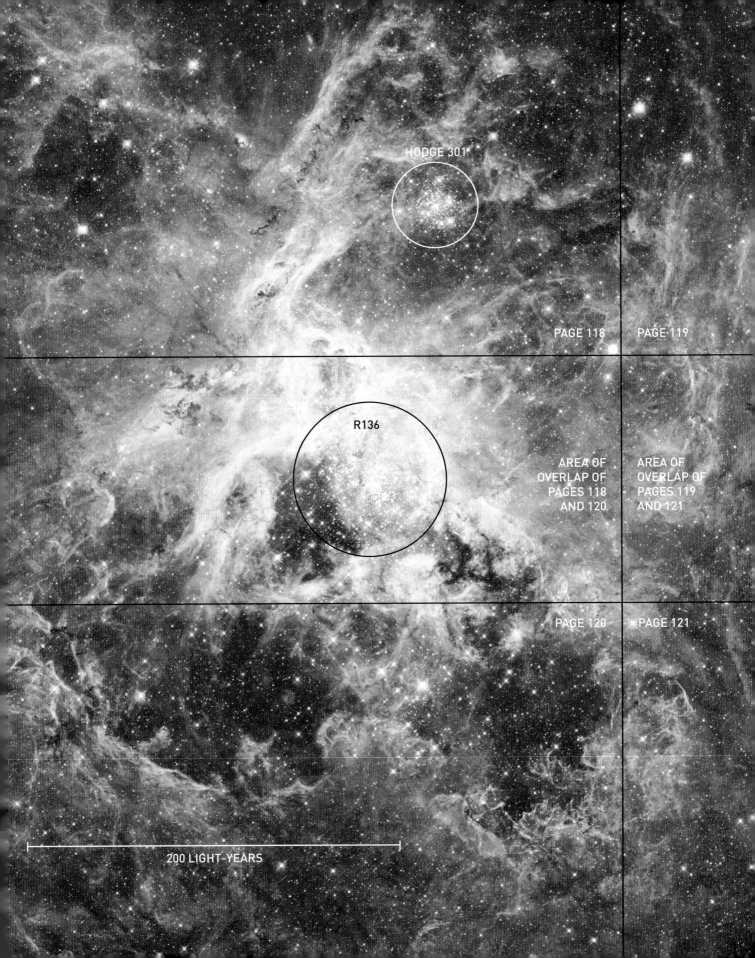

HODGE 301

R136

PAGE 118

PAGE 119

AREA OF
OVERLAP OF
PAGES 118
AND 120

AREA OF
OVERLAP OF
PAGES 119
AND 121

PAGE 120

PAGE 121

200 LIGHT-YEARS

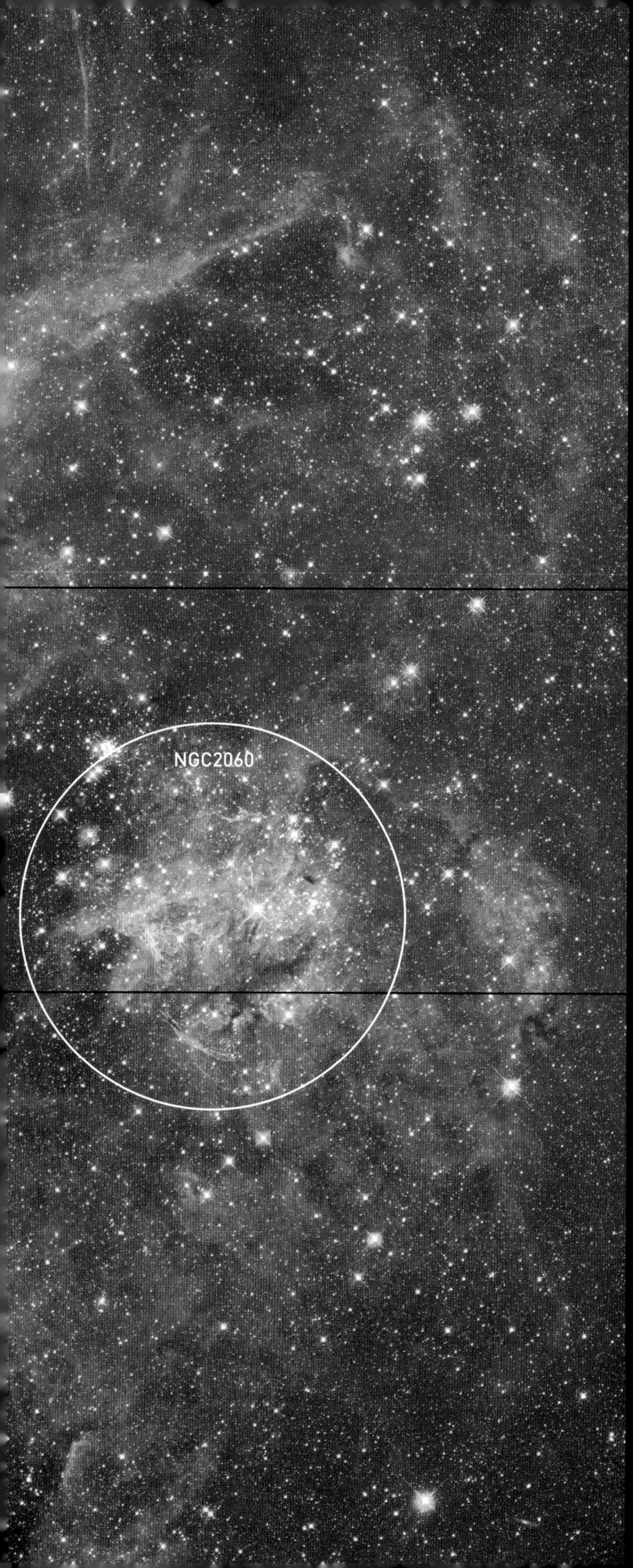

NGC2060

## STAR BIRTH IN SATELLITE GALAXIES

### 30 DORADUS

The biggest and most prolific star-forming region in or near our galaxy is 30 Doradus, which is embedded in the Tarantula Nebula. One of the largest digital mosaics ever compiled, this image comprises 15 photos taken by Hubble's Wide Field Camera 3 and 15 by the Advanced Camera for Surveys. It is shown in full-resolution detail over the following four pages. If the Tarantula Nebula were at the distance of the Orion Nebula, it would stretch from the horizon one-third of the way up to the overhead point in the night sky. Spanning roughly 650 light-years, this Hubble image reveals the stages of star birth, from embryonic stars a few thousand years old, still wrapped in cocoons of dark gas, to behemoths that die young in supernova explosions. The star clusters shown here range in age from roughly 2 million to 25 million years. 30 Doradus has been churning out stars at a furious pace for millions of years.

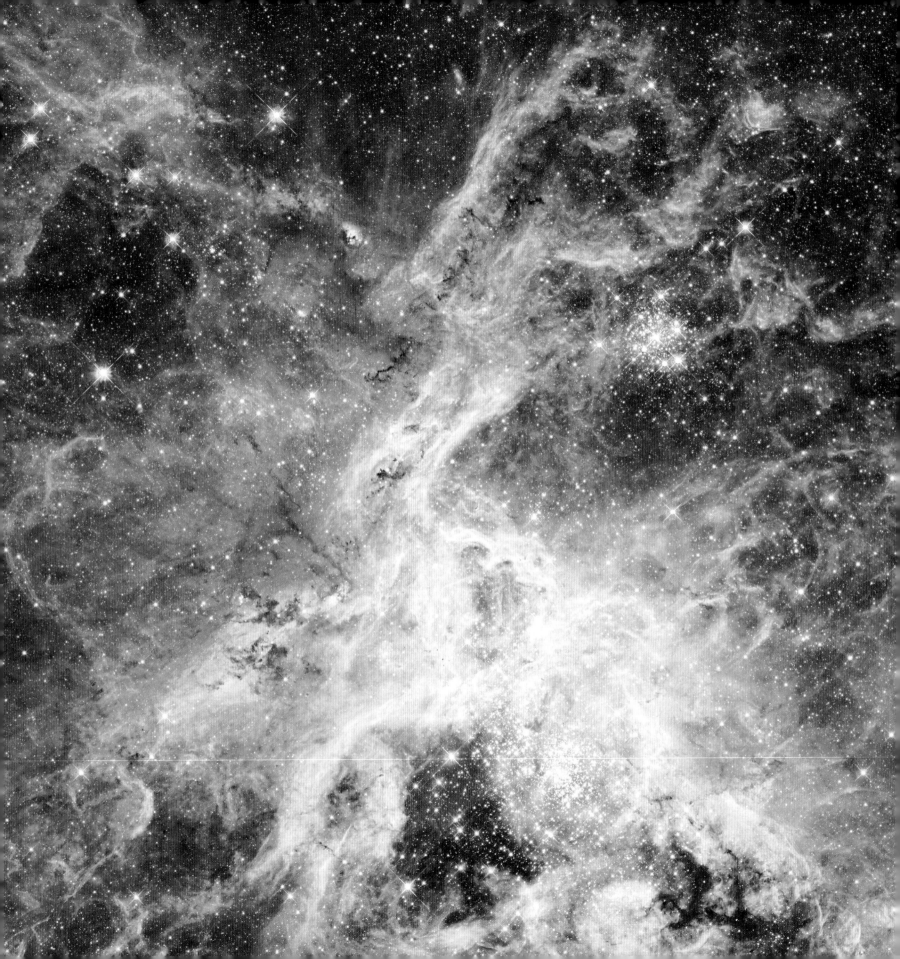

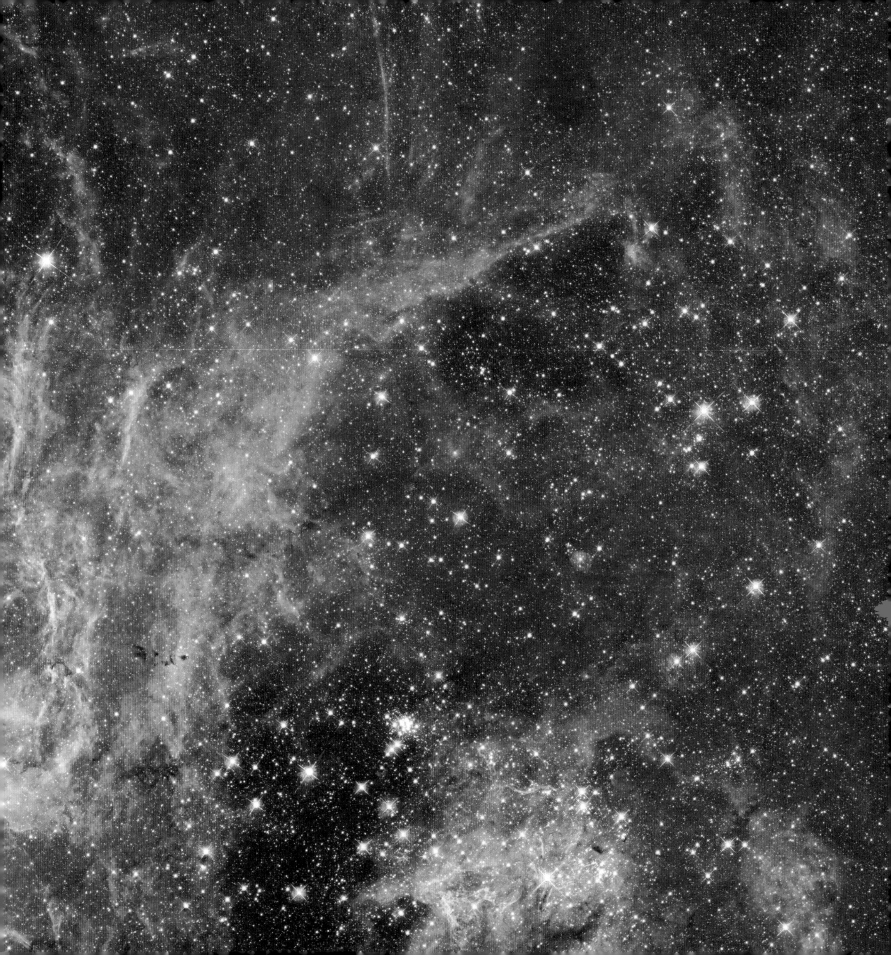

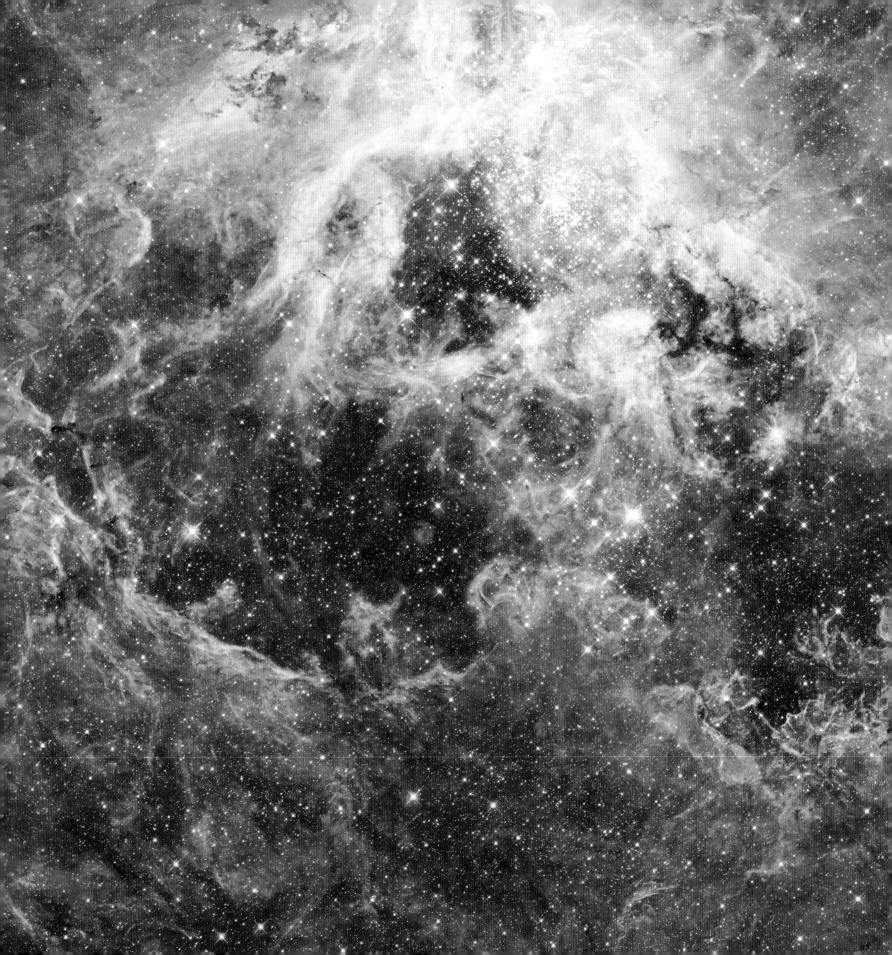

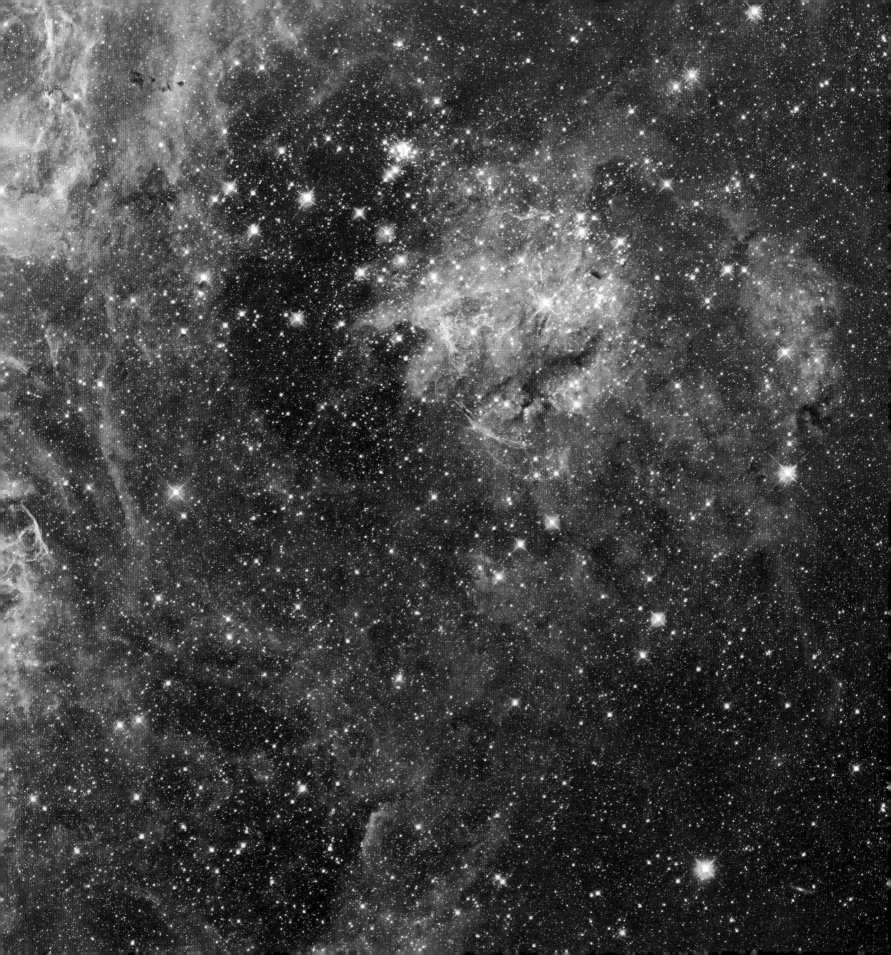

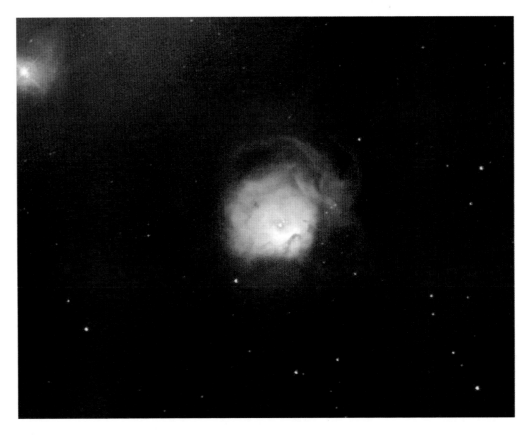

**N11A**

Resembling a delicate rose floating in space, N11A lies within a spectacular star-forming region in the Large Magellanic Cloud. It is the smallest and most compact nebula in that vicinity and represents the most recent formation of massive stars there. Shocks and strong stellar winds from the newborn massive stars in N11A's bright core have scooped out a cavity in the gas and dust.

**R136**

The brightest young star cluster in the Tarantula Nebula, R136 is only a few million years old and resides in one of the most turbulent star-birth regions in the Large Magellanic Cloud. Several of its stars are now known to be more than 100 times the mass of our Sun. These stars are destined to explode as supernovas in just a few million years. Several decades ago, astronomers were debating whether the source of the intense light at the cluster's core was a tightly bound group of stars or an unknown type of superstar over 1,000 times the Sun's mass. But the fine detail revealed by Hubble and powerful ground-based telescopes has conclusively shown that the source is an incredibly rich cluster of stars.

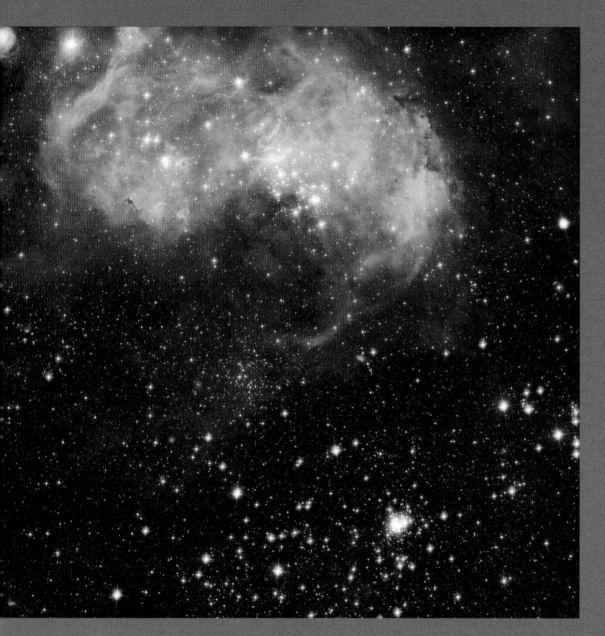

## NGC346

One of the most dynamic and intricately detailed star-forming regions in space is located inside the Small Magellanic Cloud (SMC). A striking arch cuts across NGC346, and ragged filaments give it a spinelike appearance. Several small dust globules point back toward the central cluster, like wind socks caught in a gale. At least three smaller clusters are embedded here that, combined, make up more than half of the known high-mass stars in the entire SMC. Myriad smaller, compact clusters are also visible. Some of these mini-clusters appear to be nestled in dust and nebulosity and are sites of recent or ongoing star formation. Much of the starlight from these clusters is reddened by dust-scattering blue light.

## NGC1760

This broad vista of young stars and gas clouds, in the center of the N11 star-forming complex in the Large Magellanic Cloud, is one of the most active star-formation regions seen near our galaxy.

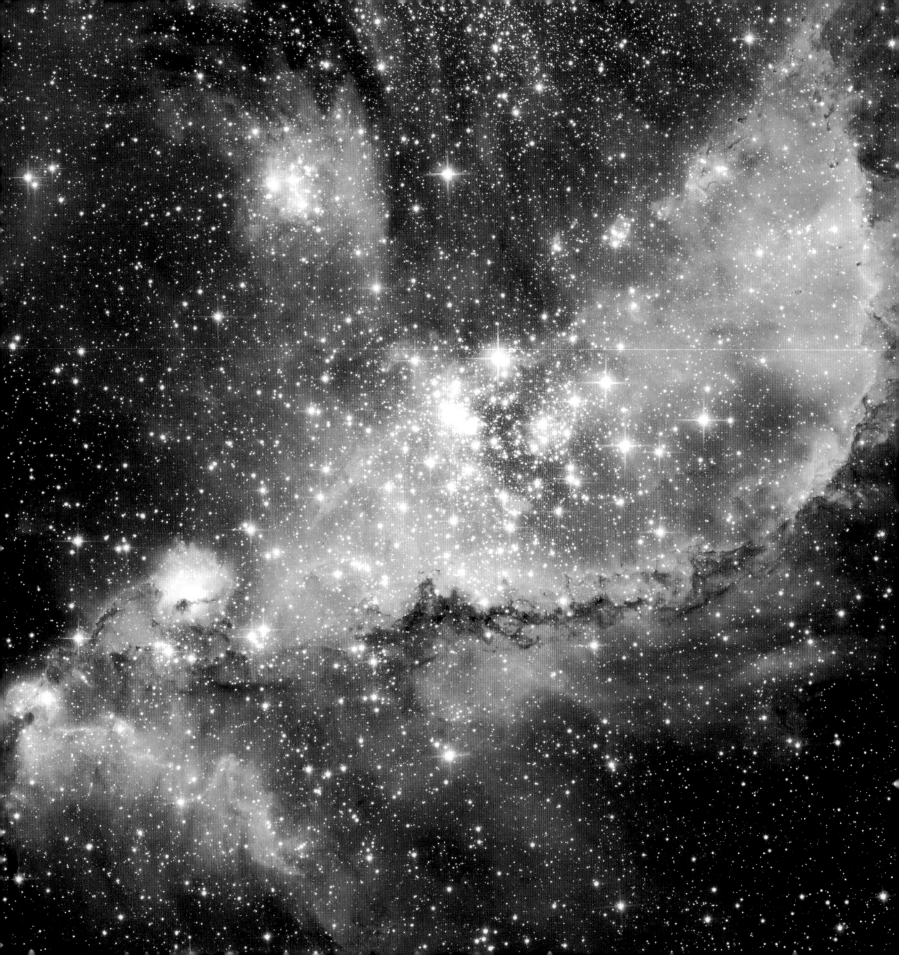

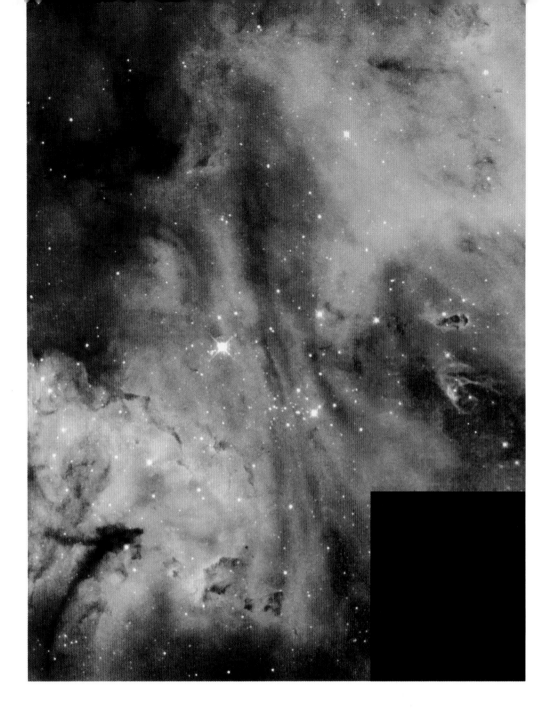

## N90

With its unusual stalagmite- and stalactite-looking features, this striking structure gives one the uncanny feel of standing at the mouth of a celestial cave. Bright blue newly formed stars are blowing a cavity in the center of a fascinating star-forming region known as N90, located in the Small Magellanic Cloud. Blazing out from the hot, young stars, high-energy radiation is sculpting the inner edge of the outer portions of the nebula, slowly eroding it away and eating into the material beyond. Dust pillars pointing toward these blue stars are a telltale sign of this eroding effect. The diffuse outer reaches of the nebula prevent the energetic outflows from streaming away from the cluster. Ridges of dust and gaseous filaments can be seen toward the upper-left part of the image and the lower-right corner. With Hubble, it is possible to trace how star formation started at the center of the cluster and propagated outward, with the youngest stars still forming today along the dust ridges.

## N180B

This particular region within the LMC, called N180B, contains some of the brightest-known star clusters. Etched against the glowing hydrogen and oxygen gases are 100-light-year-long dust streamers that run the length of the nebula, intersecting the core of the cluster near the center of the image. Also visible among the dust clouds are "elephant-trunk" stalks of dust. These dust clouds are evidence that this is still a young star-formation region.

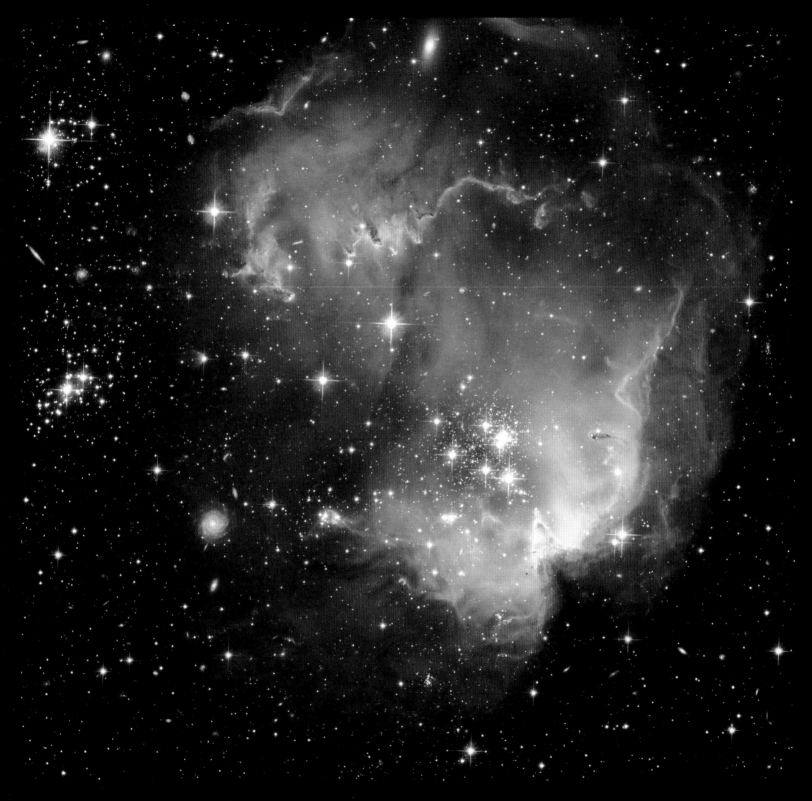

## NGC2080

Nicknamed the Ghost Head Nebula, this is one of a chain of star-forming regions in the Large Magellanic Cloud. Radiation from a core of hot, massive stars has carved a bowl-shaped cavity in the surrounding gas. The "eyes of the ghost" are two bright blobs of hot, glowing hydrogen and oxygen containing massive stars.

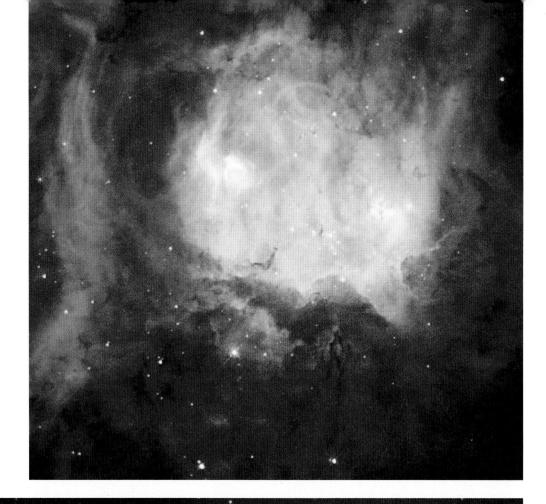

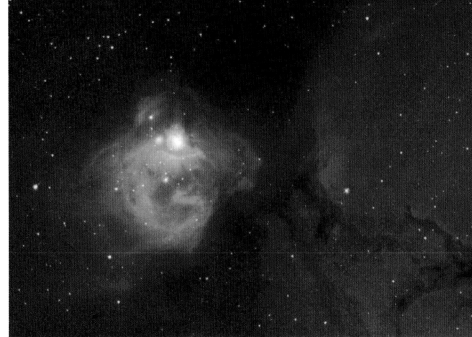

## N83B

Emerging from the shelter of their molecular cloud, young, massive, ultrabright stars are seen here as they are being born. The star at the very center of the nebula, just below the brightest region, is 200,000 times brighter than the Sun. Only 30,000 years ago, the intense light and powerful stellar winds from this star cleared away the surrounding gas to form a large bubble about 25 light-years in diameter.

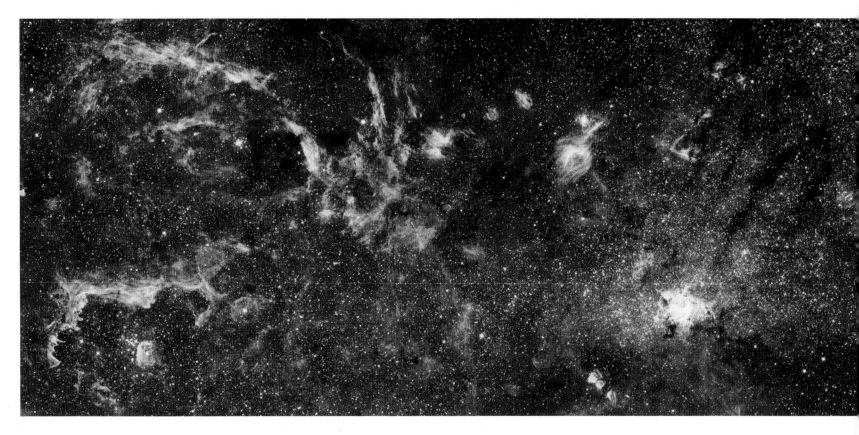

## GALACTIC CORE

The most tortuous region of star birth in our Milky Way is at the very heart of our galaxy. Monstrous molecular clouds rotate around the galactic hub. Embedded within them are immense star clusters. This composite color infrared image of the center of the Milky Way Galaxy reveals a new population of massive stars and previously unseen details in complex structures in the hot gas swirling around the central 300 light-years. The exact geometric center of the galaxy is home to a supermassive black hole nearly four million times the mass of the Sun. The galactic core is obscured in visible light by intervening dust clouds, but infrared light is able to penetrate the dust and reveal its location in the bright mass of stars and hot gas at lower right.

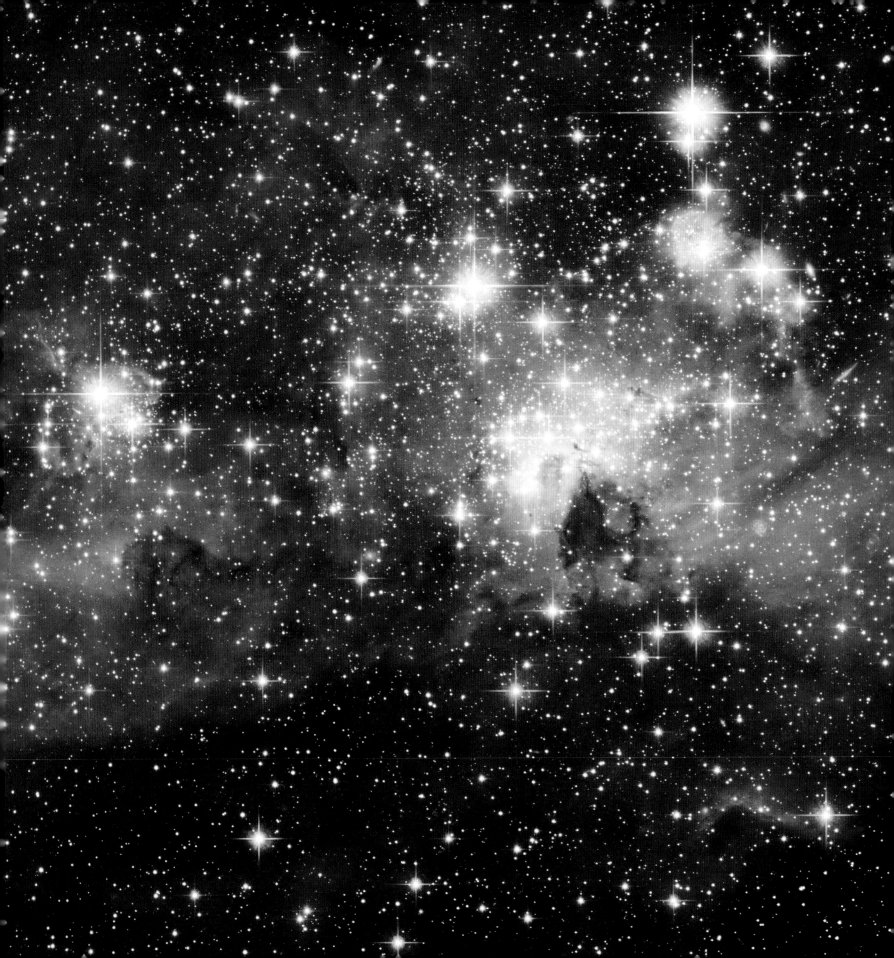

## NGC604 in M33

Right, top and bottom: This billowing cloud of hydrogen (box) is the brightest object in the Triangulum Galaxy, located 2.7 million light-years away. It radiates with the energy released by hundreds of young, bright stars. About 1,500 light-years across, this is one of the largest concentrations of glowing hydrogen in our local group of galaxies, and it's a major factory for star formation. The complex structure of NGC604, with irregular bubbles and wispy filaments alongside denser features, is being eaten away by the radiation from the young stars. The blisterlike cavities show areas of stronger erosion in the cloud. The top image was taken by a ground-based telescope; the bottom image close-up is by Hubble.

## LH95

Facing page: Spanning about 150 light-years, the star-forming region known as LH95 is at a later stage of development, when a lot of the "placental nebula" has been blown away by the actions of young stars. Some dense parts that have not been completely eroded by the stellar winds are seen as dark, dusty filaments. Usually, only the brightest, bluest, most massive stars in a star-birth region are visible, but many recently formed stars that are more yellow, dimmer and less massive are discernible as well. Also visible is a blue sheen of diffuse hydrogen gas heated by the young stars and dark dust created by stars or during super-nova explosions. Two small compact clusters of young stars, one to the right, above the center of the image, and one to the far left, complete the scene.

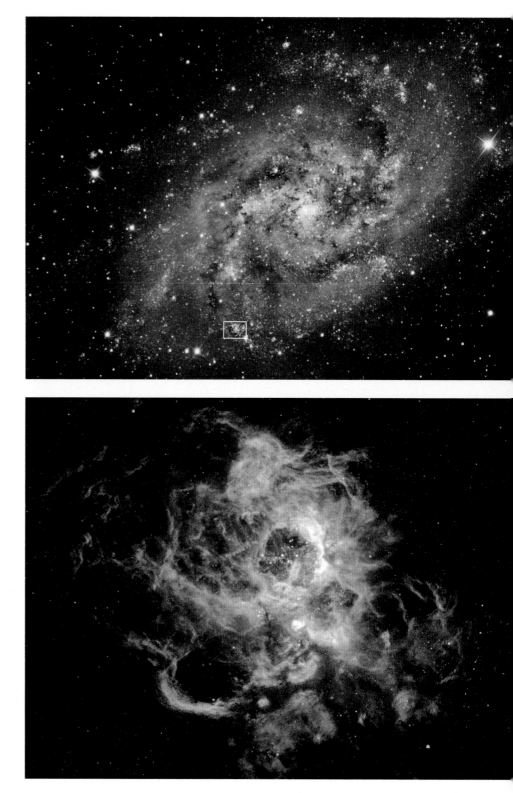

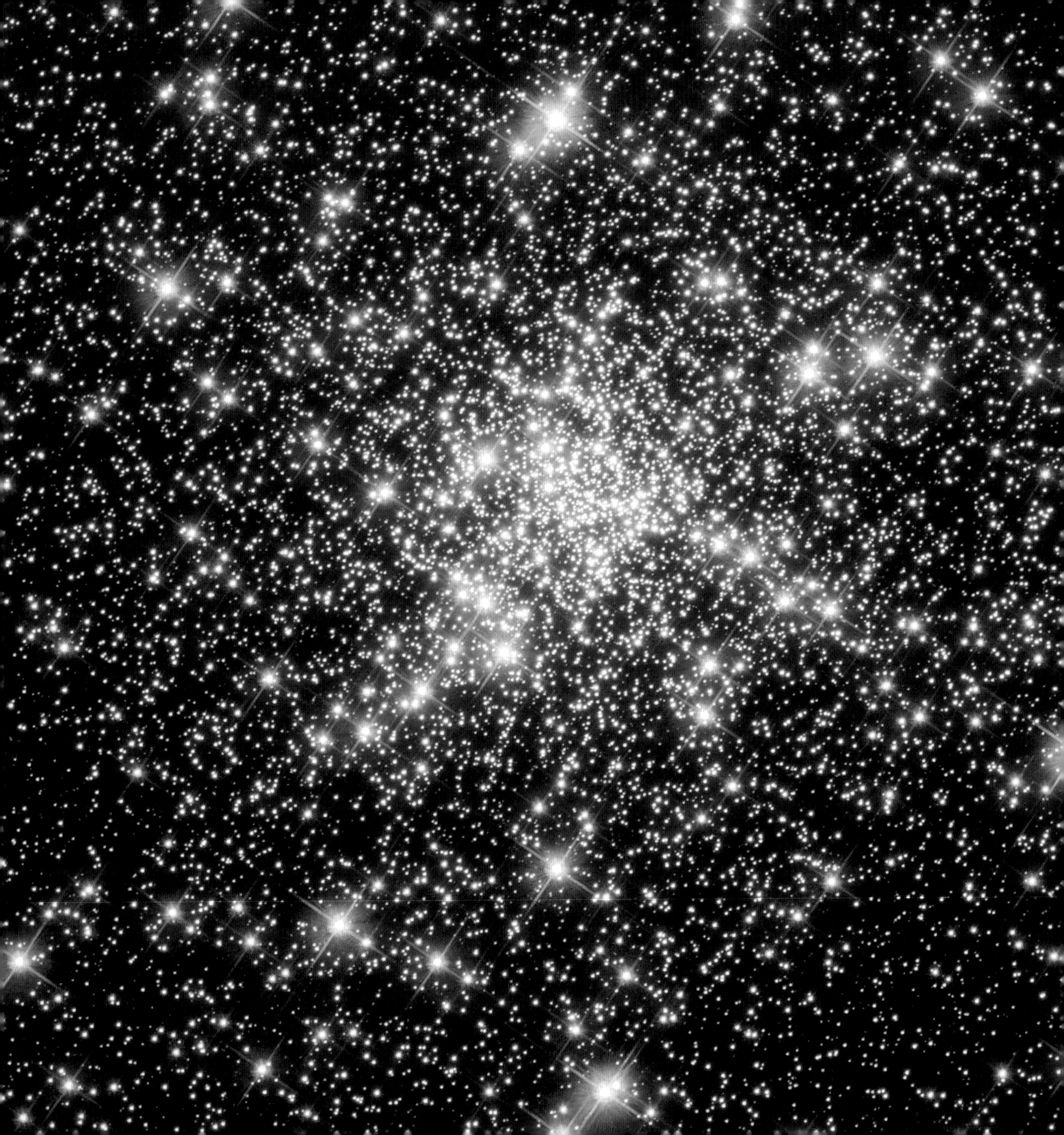

# STARRY TAPESTRY

## CHAPTER 5

Stars are the universe's basic building blocks and, in many ways, are fundamental to the existence of planets and life in the universe. Over billions of years, they have collected themselves into a hierarchy of structures, star clusters, galaxies and immense clusters of galaxies.

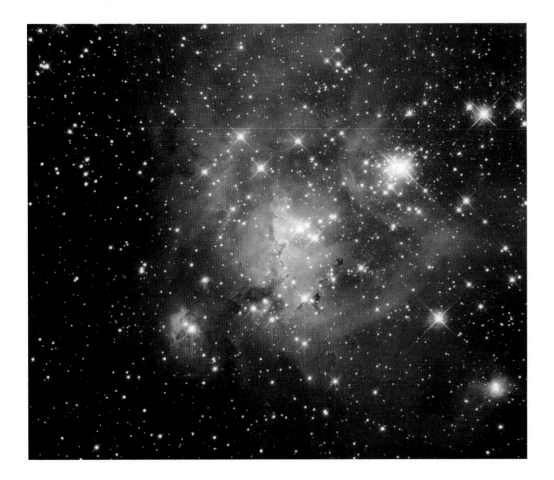

Left: Bright stellar groupings called OB associations lie in one of the largest-known star-formation regions in the Large Magellanic Cloud, a small satellite galaxy of the Milky Way. One of these, LH72 (seen here), is a group of high-mass, young blue stars embedded in a dense nebula of hydrogen gas.

Facing page: Looking like a hoard of gems fit for an emperor's collection, the globular star cluster NGC6752 is more than 11 billion years old, one of the most ancient groupings of stars known. It has been blazing for well over twice as long as our solar system has existed. NGC6752 is about 13,000 light-years distant.

Ninety percent of all stars in our galaxy are classified as main-sequence stars, which shine through nuclear fusion. The largest main-sequence stars, the blue supergiants, are 150 times the mass of the Sun and 10 million times brighter. The smallest, the low-mass red dwarfs, are about one-eighth the Sun's mass and simmer at no more than the temperature of the filament of an incandescent lightbulb. Below that mass, an object does not have a high enough internal pressure to raise the core temperature above the threshold to ignite a thermonuclear reaction and thus does not meet the most fundamental definition of a star. Such objects are known as brown dwarfs (see page 156).

A star remains on the main sequence as long as it is in perfect equilibrium, meaning gravity

Facing page: A vast spherical nest of 10 million stars about 17,000 light-years distant, the globular cluster Omega Centauri is the brightest and most massive of more than 100 globular clusters that orbit, like satellites, around the core of the Milky Way Galaxy.

Right: The enhanced colors in this deep peek into the core of Omega Centauri show stars of different temperatures. The bright red stars are aged red giants nearing the end of their lives. They are large but relatively cool. The brilliant blue stars burn nuclear fuel at extremely hot temperatures, giving them their predominantly blue color; they, too, are near the end of their life spans. Speckling the image like grains of salt are dimin - utive white stars that are like our Sun in terms of temperature and age. Finally, the profuse scattering of faint red dots throughout the image indicates burned-out stars— cooling cosmic cinders that mark a once glorious youth.

is pulling the star inward, while the radiation from fusion reactions is pushing outward. The inward and outward forces balance each other, and the star is a stable sphere. The mass of a star predetermines its longevity. Our Sun will remain a main-sequence star for 10 billion years (today, it's halfway through this span). Blue-white stars more massive than the Sun burn out in less than a billion years. The most massive blue stars burn furiously and blow themselves apart in less than 50 million years.

At the opposite extreme, cool red dwarfs burn for a trillion years or more, which is many times the 13.7-billion-year present age of the universe. If our current ideas about the probable future of the universe are correct, stars will not be around forever. It will be a lonely, dark place for a long, long time after the last star fades to black.

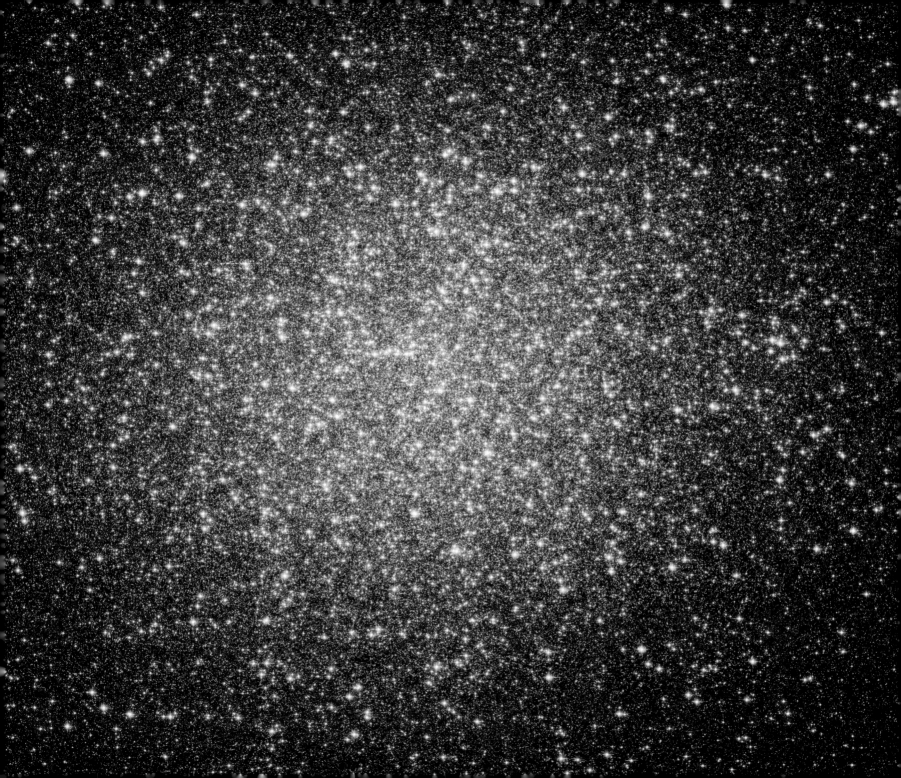

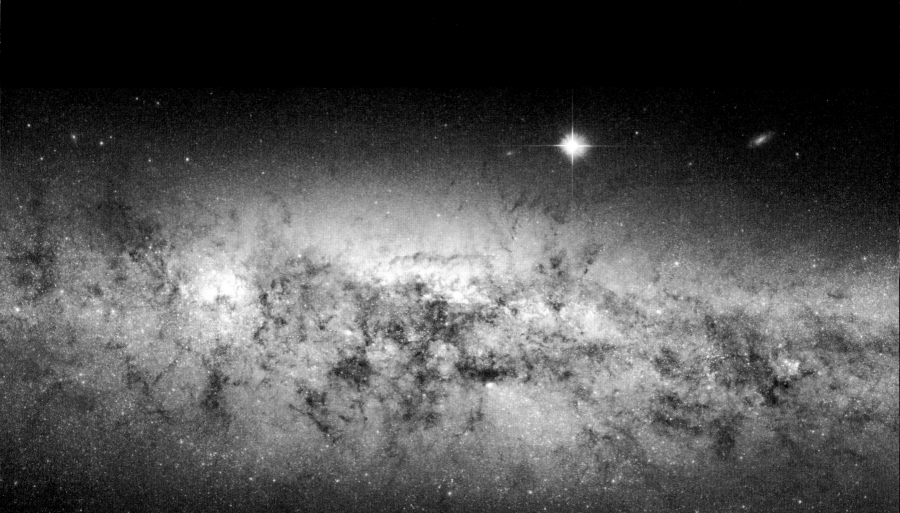

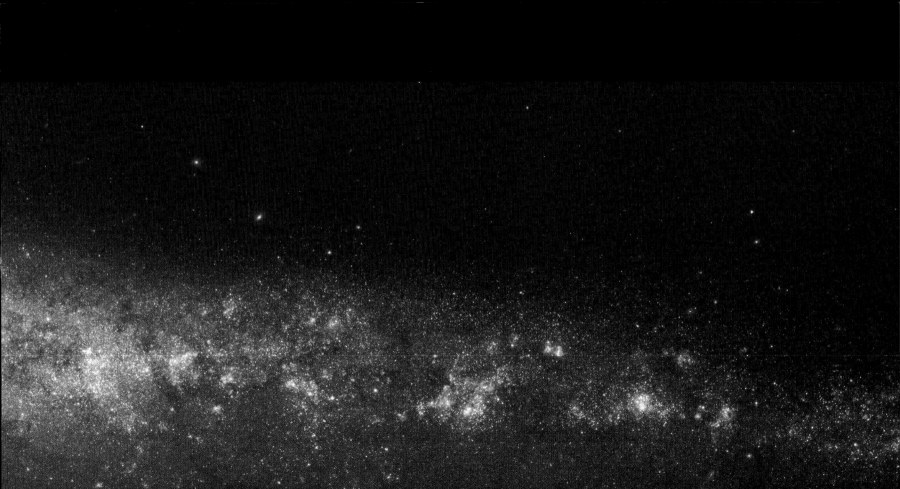

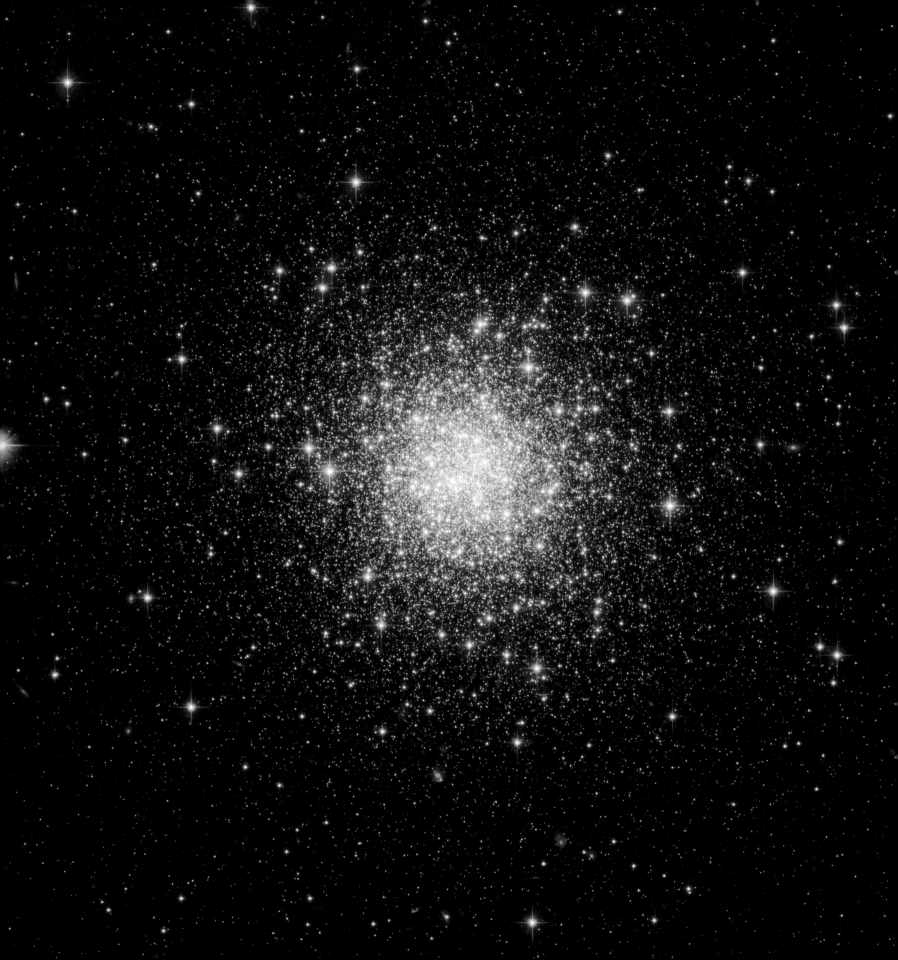

Open star clusters are groupings of young stars briefly held together by gravity. Born inside dense molecular clouds in the disk of our galaxy, most open clusters have only a short life as stellar swarms, because they are easily dismantled by external gravitational effects, especially from giant molecular clouds and other star clusters. Unlike globular clusters, which are billions of years old, open clusters will disintegrate in 100 million years. The escaped individual stars continue to orbit the Milky Way on their own. Our Sun was probably born in this way.

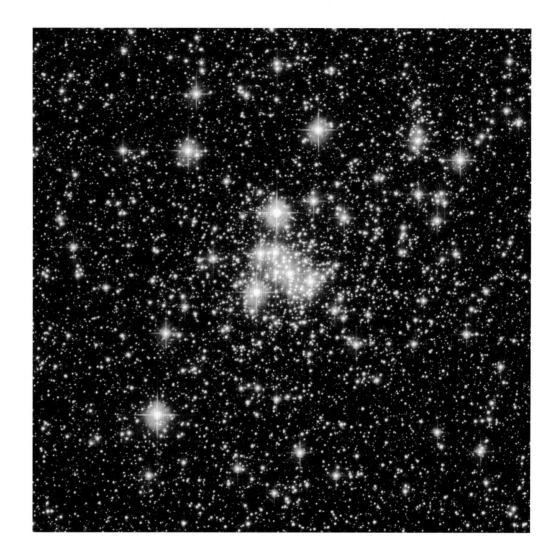

Globular star clusters are gravitationally bound spherical concentrations of 10,000 to 1,000,000 stars. At least 150 are found in the halo of our galaxy, outside of its flat spiral disk. These are the leftover building blocks of the clusters that merged to form our galaxy and therefore contain the oldest stars in the Milky Way. At the center of these clusters, stars are relatively densely packed, just light-weeks apart instead of light-years, as they are in the Sun's vicinity. As seen from the core of a globular cluster, the night sky would be so brightly decorated with stars, you could read this book by starlight alone.

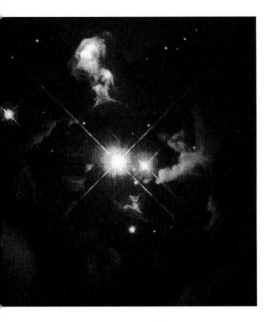

At the end of the protostar phase, the star is as hot as our Sun yet hasn't started nuclear fusion reactions. Known as a T Tauri star, it can have entangled magnetic fields and great sunspots that, together, trigger intense X-ray flares and extremely powerful stellar winds. A star remains in the T Tauri stage for about 100 million years. It then ignites nuclear fusion reactions at its core, where hydrogen nuclei combine to form helium nuclei. In the process, some of the matter is converted to energy.

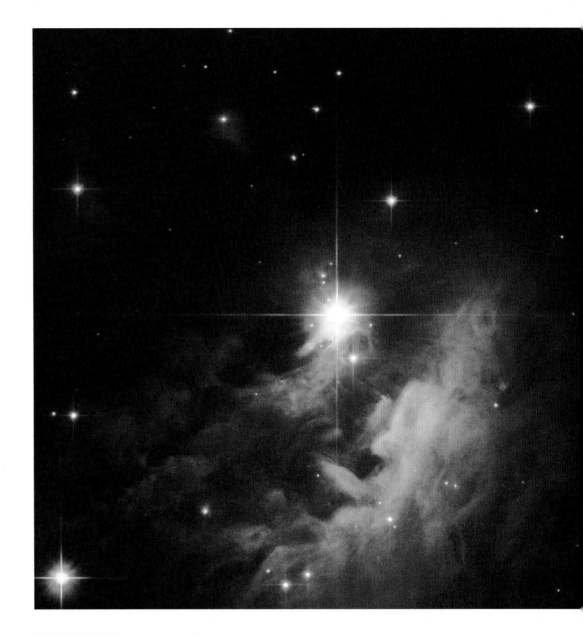

EMBRYONIC STAR

The term "protostar" is used to describe the embryonic stage of star formation. A protostar is a dense nodule of hydrogen and helium that collapses within a giant cloud of molecular hydrogen. The energy released by the protostar is created only by the gravitational compression of the gas. This is a very brief phase in a star's birth, lasting only 100,000 years.

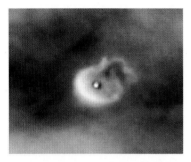

## YOUNG STELLAR OBJECTS

Fundamental to star birth is the fact that the universe likes making pancakes —or, more precisely, disks. Galaxies collapse to form disks; disks exist around black holes; so disks around stars seem natural. As the collapsing cloud shrinks, whether around a galaxy core, a black hole or a star, its rotation rate increases to conserve angular momentum. This is exactly what happens when figure skaters retract their arms and "spin up." The disk forms around the star, buoyed by centrifugal force.

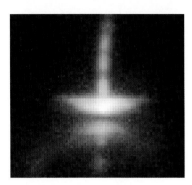

Although the centrifugal force is maximum within the disk, it slows but does not necessarily halt dust and gas accretion onto the star. The disk then begins to collapse onto the star and is clumpy and turbulent. Material from the cocoon of gas and dust around the fledgling star feeds the disk. Depending on the thickness of the disk, this material spirals onto the star, where it is heated, then fiercely ejected in billion-kilometer-long jets that herald the birth of a new star. The jets dramatically poke out of the embryonic cloud like a skyrocket trailing a plume. As clumps of gas and dust fall onto the star, the irregular flow of infalling material is evident in images of jets that reveal a string-of-pearls appearance.

## STELLAR-JETS SEQUENCE

Hubble can record how clumpy jets of material ejected from a young star change over time. Several bright regions in the clumps signify where material is colliding, heating up and glowing. The images show that a couple of bright areas on the right faded as the heated material cooled. Two regions at left, however, brightened over the 13-year span of Hubble observations, pinpointing fresh collision sites. The blobby material is speeding along at more than 300,000 kilometers per hour.

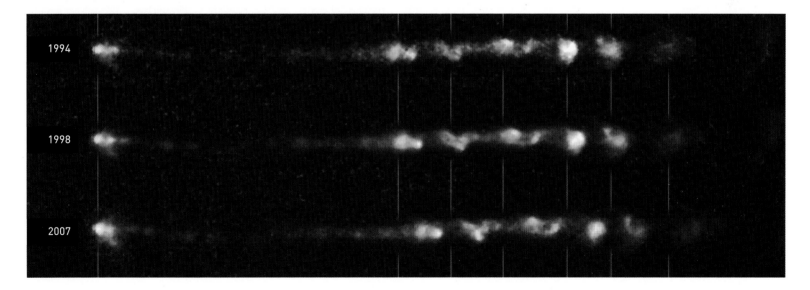

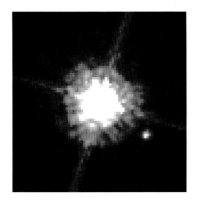

## RED DWARF/BROWN DWARF

CHXR 73, a red dwarf star, appears as a brilliant cross in this Hubble image due to the mirror-support structure in the telescope. But the key object of interest here is CHXR 73 B, the spot at lower right. At 12 times the mass of Jupiter, it is large enough to be a brown dwarf, a failed star. Unlike stars, brown dwarfs do not have quite enough mass to sustain hydrogen fusion reactions in their cores, which is what powers stars like the Sun. But their cores are warm enough that brown dwarfs shine dimly in infrared light, which Hubble detected. CHXR 73 B is 30 billion kilometers from its red dwarf sun. Red dwarfs are the most common class of star in the cosmos. Brown dwarfs may be abundant as well, but their faintness makes them exceedingly difficult to detect.

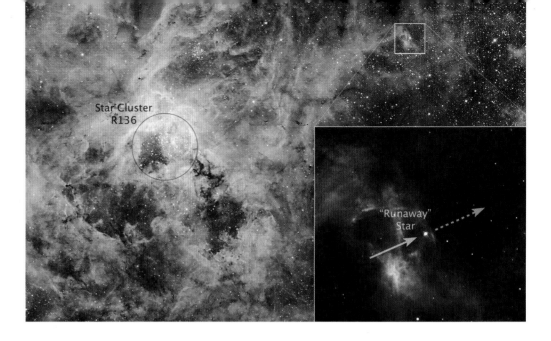

## RUNAWAY STAR

Hubble caught a blue-hot star, 90 times more massive than the Sun, hurtling across space so fast that it could make a trip from Earth to the Moon in one hour. The only way the star could have reached this speed is to have been ejected from the star cluster R136, where it was born. The small box at the upper right shows how far the star has traveled from its birthplace. One possibility is that a rogue star entered the binary system where the star lived, causing the runaway star to be gravitationally ejected from the binary system. Only a very massive star would have the gravitational energy to eject something 90 solar masses in size. This is strong circumstantial evidence to support the assumption that stars up to 150 solar masses live in the cluster.

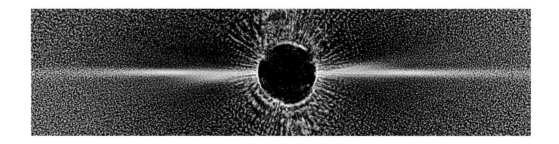

## BETA PICTORIS DISK

After a star enters the main sequence, many asteroids and planetary bodies that initially formed with the star will collide to create a surrounding dust disk. The granddaddy of these circumstellar disks is the star Beta Pictoris, located 63 light-years away in the southern constellation Pictor. Long before exoplanets (planets of other stars) were discovered, Beta Pictoris got astronomers' attention because it has an odd excess of infrared radiation for a star of its temperature. This was interpreted as the glow of a warm dust disk encircling the star. Where there's dust, astronomers reasoned, there could be planets. Newborn planets should generate dust through collisions between protoplanets. For example, our Moon may have been born out of a grazing blow between a Mars-sized embryonic planet and the newborn Earth.

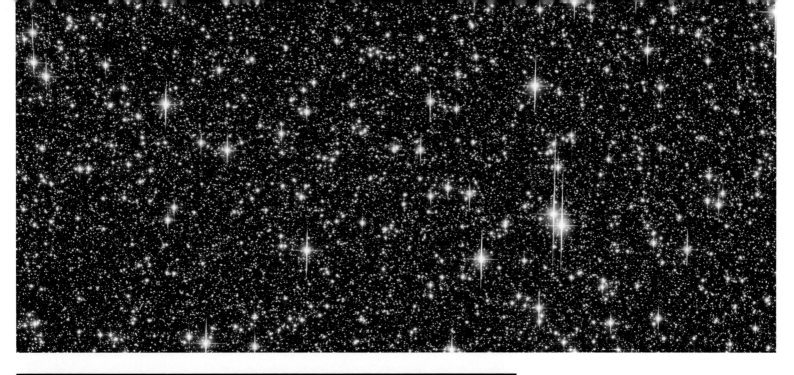

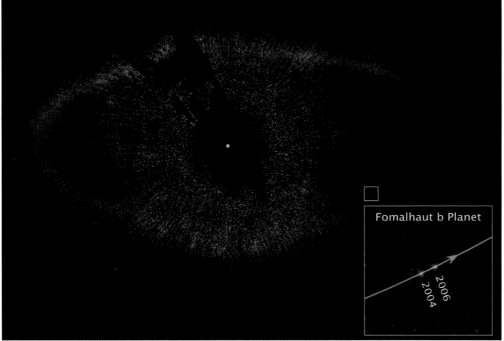

Fomalhaut b Planet

2006
2004

## FOMALHAUT

A thin ring of dust and ices, 2.3 billion kilometers in diameter, encircles the star Fomalhaut in this near-infrared Hubble image. The narrow shape of the ring offered early evidence that a planet might be gravitationally shepherding the dust. The small white box pinpoints the location of a suspected planet. The planet as detected in 2004 and again in 2006 is shown in the inset box at bottom right. It appears to be following an 872-year orbit, as predicted.

## SAGITTARIUS WINDOW

This star field of over 100,000 stars, all at least the age of the Sun, is one of Hubble's richest images. The telescope was aimed directly into the core of our galaxy. There are no younger stars here because the core stopped making stars billions of years ago. This "keyhole" view fortuitously penetrates the dust and star clouds of the Milky Way, allowing us to see through 26,000 light-years of space to the crowded center of our galaxy, in the direction of the summer constellation Sagittarius. The survey uncovered 16 stars with unconfirmed "candidate" planets orbiting them. These particular planets were detected because they are so close to their stars that they complete an orbit in just a few days. Although Hubble didn't photograph the planets, it did measure subtle dips in brightness as each planet passed in front of its parent star.

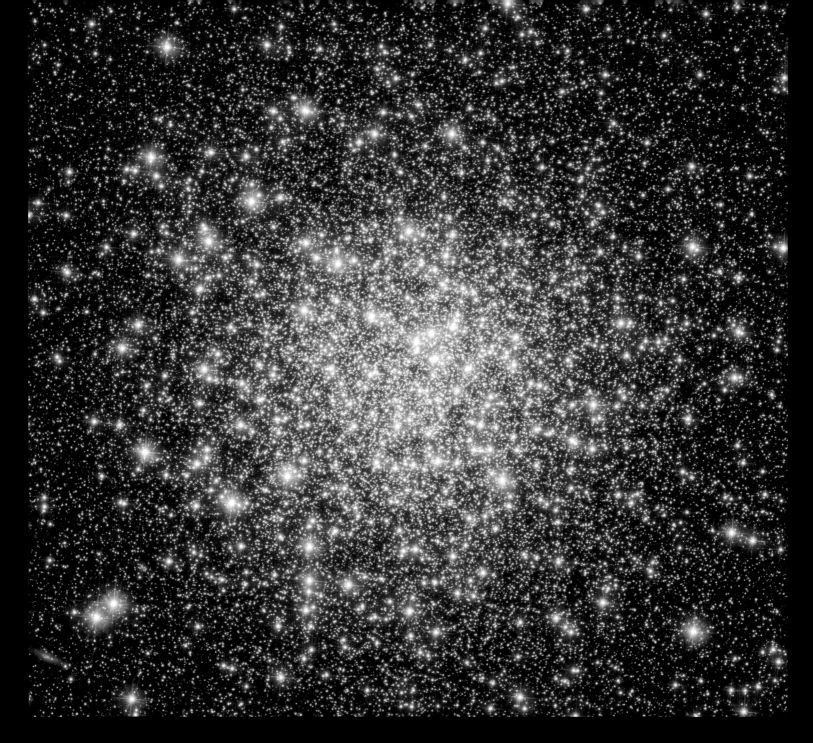

## M5

This dazzling array of stars formed more than 12 billion years ago, but there are some unexpected newcomers on the scene. Stars in globular clusters are born in the same stellar nursery and grow old together. The most massive stars age quickly, exhausting their fuel supply in less than a million years and ending their lives in spectacular supernova explosions. This process should have left the ancient cluster Messier 5 with only old low-mass stars, which, as they age and cool, become red giants. But many young blue stars, called blue stragglers, are apparent in this cluster. They were created either by stellar collisions or by the transfer of mass between binary stars. M5 lies 25,000 light-years away in the constellation Serpens.

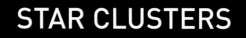

# STAR CLUSTERS

## M15

The dazzling stars in Messier 15 are all roughly 13 billion years old, making them among the most ancient objects in the universe. M15 is one of the densest globulars known, with the vast majority of the cluster's mass concentrated in the core. In the case of dense globulars like this one, astronomers think that the cluster has undergone a process called core collapse, in which gravitational interactions between stars has caused many members of the cluster to migrate toward the center. M15 is about 35,000 light-years from Earth.

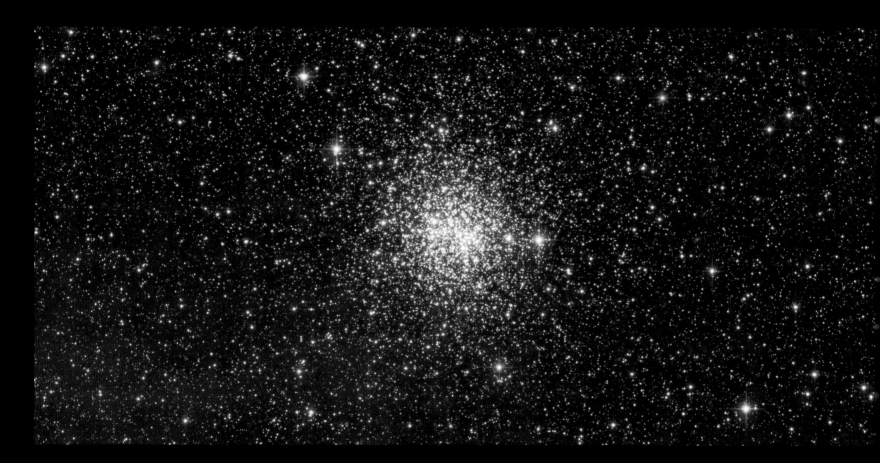

## NGC1872

A rich cluster of thousands of stars dwells in the Large Magellanic Cloud. Star clusters are usually classed as either open or globular, but NGC1872 has characteristics of both. It is as rich as a typical globular yet much younger, which is evident in the young blue stars sprinkled across it. Such intermediate clusters are common in the Large Magellanic Cloud, but not in our home galaxy.

M13

This Hubble portrait of the core of the great globular cluster Messier 13 provides an extraordinarily clear view of the hundreds of thousands of stars in the cluster's central region. M13 is 25,000 light-years away and 145 light-years in diameter.

## PALOMAR 1

One of the mysteries surrounding the faint globular star cluster called Palomar 1 is that it is probably half the age of most globular clusters, which date back to the formation of our galaxy, 12 billion years ago. This suggests that Palomar 1 was created in a different formation process. Perhaps it was the remnant core of a small galaxy that strayed too close to our Milky Way Galaxy 500 million years ago and was shredded by gravitational tidal forces. Over billions of years, our galaxy has cannibalized numerous smaller galaxies in its vicinity.

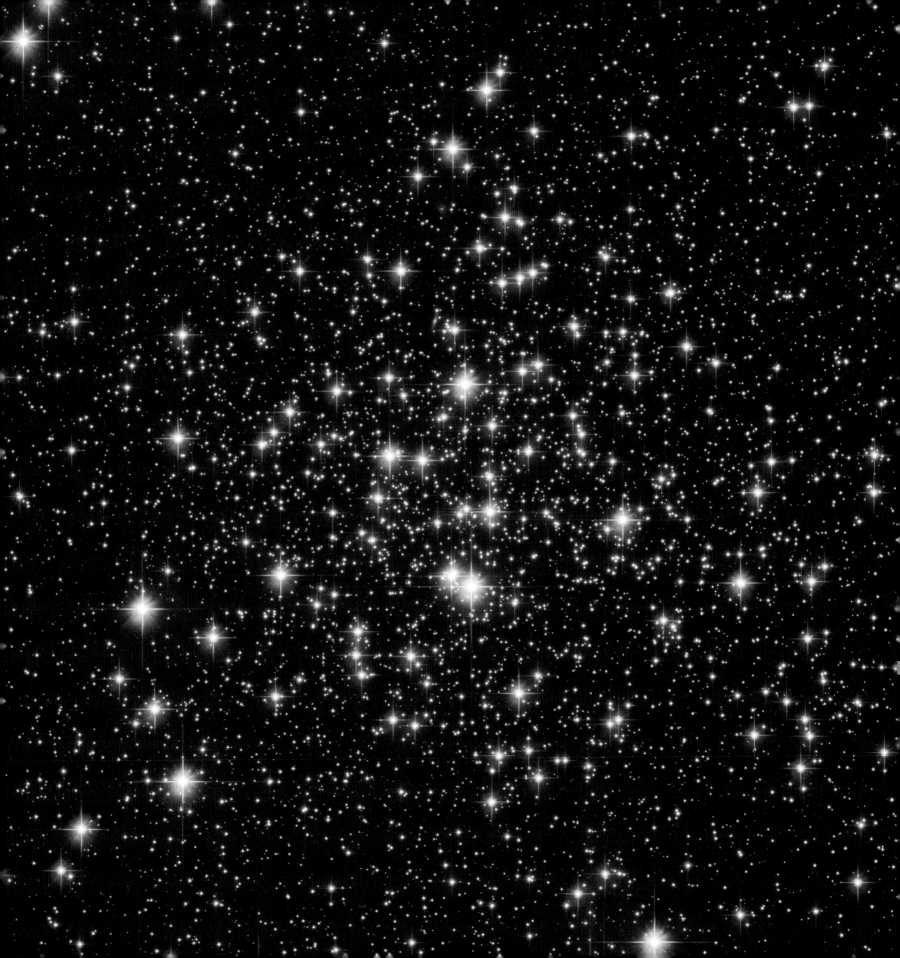

## M71

Because Messier 71 (left) is more loosely packed than other globular clusters, it was first thought to be an open cluster. But its stars are all around 12 billion years old, the same age as stars in most globulars. M71 is 13,000 light-years from Earth and 27 light-years across.

## GHOSTLY REFLECTION

The wispy tendrils of a dark interstellar cloud are made visible by the flood of light from one of the brightest stars in the Pleiades star cluster. Like a flashlight beam shining on the wall of a cave, the star's light is reflecting off the surface of pitch-black clouds of cold gas laced with dust. The blue star Merope is just outside the frame on the upper right. The Pleiades, or Seven Sisters, is an open star cluster in Taurus that is easily visible with the unaided eye.

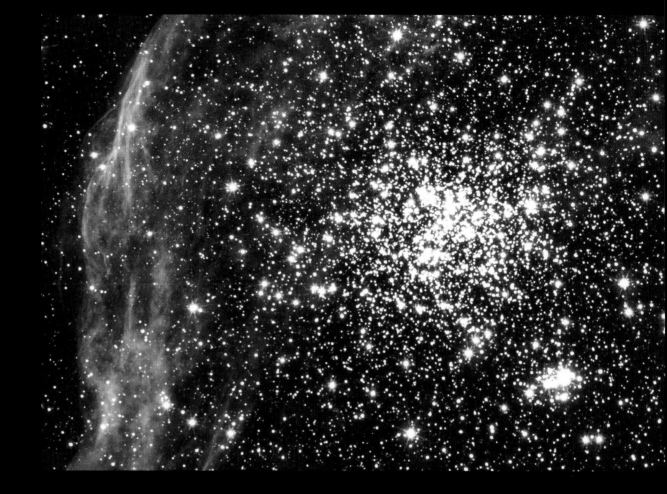

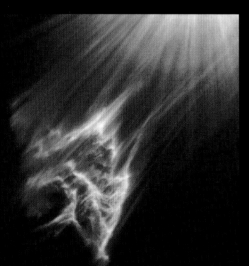

## NGC1850

A type of object unknown in our own Milky Way Galaxy, NGC1850 is a young, "globularlike" star cluster, the second brightest star cluster in the Large Magellanic Cloud. NGC1850 has two components: the main cluster in the center and a younger, smaller cluster below and to the right. The main cluster is about 50 million years old; the smaller cluster is only four million years old. Diffuse gas created by the explosion of short-lived, very massive stars surrounds NGC1850.

# HUGE STARS

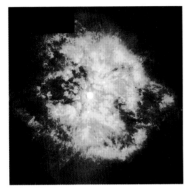

### WR124

Resembling an aerial fireworks explosion, the massive star WR124 is surrounded by hot clumps of gas being ejected into space. Vast arcs of glowing gas around the star are resolved into filamentary, chaotic substructures by Hubble. The massive central star is an extremely rare and short-lived class of super-hot star. It is going through a violent transitional phase characterized by the fierce ejection of mass. The blobs probably result from the furious stellar wind that does not flow smoothly into space.

### PISTOL STAR

Unleashing up to 10 million times the power of the Sun and big enough to fill the diameter of the Earth's orbit, this is the brightest star yet found in the galaxy. It releases as much energy in six seconds as our Sun does in one year. This infrared image reveals a bright, reddish nebula, created by massive stellar eruptions, that would nearly span the distance from the Sun to Alpha Centauri, the nearest star to the Earth's solar system. The Pistol Star can be seen only in infrared light because it lies beyond the great star cloud in Sagittarius. If the star were visible, it would go on record as the farthest star observable with the naked eye, at a whopping distance of 27,000 light-years.

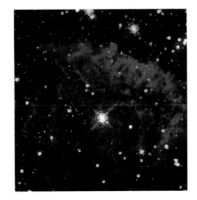

Drifting through space, the stellar grouping known as NGC2040 is made up of stars that have a common origin. A nebulous blue haze gives the cluster an ethereal look. NGC2040 is so young that it contains extraordinarily hot and massive blue stars. It is located in the Large Magellanic Cloud, a hotbed of star formation.

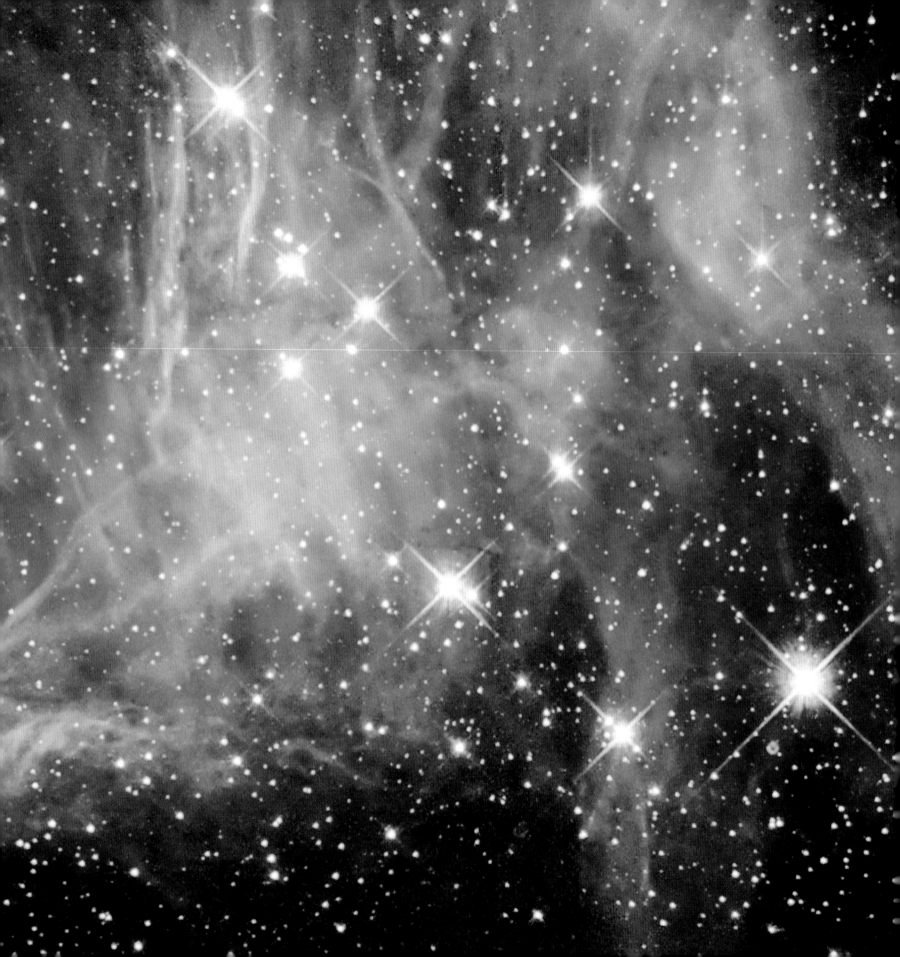

## WR25

A pair of colossal stars known as WR25 and Tr16-244 is located within the open cluster Trumpler 16, which itself is embedded within the Carina Nebula. The bluest and brightest star in the image, WR25 is the most massive of the pair. It is actually two stars that are too close together to be resolved even by Hubble. At 50 and 25 times the mass of our Sun, respectively, they whirl around each other in less than a year. Numerous much smaller red dwarf stars pepper the image.

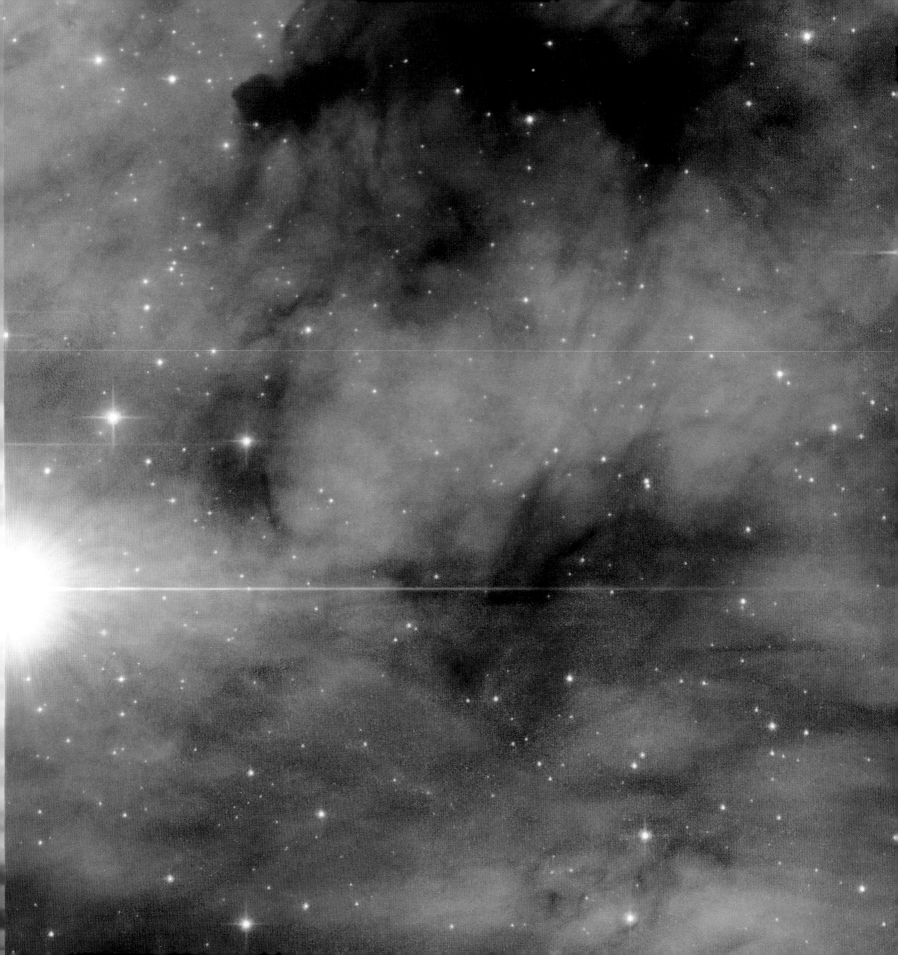

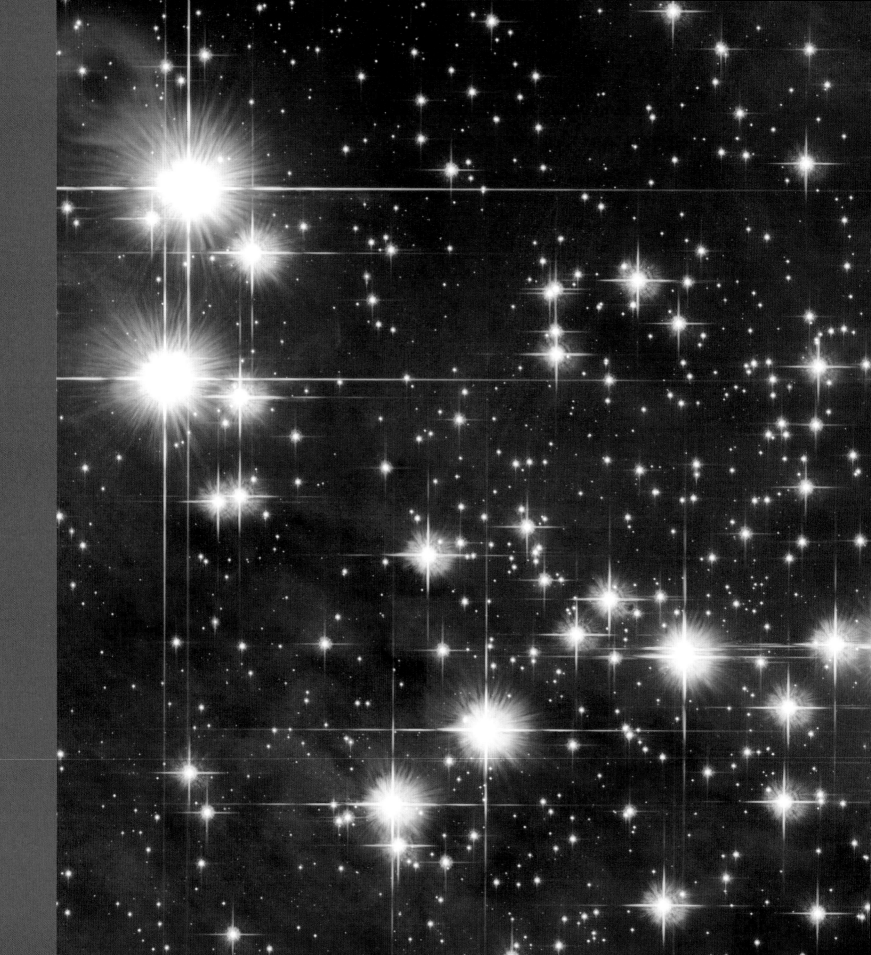

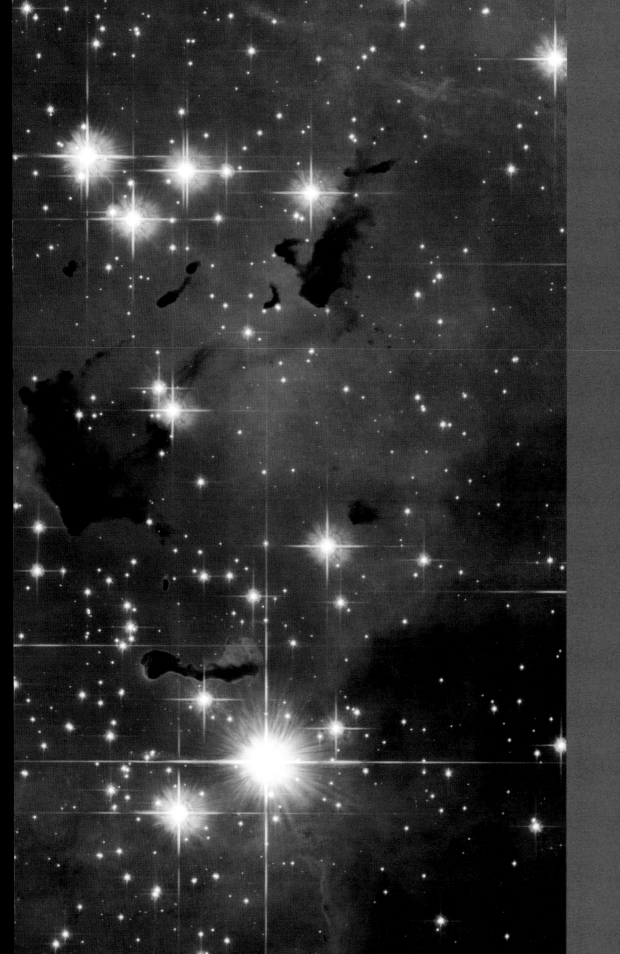

## STELLAR MATERNITY WARD

The earliest stages of star formation are fast and complex and still poorly understood. But we do know that the process starts in dark, apparently formless clouds like these. Quiescent clouds become gravitationally unstable and begin to collapse, breaking into smaller pieces. The dimensions of this collapse are mind-boggling. The gas density rises by a factor of a billion trillion. This is comparable to something the size of Lake Superior being compressed to the size of a dime.

Gravity fragments the clouds into predictable proportions of small-, medium- and high-mass stars. This is analogous to finding a few boulders, lots of rocks and even more pebbles on a beach. A very small fraction of the cloud will become blue supergiants, dozens of times to more than 100 times the mass of our Sun. A few percent will be in the range of the Sun's mass. And the vast majority of stars, the red dwarfs, will be one-quarter to one-sixth the Sun's mass. These are the most numerous stars in our galactic neighborhood.

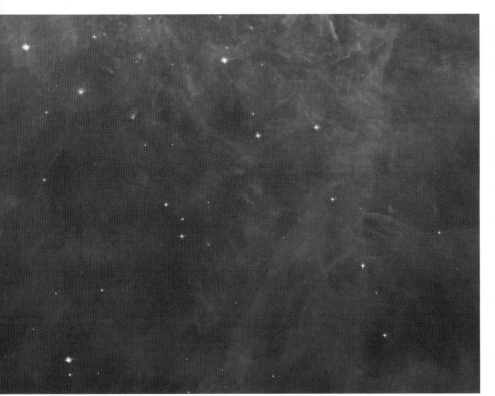

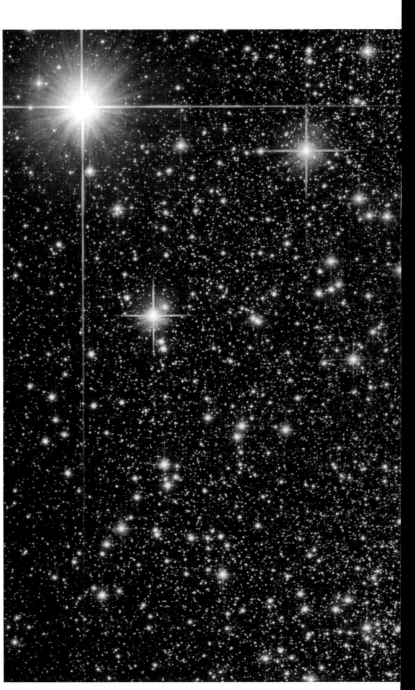

## BROWN DWARFS

The universe's underachievers
are the brown dwarfs, a class
between stars and planets. Sometimes called failed stars, they
are less than 75 to 80 times the
mass of Jupiter and are too cool
to sustain nuclear fusion. Long
theorized to exist, brown dwarfs
are exceedingly dim. Regions deep
within the Orion Nebula, as shown
in this Hubble image, reveal numerous brown dwarfs drifting through
space. Approximately one to three
times the diameter of Jupiter,
brown dwarfs are small compared
with main-sequence stars.

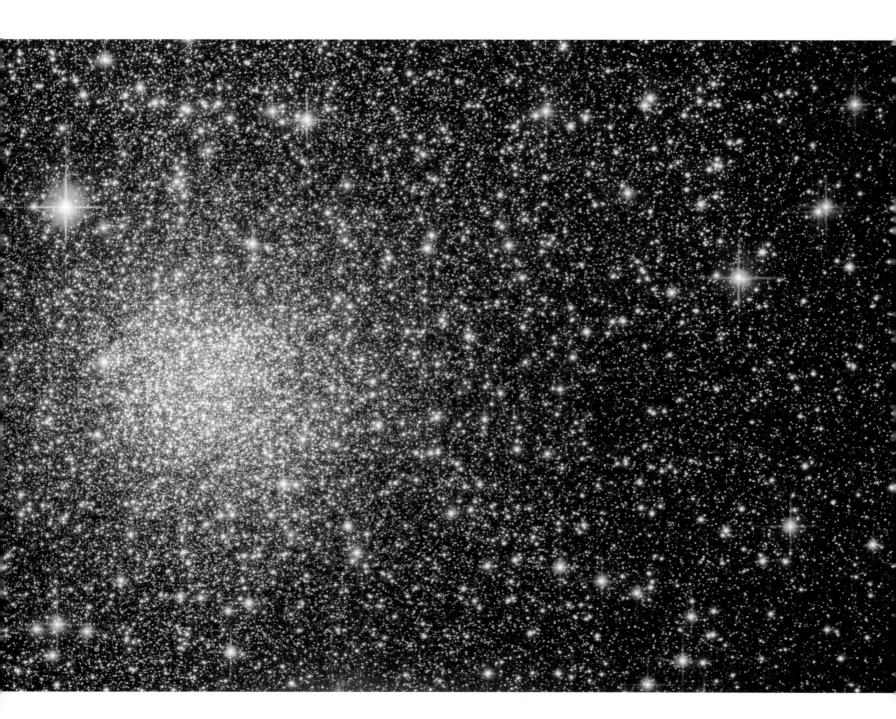

Unlike most globular clusters, Terzian 5 is embedded in the central bulge of the Milky Way Galaxy, rather than orbiting high above the galactic plane. The cluster contains a curious mix of stellar populations. Some stars are only six billion years old, while others are 12 billion years old. This suggests that the cluster is a relic of a dwarf galaxy that merged with the Milky Way during its very early days and later went through a second epoch of star formation.

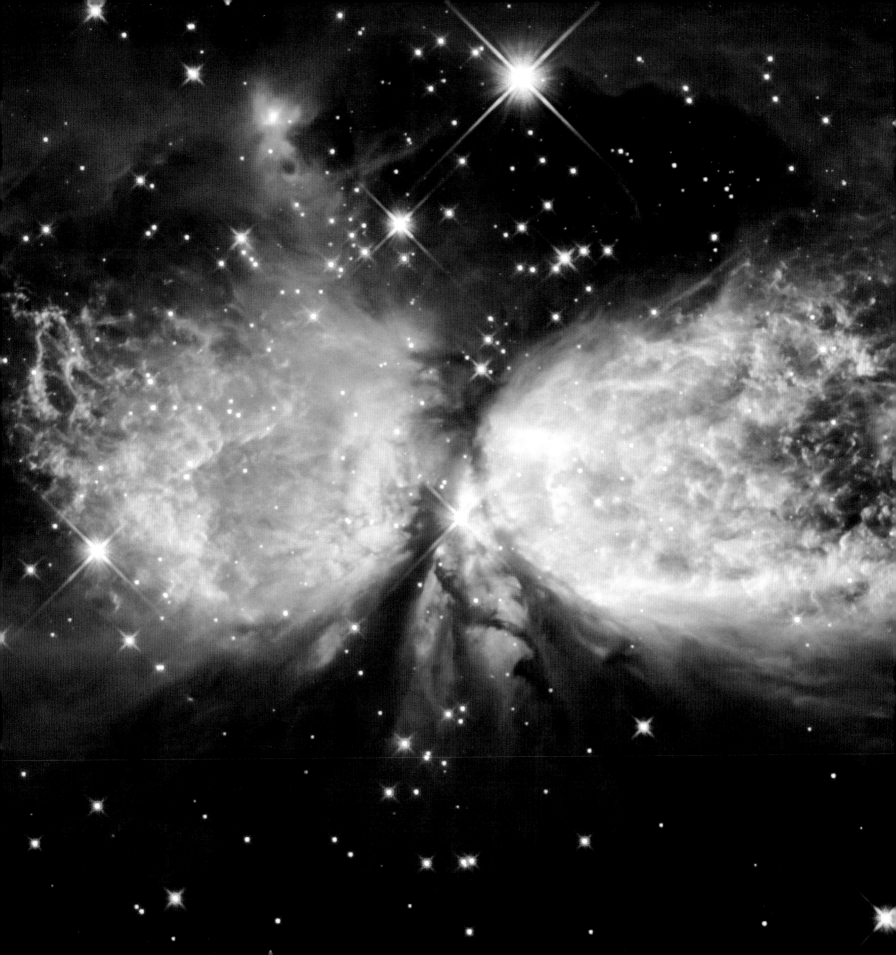

# BLAZE OF GLORY

## CHAPTER 6

American General Douglas MacArthur famously said upon his retirement in 1951: "Old soldiers never die; they just fade away." Most stars are like the soldiers in this legendary quote—when their nuclear fusion fuel is exhausted, they slowly fade to black. But as with people, there are exceptions to this scenario. Some stars, particularly the most massive ones, end with a bang or a series of violent death throes. The death of a star occasionally produces a detonation so powerful, it can be seen halfway across the universe.

The story begins with the Sun, the star we know most intimately. Every second, the Sun converts 650 million tons of hydrogen into 600 million tons of helium and, in the process, transforms 50 million tons of hydrogen into pure energy that makes our star shine.

The Sun has been pouring out this energy for 4.5 billion years. But it can't go on forever. Eventually, the Sun's fuel will run out. It's hard to imagine a day on Earth when the golden ball of the Sun no longer rises. But that day of reckoning is coming, though not for another five billion years.

How do we know this?

We wouldn't, save for the fact that we live in a galaxy which is still making stars. Stars are too long-lived for us to watch a full life cycle elapse. That's as impossible as it would be for a mayfly, which may live for only a matter of hours, to watch human growth.

The night sky presents a tapestry of the dying fireworks of expiring stars. But they do not die in vain. The heavier elements cooked up in the million-degree furnaces are blasted into space, where they are eventually incorporated into new stars and planets. The Milky Way judiciously recycles hydrogen from spent stars to make new stars, a process that has been going on for at least 10 billion years.

The end of a star's life is a dramatic spectacle that can be as eerie and enticing to watch as the birth of a star. The final years of the life of a Sun-like star are an arm-wrestle between the force of gravity perpetually attempting to crush a star under its own weight and the push back from the radiant energy pouring out from the nuclear furnace at its core. If the core throttles back on its output, gravity puts the squeeze on the star, causing it to shrink, get hotter and brighten. If the fusion reactions grow in strength, the star expands, reddens and cools.

A star's fusion engine, therefore, cannot regulate the thermostat forever. Later in a star's life, it begins to fuse hydrogen in a spherical shell surrounding an inert core of helium "ash." As the rate of fusion increases, the star expands into a red giant. The final stages of a star's existence are punctuated by relatively swift and major changes in size and temperature. The helium ash core contracts, releasing gravitational energy. Some of this energy heats up the center, while the rest heats the star's upper layers, causing them to expand and redden.

A bipolar star-forming region known as S106 looks like a soaring celestial snow angel. The outstretched "wings" of the nebula record the contrasting imprint of heat and motion against the backdrop of a colder medium. Radiation from a massive young star, IRS 4 (Infrared Source 4), is responsible for the furious activity seen in the nebula. Twin lobes of superhot gas extend outward from the central star, while a ring of dust and gas orbiting the star acts like a belt, cinching the expanding nebula into an hourglass shape. At a distance of nearly 2,000 light-years, S106 is about one light-year in width.

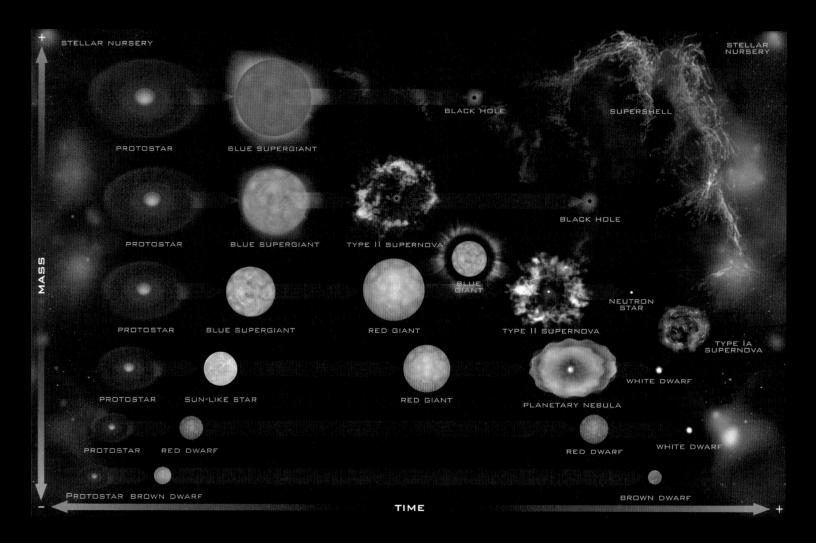

MASS

STELLAR NURSERY

STELLAR NURSERY

PROTOSTAR          BLUE SUPERGIANT

BLACK HOLE          SUPERSHELL

PROTOSTAR          BLUE SUPERGIANT          TYPE II SUPERNOVA

BLACK HOLE

BLUE GIANT

PROTOSTAR          BLUE SUPERGIANT          RED GIANT          TYPE II SUPERNOVA

NEUTRON STAR

TYPE IA SUPERNOVA

PROTOSTAR          SUN-LIKE STAR          RED GIANT          PLANETARY NEBULA

WHITE DWARF

PROTOSTAR          RED DWARF          RED DWARF          WHITE DWARF

PROTOSTAR BROWN DWARF          BROWN DWARF

TIME

The various evolutionary tracks for the life history of a star are shown in this diagram. All stars form out of collapsing clouds of gas and dust called protostellar nebulas. A star's destiny is predetermined by the mass that accumulates during its construction. The least massive stars, called red dwarfs, burn for a trillion years before depleting their fuel. A star the mass of our Sun lives for billions of years as a yellow dwarf. At the end of its life, as reserves of hydrogen fuel diminish, the star expands into a red giant. When nuclear fusion stops, the star gravitationally collapses into a white dwarf and leaves behind a spectacular planetary nebula. Stars several times more massive than the Sun shine brilliantly as blue giants but die in catastrophic supernova explosions. The residual stellar core collapses to form either a neutron star or a black hole.

The red giant star briefly contracts until it becomes so hot in its center that it triggers helium fusion at the core. Helium atoms collide to make carbon and oxygen atoms. The star burns helium for just 100 million years more.

Near the end of the star's life, its surface pulsates and shudders with seismic energy from changes in the activity of the fusion furnace. With each pulse, which lasts about a year, the surface layers expand and cool. Every time this happens, some of the stellar exterior is cast off. The hot star bleeds mass into space through a continuous "stellar wind" of gas streaming from its surface.

These dying stars slowly become embedded in a cocoon of dust of their own making. The blue-white hot fusion core contracts and grows hotter as it loses layers that are too hot to condense into dust. The stellar core is slowly exposed as the gas layers are stripped away. Hubble has photographed light from such stellar cores that makes brilliant "sunbeams" shine through the mottled clouds of dust around it. Finally, the remnant stellar core shrinks to the size of Earth and unleashes a flood of ultraviolet radiation that causes the gases in the outflow surrounding nebula to glow.

Our galaxy is littered with opulent death shrouds of such dying stars. They are called planetary nebulas because, historically, some resembled the disks of planets when viewed through a telescope. But the fact that many planetary nebulas are not spherical but are shaped more like an hourglass is compelling evidence that many of these dying stars have a companion. The companion star's orbit helps build a thick disk of dust that squeezes off material flowing from the dying star. Like a tube of toothpaste being squeezed, there is only one direction for the hot stellar wind to escape: along the star's rotation axis.

Bottom left: A bipolar hot-dog-shaped ejection of hot gases extends from a central cocoon of gas and dust in NGC6886. The embedded white dwarf is hot enough to emit a gusher of ultraviolet radiation that fluoresces the gases in this stunning planetary nebula.

Bottom right: When its central star becomes a white dwarf, the Westbrook Nebula will blossom into a fully illuminated planetary nebula. At present, it is an opaque, relatively short-lived cloud of gas ejected along the direction of the dying star's spin axis. Only a few hundred so-called protoplanetary nebulas are known in the Milky Way.

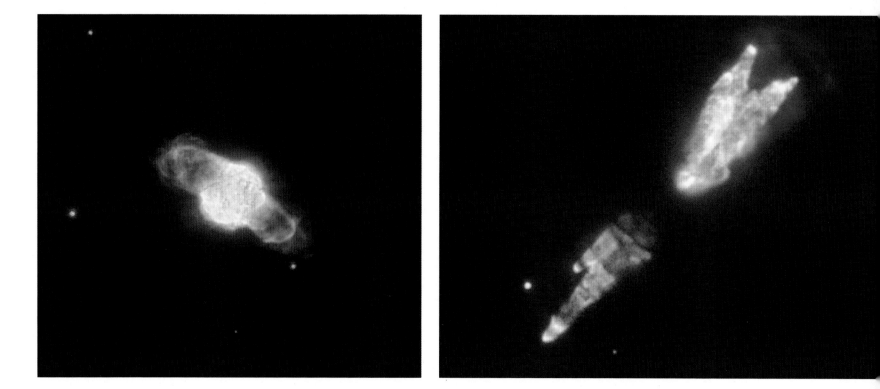

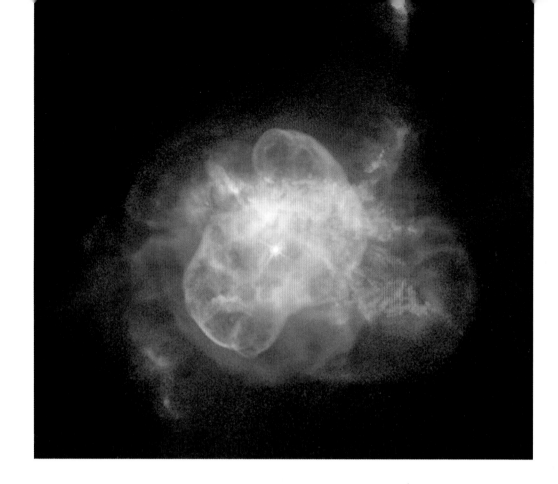

Multiple shells of material ejected by the dying star at the center of NGC6210 can be seen in this Hubble image, which shows the inner region of the planetary nebula in unprecedented detail. The central star is surrounded by a thin, bluish bubble displaying a delicate filamentary structure. The bubble is superposed on a reddish gas formation in which holes, filaments and pillars are clearly visible.

This gives birth to an expanding smoke-ring-like feature that lasts for only a few tens of thousands of years—a blink of an eye on the galactic timescale. But it spews heavier elements back into space for the creation of successive generations of stars and planets.

Hubble has assembled a dazzling collection of pictures of these stellar death shrouds that reveal surprisingly intricate glowing patterns spun into space by aging stars: pinwheels, lawn-sprinkler-style jets, elegant goblet shapes and even some that look like a rocket engine's exhaust. These incandescent sculptures are forcing astronomers to rethink stellar evolution. In particular, the patterns may be woven by an aging star's interaction with unseen companions, such as planets, brown dwarfs or smaller stars.

Hubble has also uncovered "doughnuts" of dust girdling a star that pinch outflowing gas. These may be linked to the influences of an invisible companion, such as another star or a massive planet. This creates jets of high-speed particles that shoot out in opposite directions from the star and plow through surrounding gas, like the stream of water from a garden hose hitting a sandpile. Hubble sees remarkably sharp inner bubbles of glowing gas—like a balloon inside a balloon—blown out by the faster-moving hotter gas that appears later in the star's demise.

Strange, glowing "red blobs" positioned along the edge of some planetary nebulas may be chunks of older gas caught in the stellar gale of hot flowing material from the dying star. Other planetaries display pinwheel patterns formed by a symmetrical ejection of material, creating intricate structures that are mirrored on the opposite side of the star.

It remains a puzzle, though, how these nebulas acquire their complex shapes and symmetries.

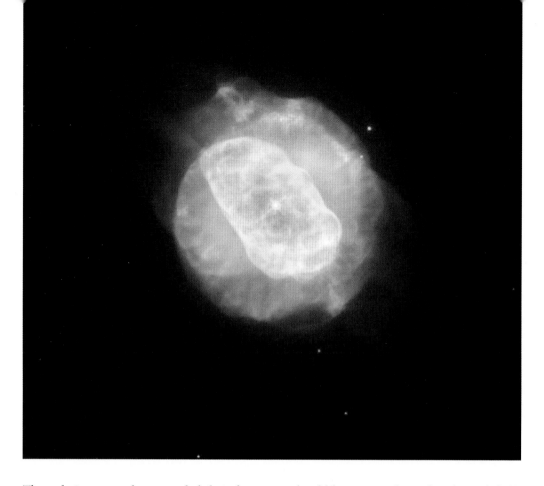

Like a balloon within a balloon, planetary nebula NGC5882 has an elongated inner shell of gas and a fainter outer shell that surrounds it. Hubble's sharp view reveals the intricate knots, filaments and bubbles within these shells.

The red giant stars that preceded their formation should have ejected simple spherical shells of gas. Hubble's ability to see very fine structural details is enabling astronomers to look for clues to this cosmic enigma.

A far more violent fate awaits stars that are over seven times the mass of the Sun. These stars burn their fuel at a blindingly fast rate, blazing 1,000 times brighter than the Sun. And even though they have bulked up on much more hydrogen fuel than the Sun's initial supply, it is largely burned off in barely 100 million years.

Then the fireworks begin.

Late in its life, a massive star has an "onion-skin" structure of concentric shells of nuclear fusion. The outermost shell converts hydrogen into helium; the next layer down, helium into carbon; the level below that, carbon into oxygen; and so on. These stars are immense factories for creating elements that will later be used as building blocks for other stars, planets and life forms in our galaxy.

But the superstar forms a cold heart of iron. And this element cannot sustain nuclear fusion. No longer supported by the release of nuclear energy, the exhausted star's massive core implodes. The shock wave rumbles up through the star and smacks head-on into an avalanche of infalling material from the star's outer layers. The resulting blast wave rips the star to shreds. It unleashes as much energy in a few weeks as the Sun does over 10 billion years.

The result of the collapse may leave behind a rapidly rotating neutron star that can be observed many years later as a radio pulsar. Or, if the core is especially massive, it will become a black hole.

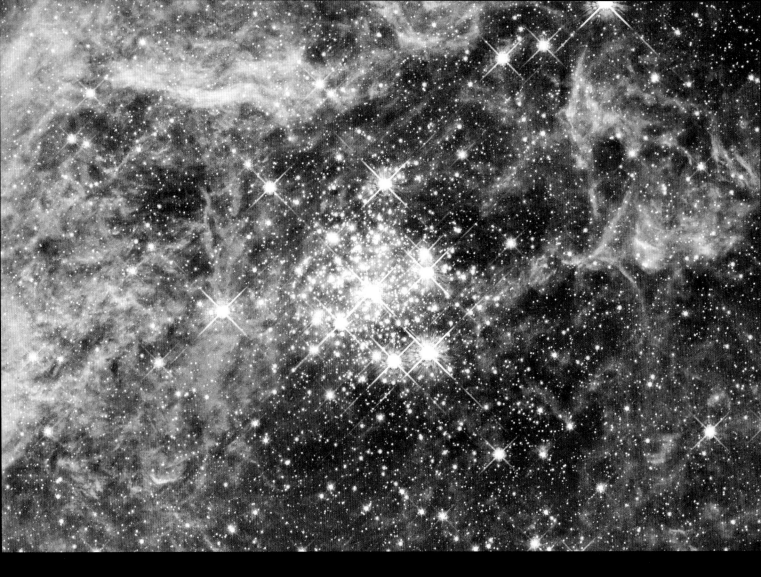

Supernovas can explode like a string of firecrackers even in a young cluster of stars, because the most massive stars in the cluster burn out in just a few million years, triggering a titanic explosion. Both of the young clusters shown on these two pages are 170,000 light-years away in the Large Magellanic Cloud. Hodge 301, above, is only 25 million years old, yet some of its aging red supergiant stars have already ended their lives as supernovas. The star cluster NGC2060, facing page, contains a supernova that exploded about 10,000 years ago, blowing out gas surrounding the cluster.

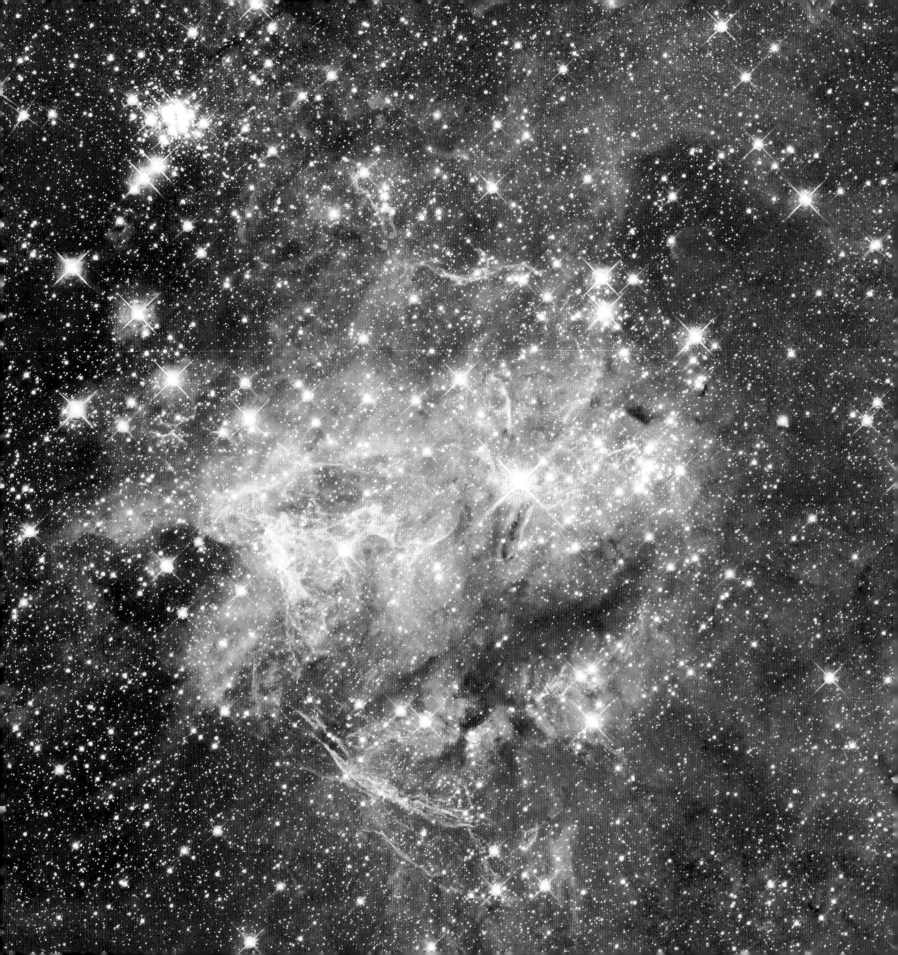

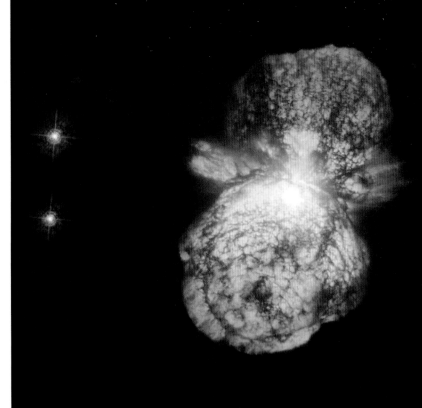

## VY CANIS MAJORIS

Below: This red giant star is 40 times the mass of our Sun and 500,000 times brighter. Its eruptions have formed loops, arcs and knots of material that are hurtling into space in all different directions. They were produced by localized eruptions from active regions on the star's surface. At every outburst, the star loses about 10 times more mass than its normal rate. If the bloated VY Canis Majoris were to replace the Sun, its surface would extend to the orbit of Saturn. The arcs and knots represent massive outflows of gas probably ejected from large star spots or convective cells on the surface, analogous to the sunspots and prominences associated with the Sun's magnetic fields but on a vastly larger scale.

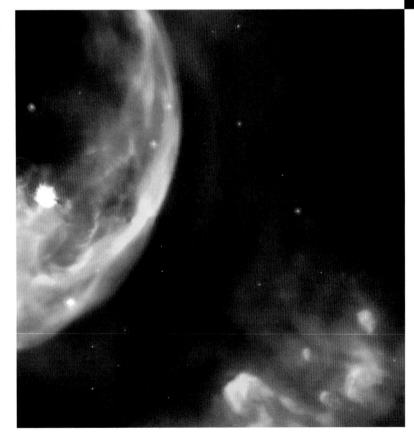

## ETA CARINAE

Above: During the early 1800s, observers in the southern hemisphere noticed that Eta Carinae was gradually increasing in brightness. By 1843, it ranked as the second brightest star in the night sky. It then dimmed over the next half-century, fading below naked-eye visibility by 1900. Eta Carinae had released as much visible light as a supernova yet survived the event. This slow-motion explosion produced two lobes and a large, thin disk, or waist, presumably around the star's equator. The figure-eight structure is laced with dust, giving it a filamentary appearance. Eta Carinae is likely a double-star system, with the primary star about 100 times our Sun's mass and a companion half that size. The larger star could explode any day, destroying its companion. Since Eta Carinae is 8,000 light-years away, it may have already exploded, sending a blast of light and other radiation heading in our direction.

## NGC7635

This remarkably spherical six-light-year-diameter "bubble" marks the boundary between an intense stellar wind and the more quiescent interior of the nebula. The beefy central star, which is 40 times the Sun's mass, has stellar-wind hurricanes moving outward at more than six million kilometers per hour. The bubble's surface marks the leading edge of the wind's gust front. As the shell expands, it slams into mottled regions of cold gas, which slows the expansion by differing amounts and creates the bubble's rippled appearance. Located in the constellation Cassiopeia, the Bubble Nebula is 7,000 light-years away.

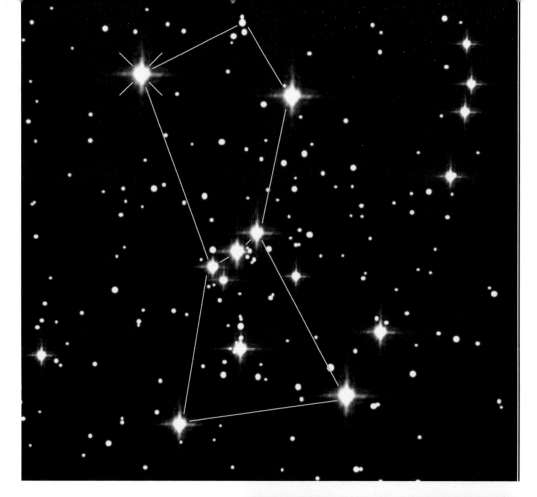

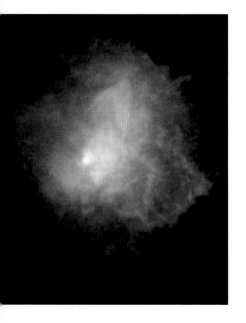

## BETELGEUSE

Stars are so far away that they remain mere pinpoints of light in telescopic images, even in Hubble's crisp views. One exception is the red supergiant star Betelgeuse, located 600 light-years away, in the constellation Orion. The bloated, aging star has expanded to beyond the diameter of Jupiter's orbit. If our Sun were reduced to the size of a tennis ball, Betelgeuse, on the same scale, would fill a sports arena. Hubble resolved the disk of the mighty star, revealing a huge ultra-violet atmosphere with a mysterious hot spot on the surface of the stellar behe-

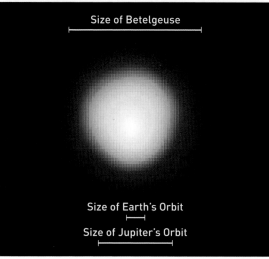

Size of Betelgeuse

Size of Earth's Orbit

Size of Jupiter's Orbit

moth. Twice the diameter of the Earth's orbit, the enormous bright spot is at least 1,600 degrees C hotter than the surface of Betelgeuse. This is a huge convection cell where hot material from the star's interior is bubbling up to the surface. Subsequent studies with other telescopes show that the star has a mottled surface and huge plumes of ejected gas.

# AGING STARS

## NGC3918

A red giant star lies at the center of this immense cloud of gas. During the star's final convulsive phase, as its nuclear furnace becomes unstable, huge clouds of gas are ejected from the surface of the star. Intense ultraviolet radiation from the tiny remnant star then causes the surrounding gas to fluoresce. With its bright inner shell of gas and a more diffuse outer shell that extends far from the nebula, NGC3918 has a distinctive eyelike shape that looks as if it could be the result of two separate ejections of gas. Actually, the shells formed at the same time but are being blown from the star at different speeds. The powerful jets of gas emerging from the ends of the large structure are traveling at speeds of up to 320,000 kilometers per hour.

## MINKOWSKI 2-9

This is a striking example of a "butterfly," or bipolar, planetary nebula. If Minkowski 2-9 is sliced in half vertically, each side appears much like the exhaust from a jet engine. The central star is really a binary system. The gravity of one star pulls weakly bound gas from the surface of the other and flings it into a thin, dense disk that surrounds both stars and extends well into space. The disk measures approximately 10 times the diameter of Neptune's orbit and can actually be seen in Hubble pictures.

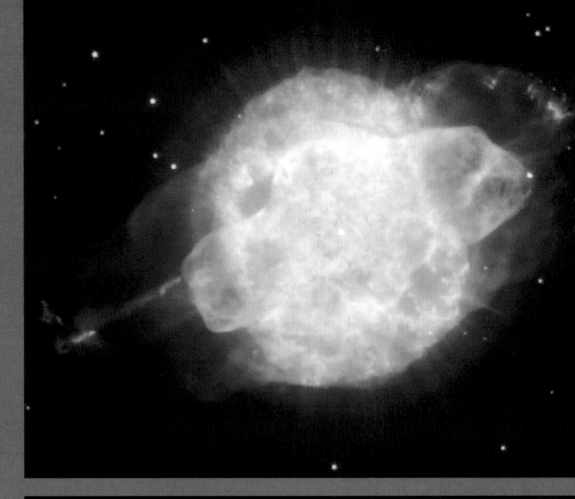

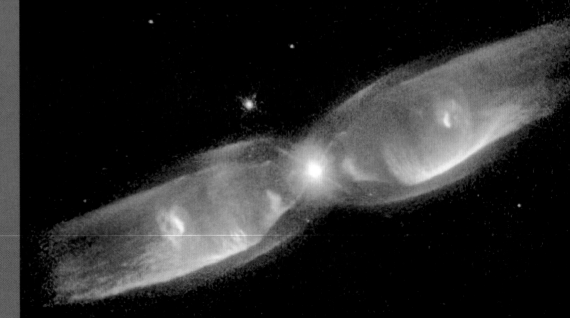

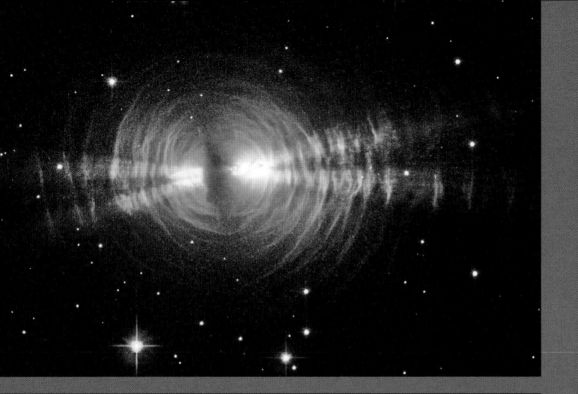

## EGG NEBULA

Concentric dust layers extend over one-tenth of a light-year from this dying sun. Running almost vertically through the image, a thick dust belt blocks the light of the central star. Twin beams of light radiate from the hidden star, illuminating the pitch-black dust like a flashlight shining in a smoky room. The nebula was photographed through polarizing filters to measure how the dust reflects light.

## RED RECTANGLE

This striking ladderlike structure, nicknamed Stairway to Heaven, surrounds a dying star. Cataloged as HD 44179, this nebula is more commonly called the Red Rectangle because of its unique shape and color. What looks like the rungs of a ladder are projections of gas cones, like a series of nested wine glasses. The outflows are ejected from the star in two opposing directions. The "rungs" may have arisen in episodes of mass ejection that occur every few hundred years and could represent a series of "smoke rings," seen almost exactly edge-on from our vantage point. The hourglass shape of the nebula suggests that the central star is actually a close pair of stars orbiting each other with a period of about 10 months. The stirring motion of the pair has ejected a dust disk that obscures our view of the binary.

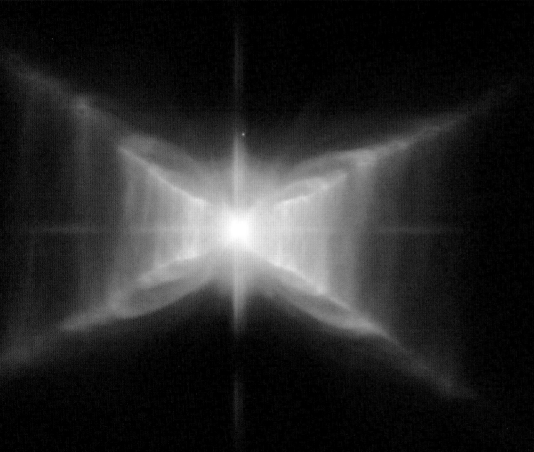

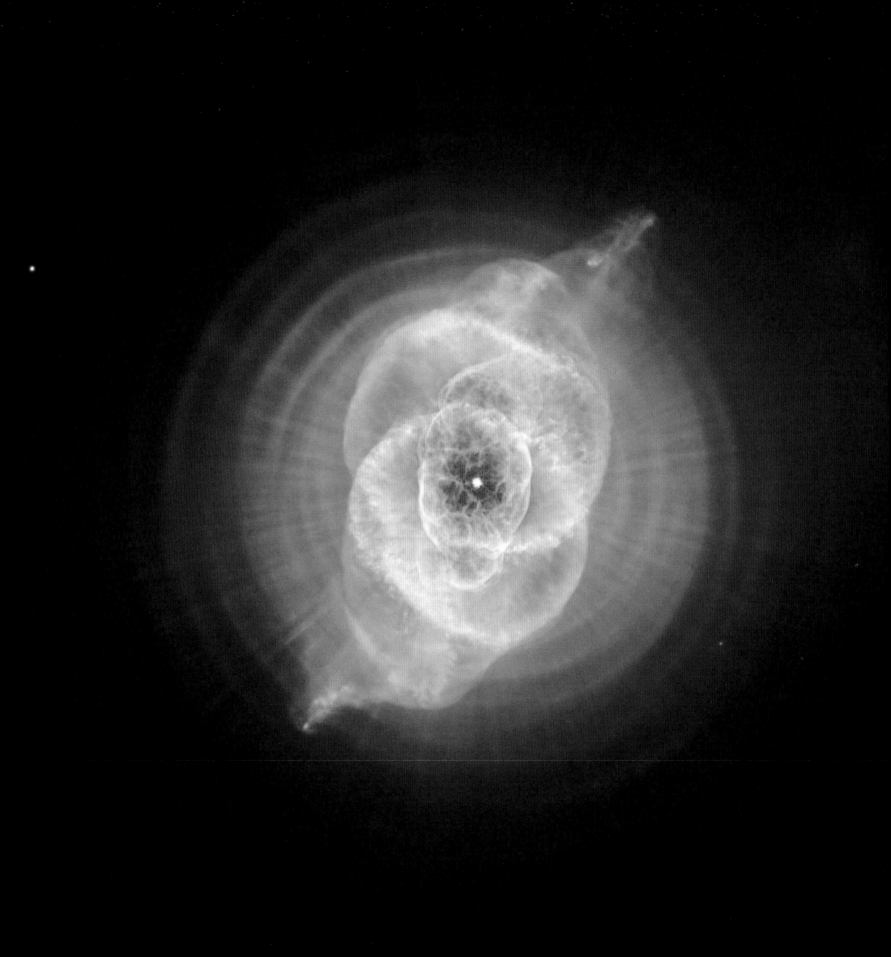

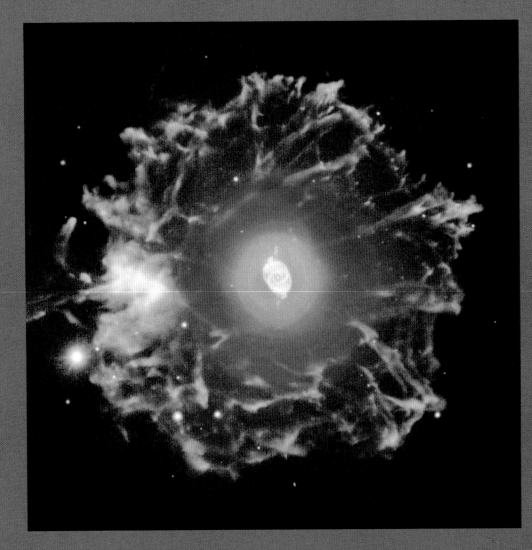

CAT'S EYE NEBULA

Among the first planetary nebulas to be discovered, the Cat's Eye was the very first to be recognized as a bubble of hot gas. Today, two centuries later, it remains one of the most complex-looking planetary nebulas known. As can be seen in the image on the facing page, Hubble resolves concentric gas shells, jets of high-speed gas and unusual shock-induced knots of gas. It also reveals a faint bull's-eye pattern of concentric dust shells. These were ejected by the star in a series of pulses before it collapsed down to a white dwarf. The pulses, which occurred at a rate of one every 1,500 years or so, created dust shells, each of which contains as much mass as all the planets in our solar system combined. The image above, a wider view taken by a ground-based telescope, shows that the core region photographed by Hubble is surrounded by an outer shell ejected millennia ago.

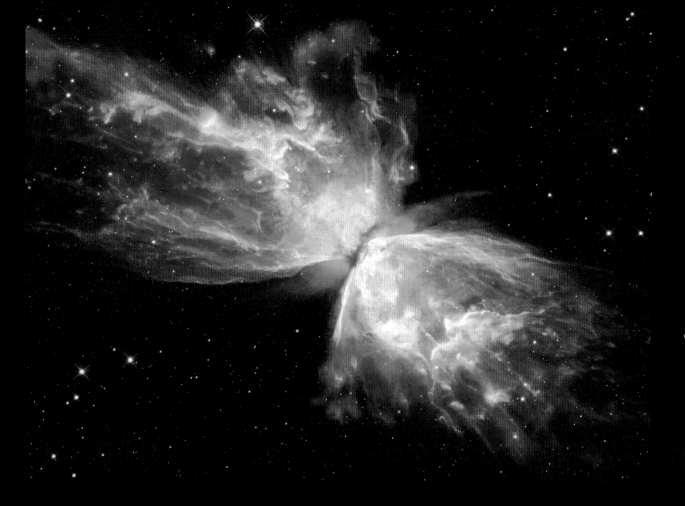

## BUTTERFLY NEBULA

Roiling bubbles of gas heated to more than 20,000 degrees C and speeding into space at over one million kilometers per hour form these ethereal "butterfly wings." This is a particularly volcanic planetary nebula because the dying star at the center is about five times the mass of the Sun. The glowing gas is the star's outer layer, which was expelled roughly 2,000 years ago. The Butterfly's "wingspan" measures more than two light-years. The central star itself cannot be seen because it is hidden within a dark, doughnut-shaped ring of dust. This thick dust belt pinches the nebula in the center, constricting the star's outflow and creating the classic bipolar, or hourglass, shape displayed by some planetary nebulas. This is strong circumstantial evidence for a double-star system at the center.

## HELIX NEBULA

In one of the most striking views of a planetary nebula, Hubble resolves a fine web of radial filamentary "bicycle-spoke" features embedded in the red and blue gas ring. Looks can be deceiving in the Helix Nebula. Although it looks like a bubble, it is actually a face-on tunnel, or open-ended barrel, of glowing gases. This fluorescing tube is pointed nearly directly at Earth, hence the bubble appearance. Thousands of cometlike filaments along the inner rim of the nebula point back toward the central star, which is a small but superhot white dwarf. These tentacles formed when a hot stellar wind of gas plowed into colder shells of dust and gas previously ejected by the doomed star. Nearly three light-years across, the huge nebula is approximately three-quarters of the distance between our Sun and the nearest star.

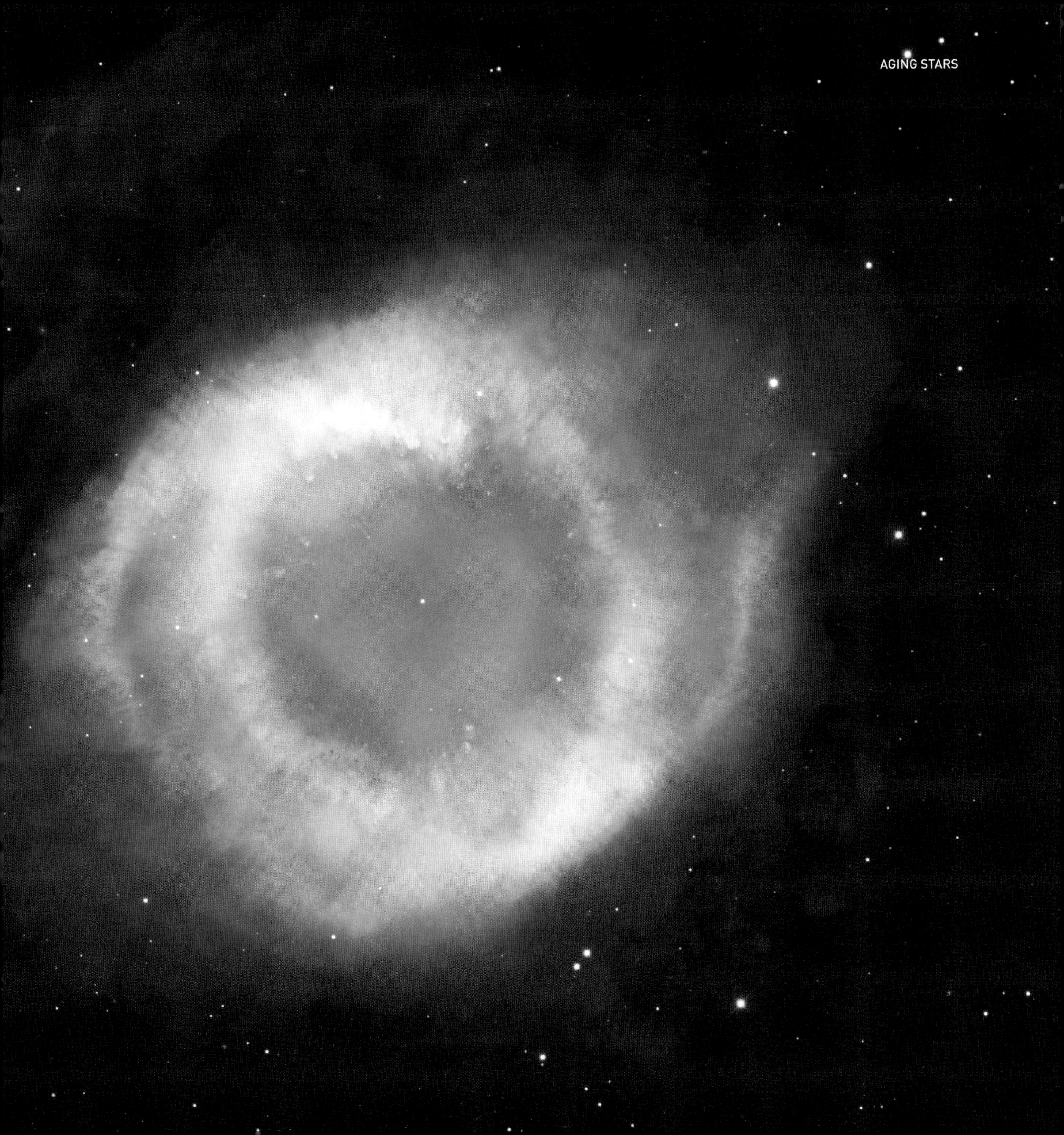

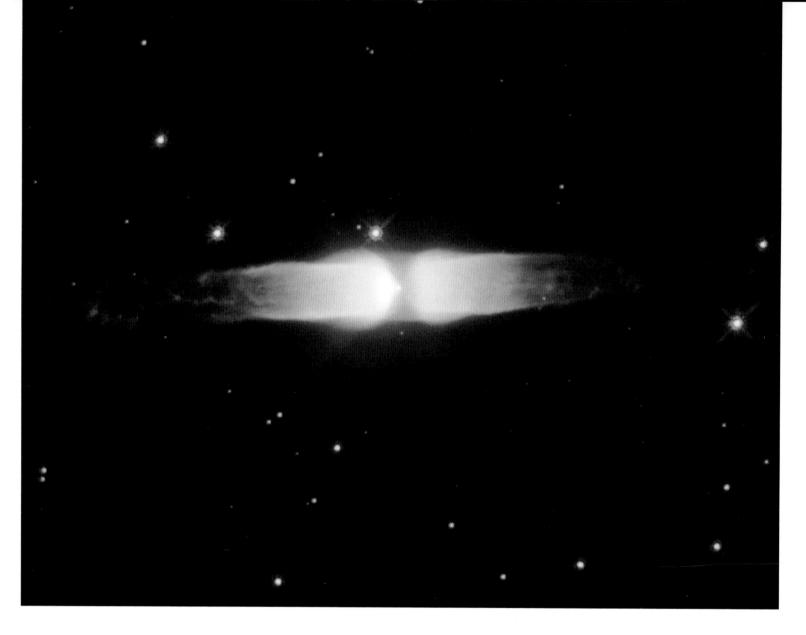

## HENIZE 3-401

The young planetary nebula Henize 3-401 looks like a pair of jet engines. The two very long, cylindrical outflows have intricate threadlike structures and tattered ends. The central star peeks out from behind a dark disk of dust presumably formed by binary stars at the center. The disk, perhaps combined with the action of powerful magnetic fields around the star, funnels the gas like a rocket exhaust.

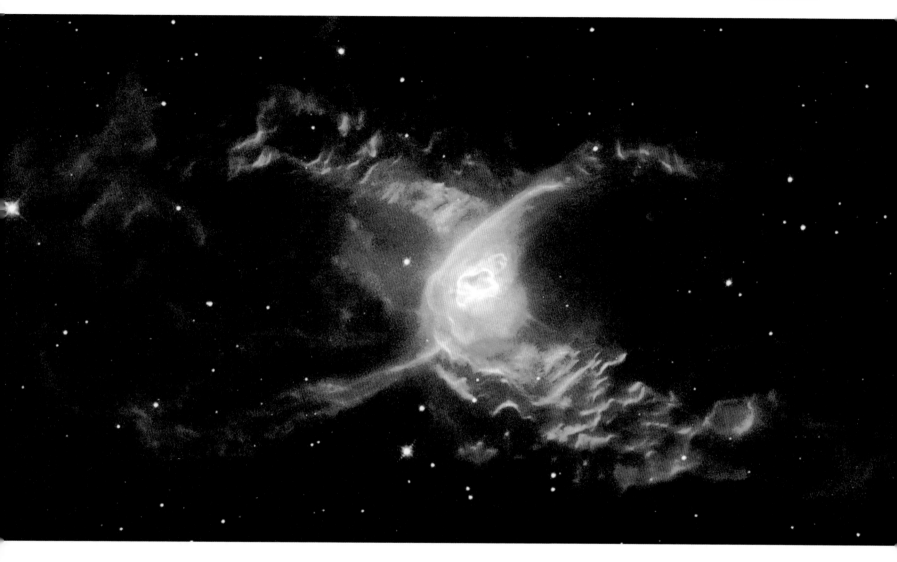

SPIDER NEBULA

Powerful stellar winds blasted off the central star create a wavelike pattern that is sculpting the four-legged Spider Nebula. The gale of charged particles is blowing at about 16 million kilometers per hour. The waves themselves move outward at a slower rate of approximately 965,000 kilometers per hour. The gas walls of the two lobed structures are not at all smooth but display complex ripples. The central white dwarf is at least 300,000 degrees C, making it one of the hottest stars known. The nebula is located 3,000 light-years away, in the constellation Sagittarius.

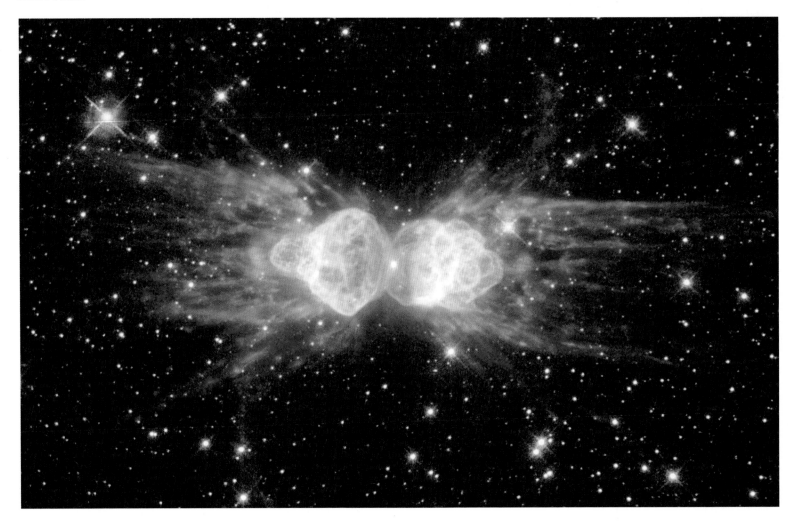

### ANT NEBULA

Designated Menzel 3 (Mz 3), this member of Hubble's celestial menagerie resembles the head and thorax of a garden ant. The central star in Mz 3 might have a closely orbiting companion that is exerting strong gravitational tidal forces which are shaping the outflowing gas. The very massive young star Eta Carinae shows a similar outflow pattern to that of Mz 3.

### ESKIMO NEBULA

Designated NGC2392, this planetary nebula is dubbed the Eskimo Nebula because, as seen in ground-based telescopes, it resembles a face peeking out of a furry parka. Through Hubble's eyes, the "furry" features look like giant comets radially pointing away from the central star, like the spokes of a wheel. The clumps that form the "comet heads" all seem to be located at a similar distance from the star. This effect is caused by hot, rarefied gas slamming into the cooler, denser gas where the fingerlike features will form. It is a characteristic signature of many planetary nebulas.

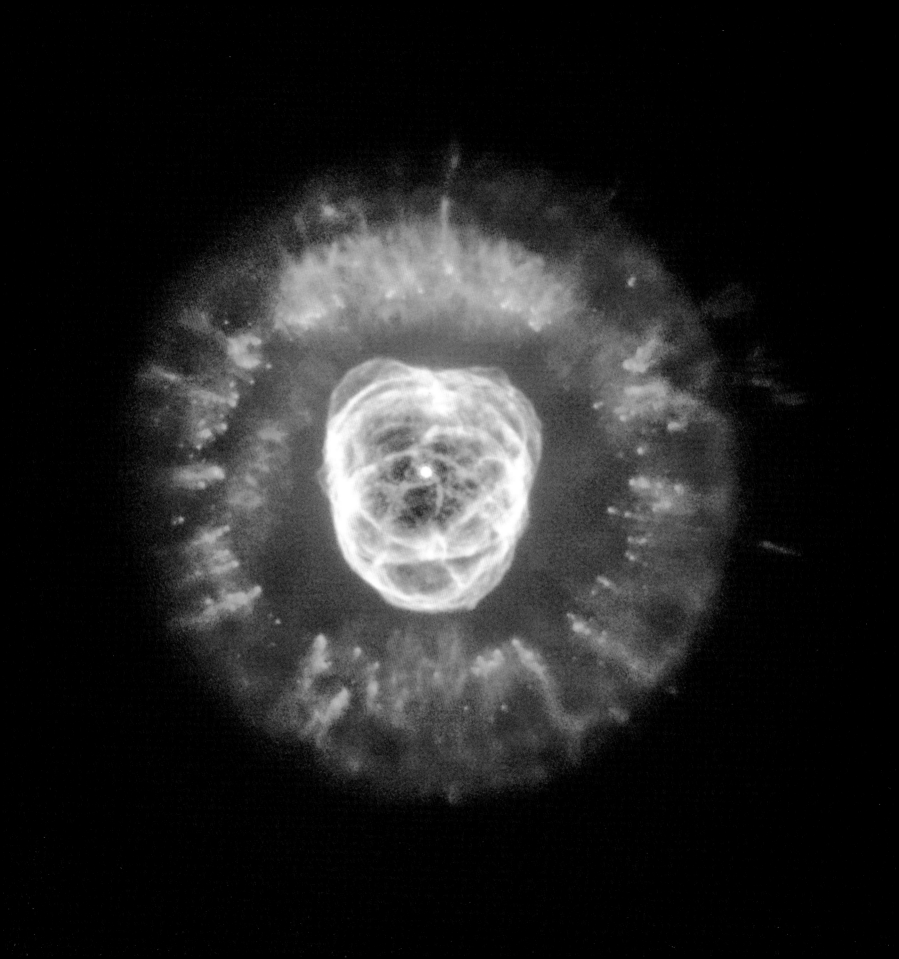

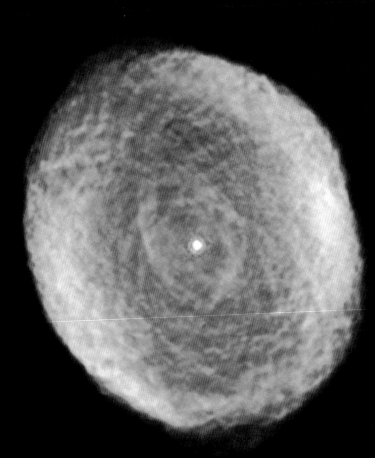

IC418

This is one of the few planetary nebulas photographed by Hubble
that has a spherical shape, which would suggest that a single
star rather than a binary pair ejected the gases that are shaping
the nebula. The inner blue bubble indicates that there's been more
than one outburst. The processes forming the delicate-looking
filaments remain a mystery.

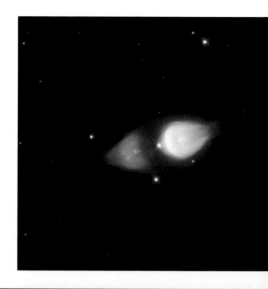

## MINKOWSKI 92

Right: Two vast onion-shaped structures on either side of an aging star give it an uncanny appearance. Intense pulsations on the surface of the dying star at the center of the nebula are puffing out material. The "onions" are made visible from the light of the central star. In a few thousand years, as the star becomes hotter, its ultraviolet radiation will light up the surrounding gas from within, causing it to glow.

## IRAS 13208-6020

This object has a clear bipolar form, with two very similar outflows of material streaming in opposite directions and a dusty ring encircling the star.

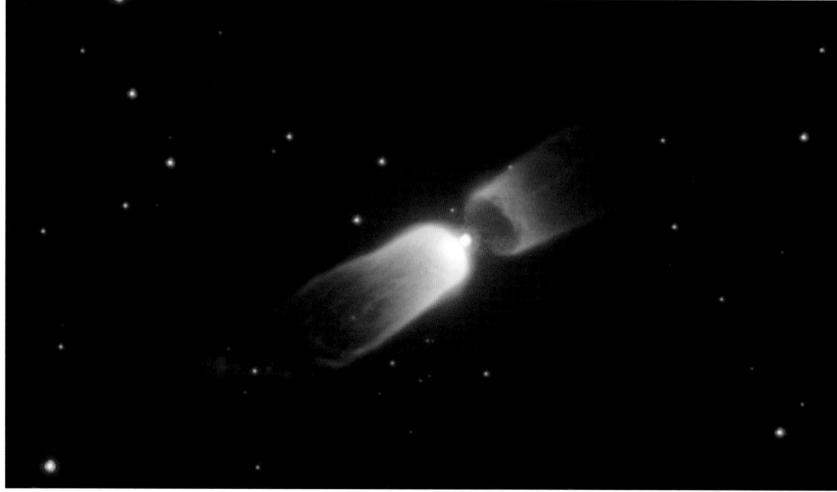

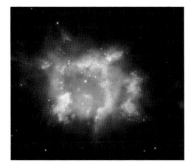

NGC6326

Above: This planetary nebula looks like a fireworks pinwheel. Sometimes a star's ejected gas creates elegantly symmetric patterns, but NGC6326 looks like a paint splat. This object is located 11,000 light-years away, in the constellation Ara.

IRAS 19475+3119

Top right:  A cloud of ejected material from the central star in this eerie, almost angelic-looking nebula shines by reflecting the star's brilliant light. Jets from the star are creating symmetrical hollow lobes. The nebula lies 15,000 light-years away, in the constellation Cygnus.

IC4634

Right: Two S-shaped ejections from a dying star make this nebula look like a mini-galaxy. These patterns suggest that there were distinct, separate waves of thrown-off gases. One is farther from the central star and, therefore, was spewed first, followed by a more recently ejected tide of matter that formed the tighter S-shape. The nebula is 7,500 light-years away, in the constellation Ophiuchus.

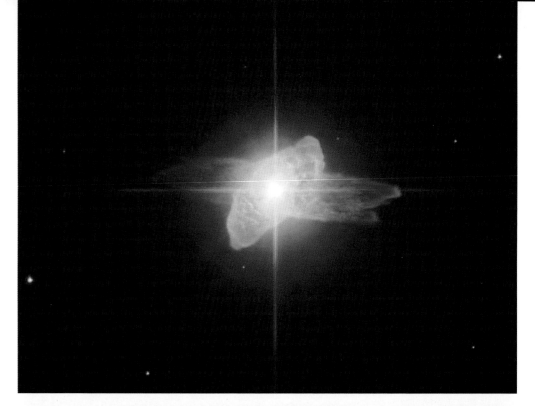

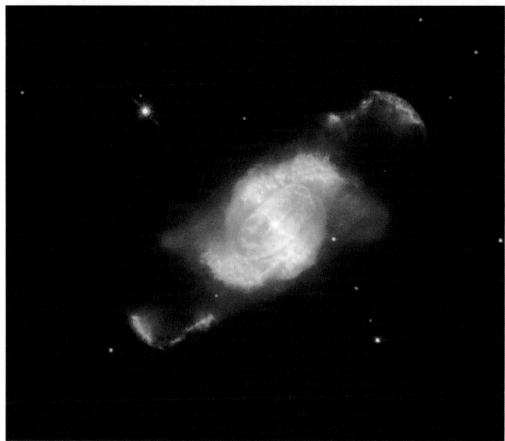

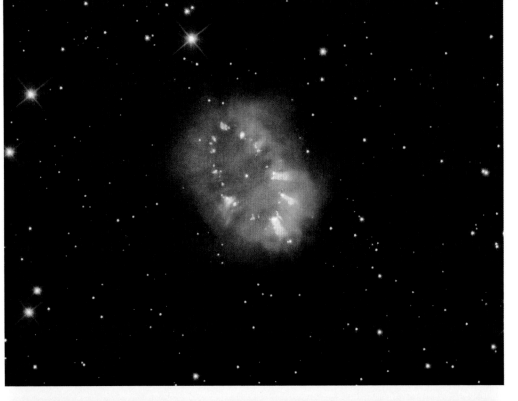

### NECKLACE NEBULA

This dazzling ring of gas, measuring 20 trillion kilometers across, is dotted with dense knots of gas that resemble jewels in a necklace. A pair of stars orbiting very close together produced the nebula. One of the aging stars ballooned to the point where it enveloped its companion, which continued orbiting inside the larger star, causing it to spin so fast that much of its gaseous envelope expanded into space. Because of centrifugal force, most of the gas escaped along the star's equator, producing the ring. The embedded bright knots are the densest gas clumps in the ring. The stars are furiously whirling around each other, completing an orbit in a little more than a day.

### IRAS 22036+5306

The cast-off material encircling this dying star could be the remnants of comets and other small, rocky bodies. The dark dust ring probably formed through the interactions of a pair of stars in the center, in which the dying star absorbed its companion. Twin jets spout from the star's poles, hurtling outward at more than 600,000 kilometers per hour. The nebula is 6,500 light-years away, in the constellation Cepheus.

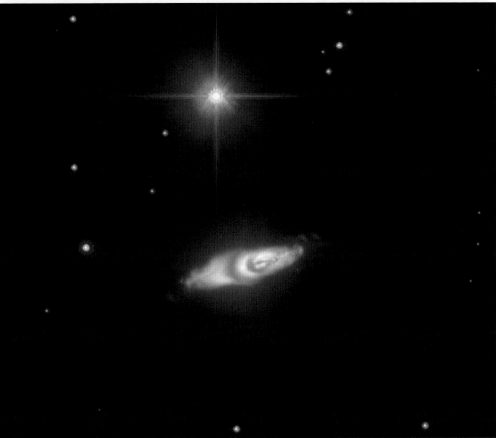

# SUPERNOVA REMNANTS: CELESTIAL TOMBSTONES

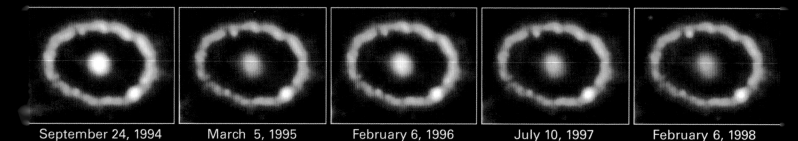

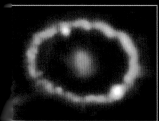 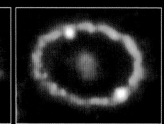 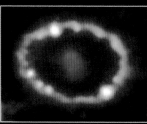 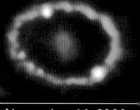

| September 24, 1994 | March 5, 1995 | February 6, 1996 | July 10, 1997 | February 6, 1998 |
|---|---|---|---|---|

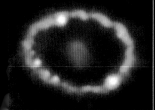 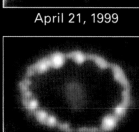 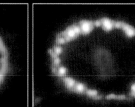 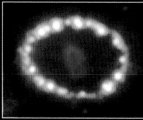 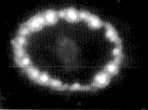

| January 8, 1999 | April 21, 1999 | February 2, 2000 | June 16, 2000 | November 14, 2000 |
|---|---|---|---|---|

| March 23, 2001 | December 7, 2001 | January 5, 2003 | August 12, 2003 | November 28, 2003 |
|---|---|---|---|---|

## SN 1987A

On February 23, 1987, the brightest supernova seen from Earth in four centuries lit up the southern sky. The doomed star exploded 169,000 light-years away, in the Large Magellanic Cloud. Although Hubble was not operating in 1987, its pictures a few years later revealed a glowing ring of gas, 10 trillion kilometers in diameter, encircling the supernova remnant. The ring is a relic of the hydrogen-rich stellar envelope that was ejected, in the form of a gentle "stellar wind," by the progenitor, which may have expanded and engulfed a smaller companion star 10,000 years before the explosion. This would have spun up the rotation rate of the giant star, causing it to shed material most efficiently along its equator. In the first few hours following the supernova, the ring glowed as it was flooded with ultra-violet radiation. When the ring began to fade, astronomers witnessed a tidal wave of kinetic energy as the supernova blast slammed into it. Shock waves resulting from the impact of the ejecta brightened 30 to 40 pearl-like "hot spots" in the ring. These blobs are growing and merging together to form a radiant circle. Over the years, astronomers have observed a long, dumbbell-shaped structure at the center of the ring consisting of two blobs of debris speeding apart at nearly 10 million kilometers per hour.

## N132D

Intricate wisps of glowing gas float amid myriad stars in this view of a 3,000-year-old supernova remnant The titanic explosion took place in the Large Magellanic Cloud. As the expanding supersonic shock wave from the supernova impacted the thin interstellar gas that pervades the galaxy, it sculpted the complex structure of N132D.

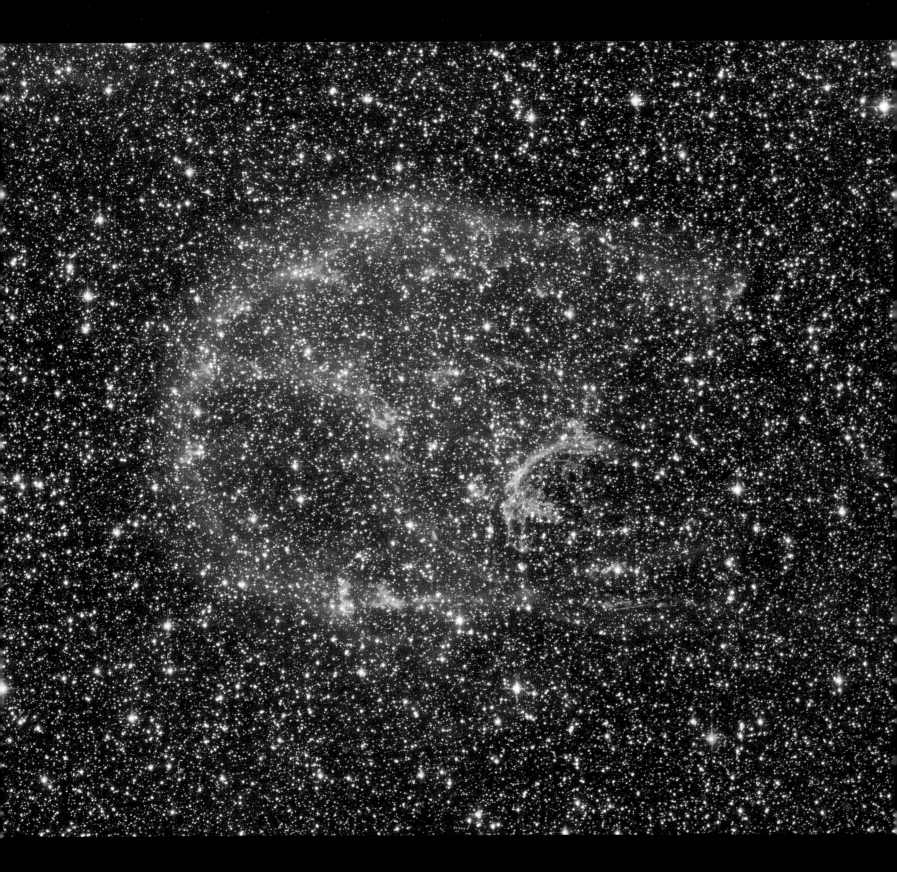

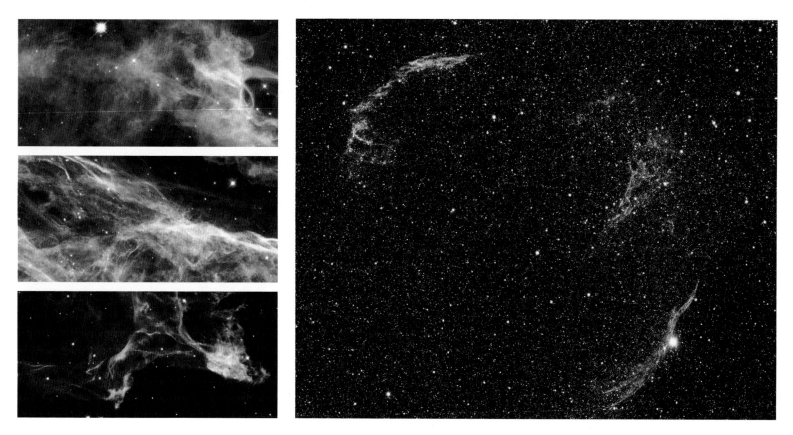

## CYGNUS LOOP

The three images above show small portions of the gigantic Cygnus Loop, seen in the wide-angle image, above right, taken by an amateur astronomer. The 15,000-year-old supernova remnant covers a region of sky six times the diameter of the full Moon. The supernova blast wave has recently slammed into a cloud of denser interstellar gas, creating shock waves within the cloud that are heating the gas and causing it to glow. Blue shows emission from oxygen, red from sulfur and green from hydrogen.

## CASSIOPEIA A

The annihilation of a star in the late 17th century created the supernova remnant Cassiopeia A, a broken 10-light-year-diameter ring of filamentary and clumpy stellar ejecta. Mysteriously, no bright star that would indicate the supernova explosion was reported by observers of the time. Huge swirls of debris glow with the heat generated by the passage of a shock wave from the blast. Cassiopeia A is expanding at nearly 50 million kilometers per hour.

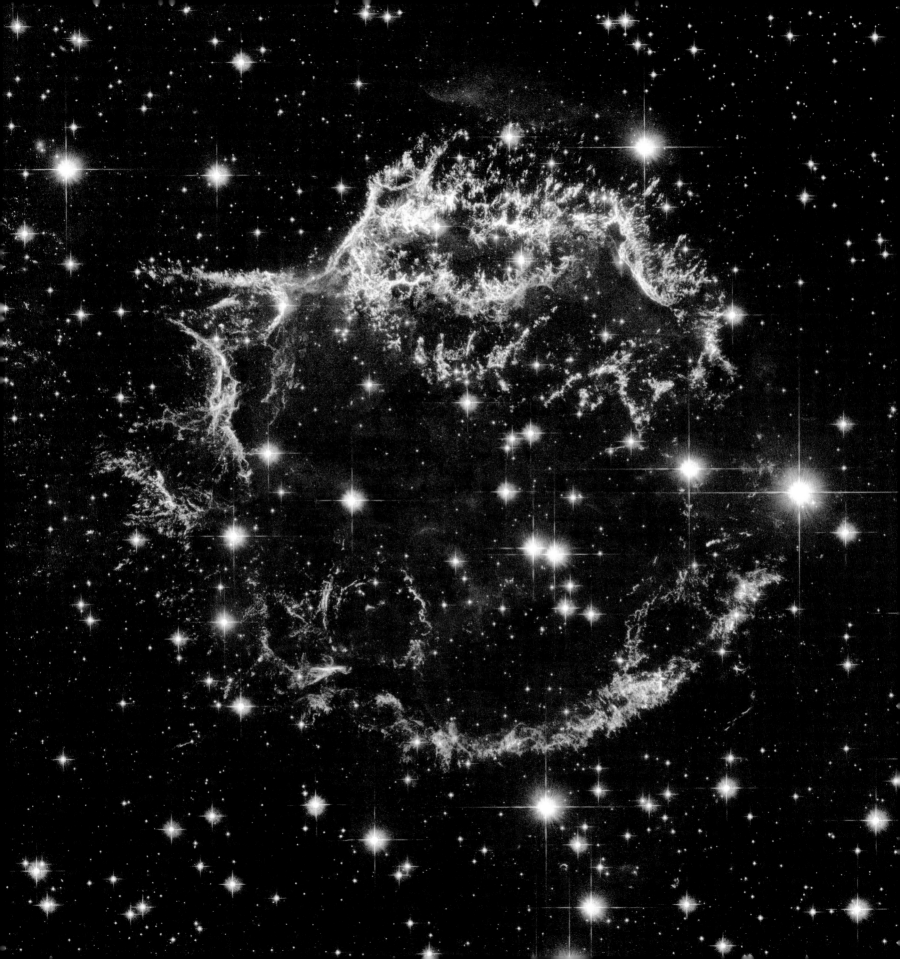

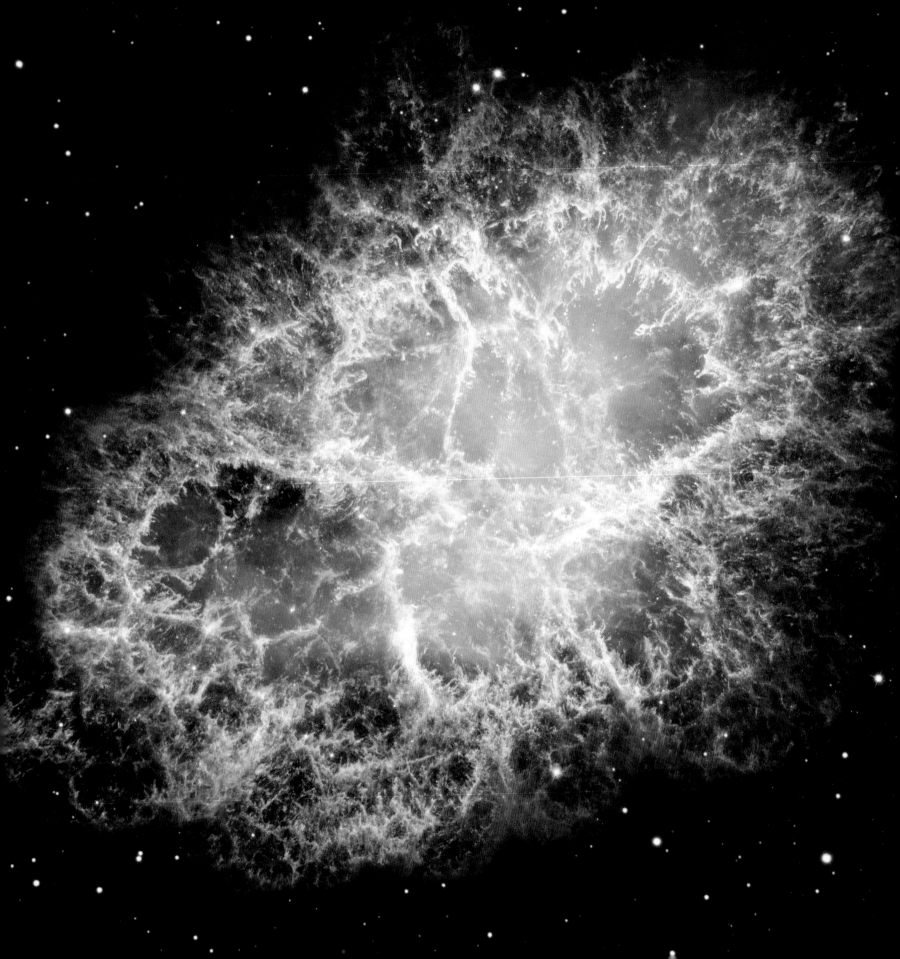

## CRAB NEBULA

The most photogenic of all the known supernova remnants was initially observed in the mid-1700s and soon became the first object in Charles Messier's catalog of nebulas. By 1900, astronomers had measured the expansion rate of the Crab Nebula and calculated that the explosion must have occurred approximately 900 years earlier. This corresponds to Chinese records of a bright star appearing in the daytime sky in 1054. Since then, the nebula has expanded to a celestial cloud six light-years across. The filaments are the tattered remains of the former star and consist mostly of hydrogen, with traces of sulfur and oxygen. In 1968, an energetic central object—the remnant core of the progenitor star—was found to be emitting radio pulses at 30 times per second, which meant the object must be rotating that fast and shooting out beams of energy. Only an ultradense neutron star could spin that fast and still remain intact. Called a pulsar, the flashing neutron star is barely visible in this Hubble image. Hubble photographed a striking pattern of concentric magnetic rings encircling the pulsar (see page 188). The ring-like structures are emitting X-rays. The high-energy particles slam into the nebular material, powering the eerie bluish glow in the interior.

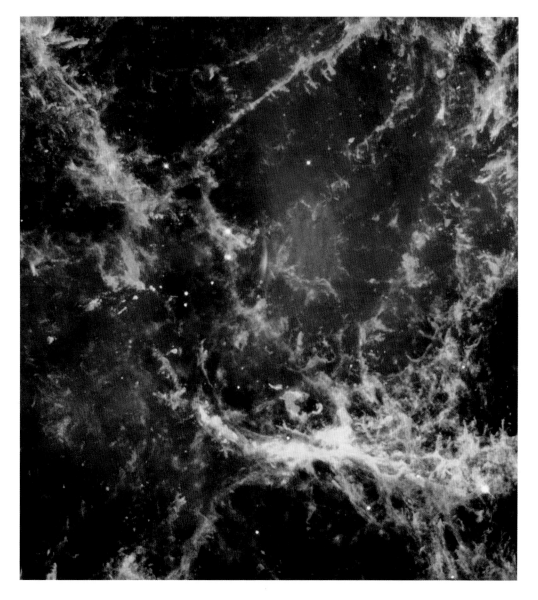

## CRAB CLOSE-UP

Material expelled from the outer layers of the Crab Nebula progenitor star during the supernova explosion formed this colorful network of filaments. While they appear to be close to the pulsar, the yellowish green filaments toward the bottom of the image are actually closer to us and are approaching at 1.6 million kilometers per hour. The orange and pink filaments toward the top of the picture include material that is rushing away from us from behind the pulsar at similar speeds. The Crab pulsar is the lower of the pair of bright stars just to the left of center.

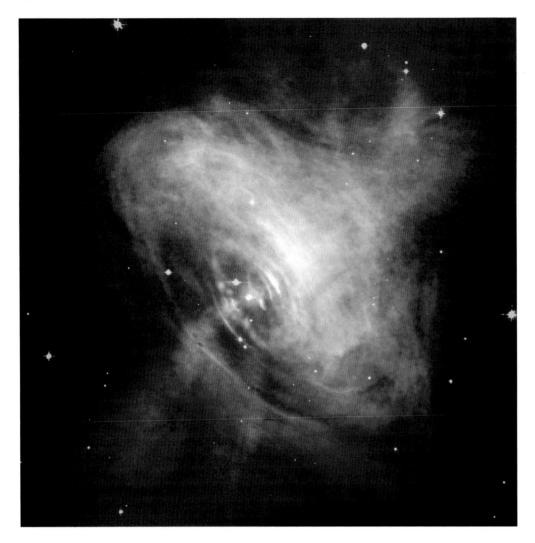

Here is a small portion of a filament on the edge of an expanding "bubble" from a supernova that exploded in the year 1006. The twisting ribbon of light corresponds to locations where the blast wave from the supernova is slamming into very tenuous surrounding gas at 10 million kilometers per hour. The hydrogen gas heated by this fast shock wave emits radiation in visible light. The bright edges within the ribbon are regions where the shock wave is exactly edge-on to our line of sight.

CRAB PULSAR

This close-up Hubble view of the region around the Crab pulsar shows bright wisps moving outward at half the speed of light to form an expanding ring that is visible in both Chandra X-ray Observatory images and Hubble optical pictures (images from both telescopes are combined here). These wisps appear to originate from a shock wave that shows up as an inner X-ray ring. The ring consists of about two dozen knots that form, brighten and fade, jitter around and occasionally undergo outbursts which give rise to expanding clouds of particles but remain in roughly the same location. A turbulent pair of jets shoots out in a direction perpendicular to the inner and outer rings.

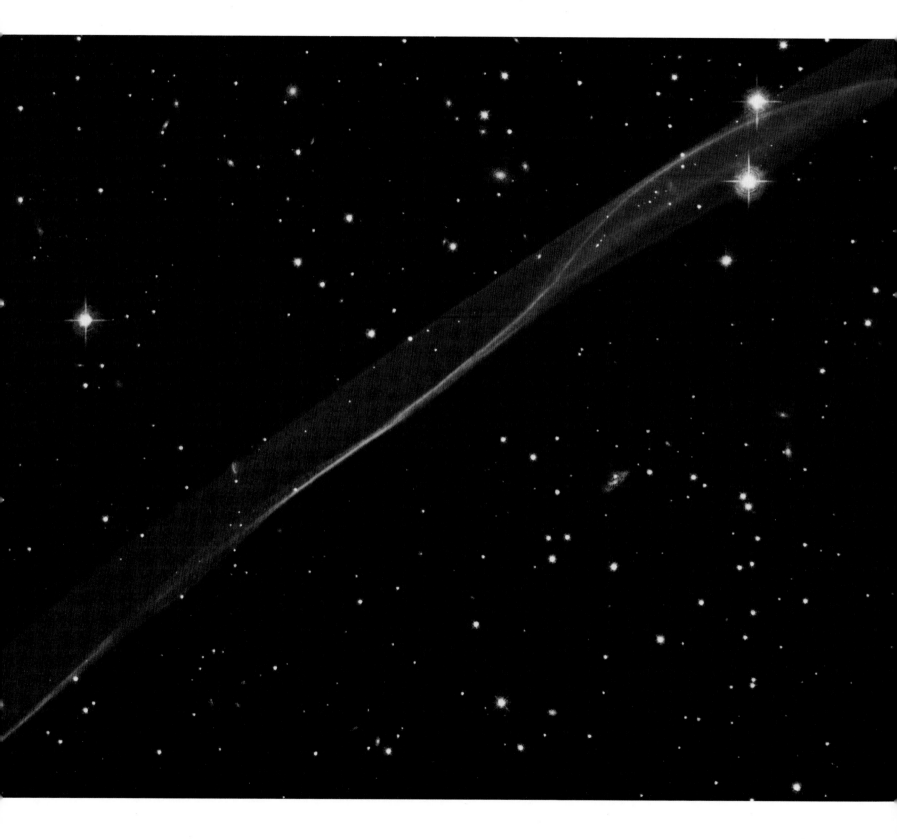

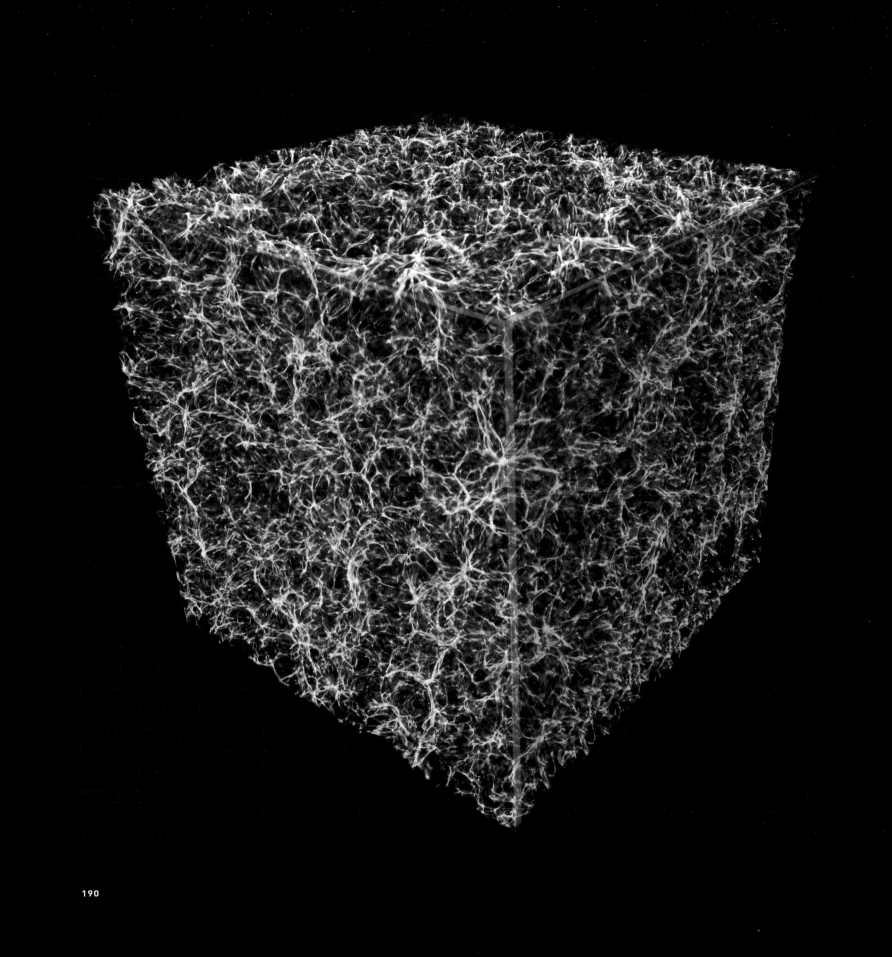

# HUBBLE'S INVISIBLE UNIVERSE

## CHAPTER 7

As spectacular as it all looks in Hubble's views, the universe remains largely hidden from us. Its 10 billion trillion stars are the only truly luminous form of matter in the cosmos. They are essentially lights draped over an unseen structure. All the stars and galaxies represent only a fraction of the mass of the entire assembly. The rest of the scaffolding is made up of a mysterious substance called dark matter.

Although astronomers have known about dark matter for more than half a century, we still don't have a clue what it is. It's not just invisible because it's dark. It is some unknown form of matter that is fundamentally different from the atoms that make stars, planets and people. Physicists hope to someday be able to create a particle of dark matter in a subatomic particle accelerator collision.

Dark matter comprises 25 percent of the universe. Not knowing the nature of dark matter is as unnerving to astrophysicsts as it would be to the rest of us if we did not understand the nature of water, despite the fact that it covers three-quarters of the Earth's surface.

More important, astronomers do realize that dark matter is the scaffolding of the universe. The primeval hydrogen gas that condensed to forge stars gravitationally collected along great filaments of dark matter which coagulated not long after the big bang. This can be demonstrated in computer simulations of the growth of the early universe under the spell of dark matter, which show the universe transitioning from a smooth distribution of matter into a spongelike structure of long filaments. The simulations look a lot like the filamentary structure of galaxies we actually see out there.

Dark matter is aloof from normal matter and doesn't interact with it. But because it has mass, dark matter has gravity, just as normal matter does. Therefore, its ghostly influence can be precisely measured. It's a little like being in a room with H. G. Wells' fictional "invisible man."

Without the gravity of dark matter, the Milky Way's 100 billion stars would not remain in a stable pancake-shaped disk but would fly off into deep space. Our galaxy is embedded in a huge cloud of dark matter that is at least 10 times as massive as all the stars in our galaxy. Likewise, entire clusters of galaxies would disperse without the gravitational glue of vast quantities of dark matter holding them together.

Some theoreticians have suggested that rather than coming from an invisible form of matter, the gravity attributed to dark matter might actually be gravity leaking into our universe from a parallel universe. And therefore, there is no dark matter. Although it is omnipresent, gravity is the weakest of the four forces in nature. One theoretical explanation is that it is anemic because it leaks into other dimensions.

But Hubble has provided direct observational evidence for the reality of dark matter by map-

A computer simulation of a trillion-cubic-light-year cube of our universe shows the structure of dark matter—the invisible scaffolding that defines the distribution of galaxies and galaxy clusters. Although omnipresent and fundamentally important, dark matter remains largely mysterous despite half a century of investigation.

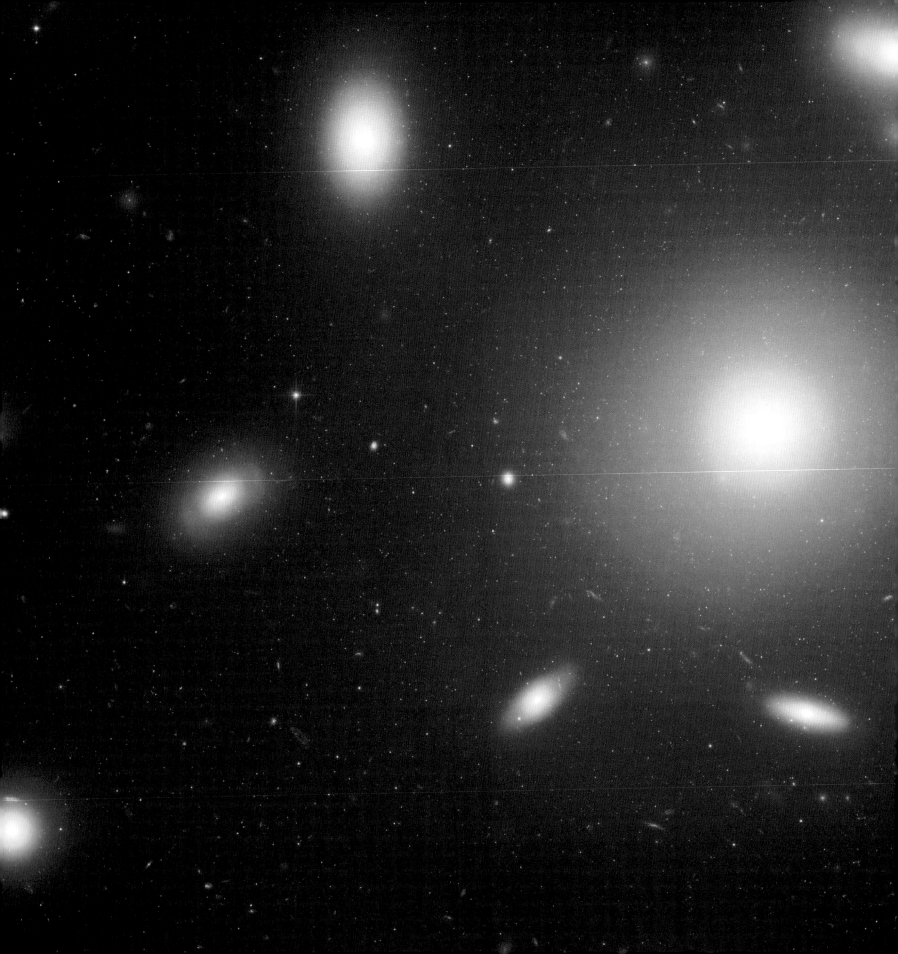

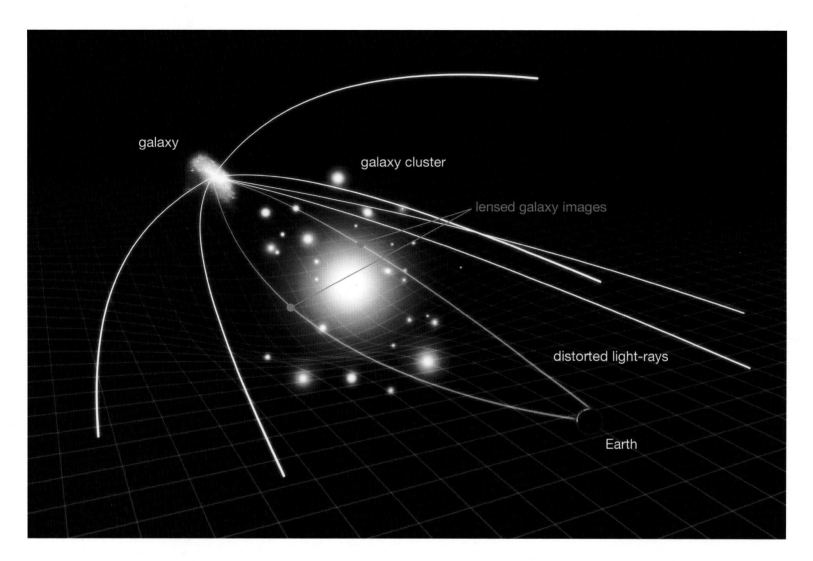

galaxy

galaxy cluster

lensed galaxy images

distorted light-rays

Earth

ping its distribution in space, rather than simply measuring the amount of its ghostly gravitational pull. This mapping of dark matter's distribution in space and time shows how galaxies have grown and clustered over billions of years. Tracing the growth of clustering under the influence of dark matter may eventually also shed light on dark energy, a repulsive form of gravity that influences how quickly dark matter clumps over time.

## SEEING DARK MATTER

Hubble has made exquisite maps of dark matter because it can see its ghostly gravitational foot-print on the universe. The phenomenon is called gravitational lensing, and it is nature's largest optical illusion.

In his theory of general relativity, Albert Einstein first hypothesized that gravity warps the fabric of space. In other words, gravity is not a force that reaches out and grabs objects. For

Above: In an effect known as gravi-tational lensing, light reaching the Earth from a distant galaxy that is exactly behind a nearer galaxy is warped, or "lensed," by the gravity of the intervening galaxy.

Facing page: The existence of dark matter pervading galaxy clusters was first suspected in studies of the Coma Galaxy Cluster in the late 1930s, but its importance was not fully appreciated until the 1970s.

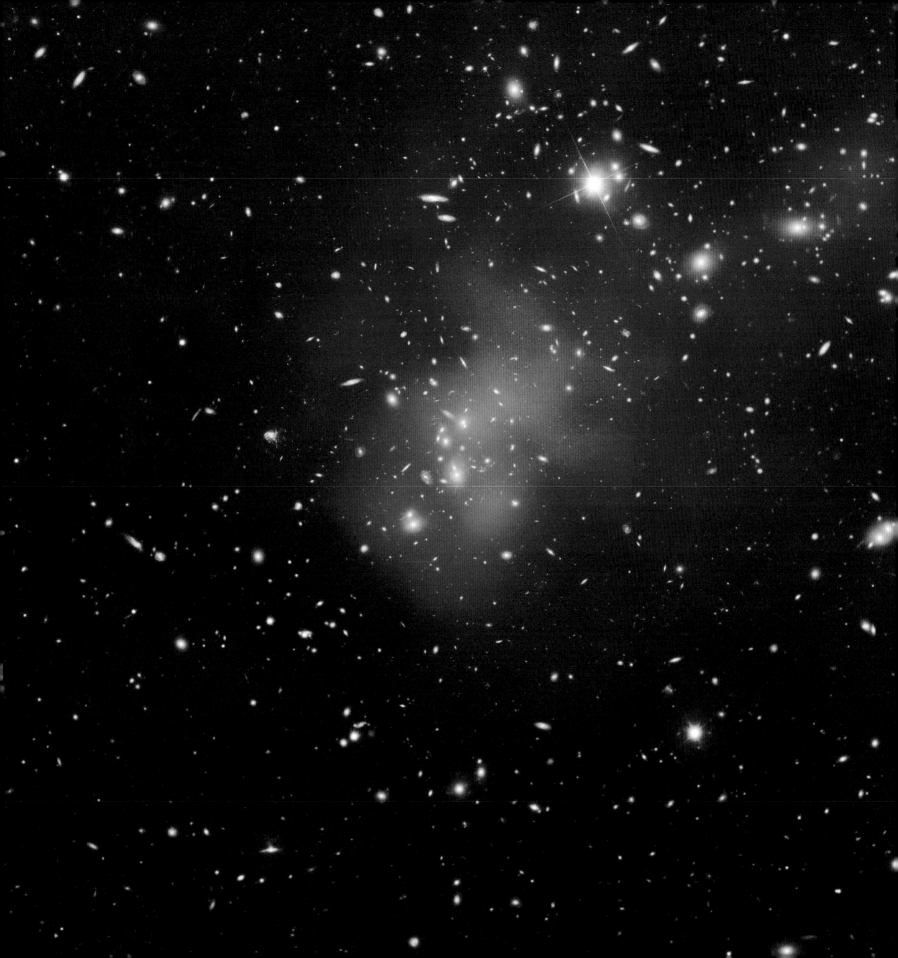

example, Earth doesn't grab on to the Moon. Rather, gravity curves space and stretches time. Warped space will change the direction and motion of any object. The Moon follows a nearly circular path about Earth because it is gliding along curved space, like a penny tossed into a concave plate at a carnival game that slides off in a different direction.

A beam of light will also be deflected if it passes through warped space. Space can bend light in the same way a magnifying glass does. Einstein realized that if a massive foreground object warps space, whether it is the Sun, a black hole or an entire cluster of stars, the light from more distant background objects will be distorted, made brighter and magnified. He hypothesized that telescopes should be able to see this effect in space, but he thought the distortion would be so small, the effects of gravitational lensing would not be easily observable.

The first direct observations of gravitational lensing in the deep universe were made in 1979, when astronomers confirmed that a distant quasar's light was split into two images by a foreground galaxy. In the 11 years between that discovery and Hubble's launch, ground-based telescopes found only about a dozen examples of gravitational lenses. But because Hubble can see objects 10 times sharper and several times fainter than ground-based telescopes, it quickly discovered a lot of small examples of gravitational lensing and uncovered a number of multilensed sources inside a typical foreground galaxy cluster.

Hubble's crisp vision is ideal for teasing out evidence for lensing all over the universe. In fact, the phenomenon of lensing is a striking example of just how crowded the universe is. Foreground galaxy clusters overlap background galaxies. In regions of space where many galaxies bunch together in huge clusters and superclusters, the result is a funhouse-mirror distortion.

There is so much dark matter in the universe that it ripples the fabric of space like a textured shower curtain. Gravity from normal matter does this too, but dark matter's effects are much more visible.

How do we know if the light from a distant galaxy has been warped? If the galaxy aligns perfectly behind a foreground object, it forms a ring of light around the object, called an Einstein ring. But such alignments are extremely rare; instead, pieces of the background galaxies typically appear on either side of the lensing body.

This presents an ideal method for precisely measuring the distribution of dark matter in a cluster of galaxies. Hubble's sharpness can pick up very small lensing artifacts embedded in a galaxy cluster. This is called microlensing. For astronomers, assembling the multi-imaged galaxy pieces to map out how the dark matter is distributed is like working on a gigantic jigsaw puzzle. Dark matter can also yield insights into the even more mysterious dark energy.

A property of space, dark energy fights against the gravitational pull of dark matter. It pushes galaxies apart from one another by stretching the space between them, thereby suppressing the formation of giant galaxy clusters. One way astronomers can probe this primeval tug-of-war is through mapping the distribution of dark matter in clusters.

Galaxy clusters had to have started building themselves very early in the universe. The clock was running out for large-scale assembly, because the more the universe expands, the more space there is between galaxies and hence the more dark energy there is. So the giant clusters we see today would have had to assemble galaxies quickly after the big bang in order to build the galaxies into the number of clusters we see today.

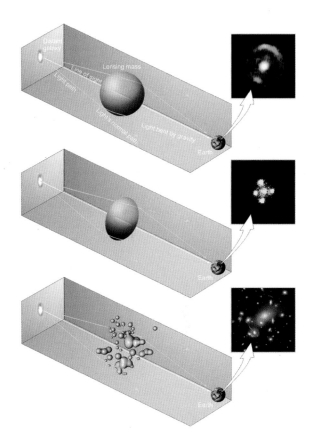

Above: Gravitational lenses produce different-shaped images depending on the shape of the lensing body. If the lens is spherical, the image appears as an Einstein ring (top); if the lens is elongated, the image is an Einstein cross (middle), which appears split into four distinct images; and if the lens is a galaxy cluster, like Abell 2218, then arcs of light are formed (bottom).

Facing page: In this view of the galaxy cluster Abell 2744, concentrations of dark matter are colored blue and hot intergalactic gas is colored red.

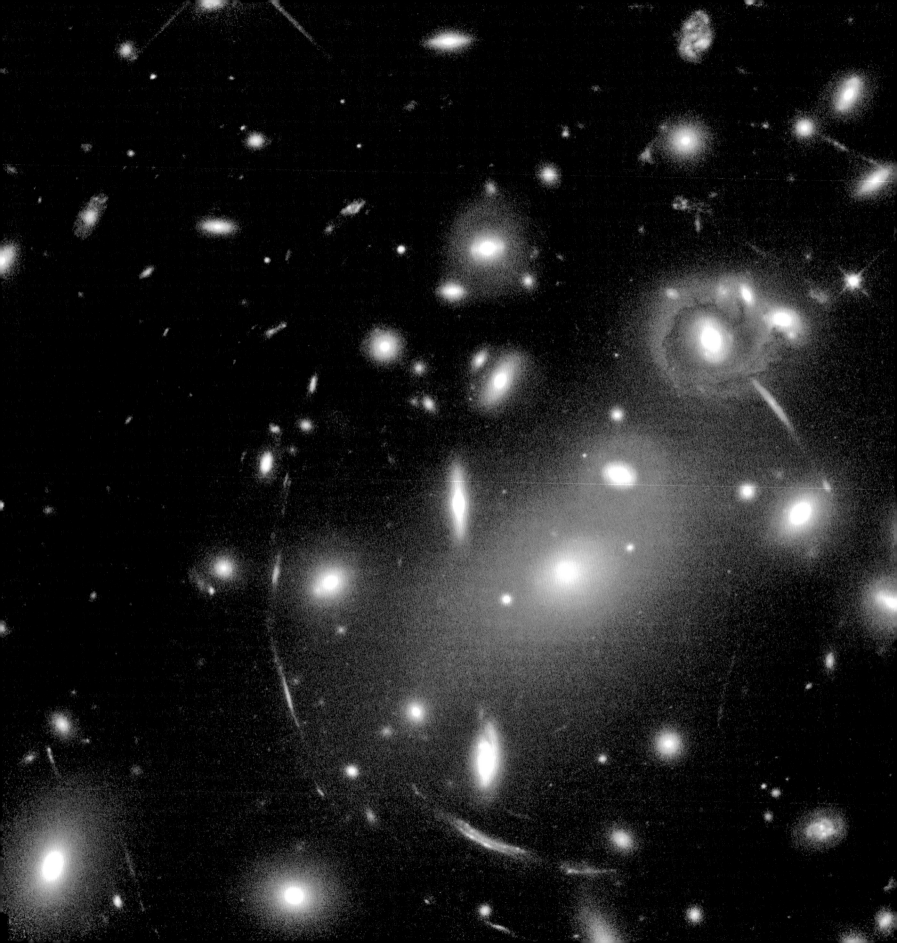

GALAXY CLUSTER ABELL 2218
The gravitational field of the cluster of galaxies known as Abell 2218 acts like a giant lens, magnifying the light of more distant galaxies far behind it. The lens effect makes the remote galaxies appear as curving arcs.

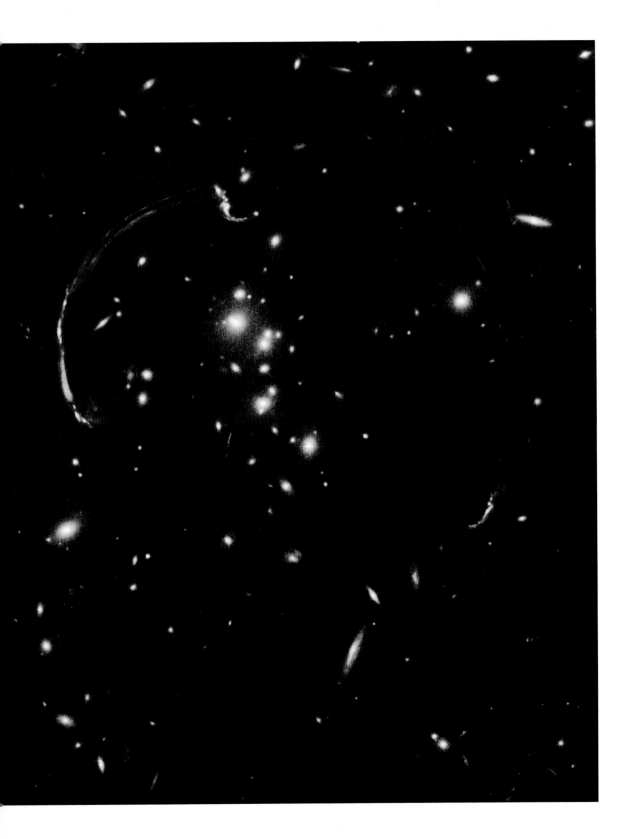

## ZOOM LENS MAGNIFIES GALAXY

One of the most striking examples of gravitational lensing is a nearly 90-degree arc of light in the galaxy cluster RCS2 032727-132623. In 2006, a team of astronomers using the Very Large Telescope in Chile measured the arc's distance and found that it is three times farther away than the galaxy cluster in which it resides. The arc is a greatly stretched-out image of a galaxy that existed when the universe was half its present age. Because of gravitational lensing, the galaxy appears 20 times larger and many times brighter than any previously discovered lensed galaxies. As is typical of gravitational lenses, the distorted image of the galaxy is repeated several times in the foreground lensing cluster. Each element was carefully studied, then used to reconstruct a model of the remote galaxy. For astronomers, determining what the galaxy actually looks like is akin to piecing together a vast Humpty Dumpty again, minus the distortion caused by the cluster's gravity. In the reconstruction, what stands out are regions of vigorous star formation that glow like bright Christmas tree bulbs in a wreath. These are much brighter than any star-formation region in our Milky Way Galaxy.

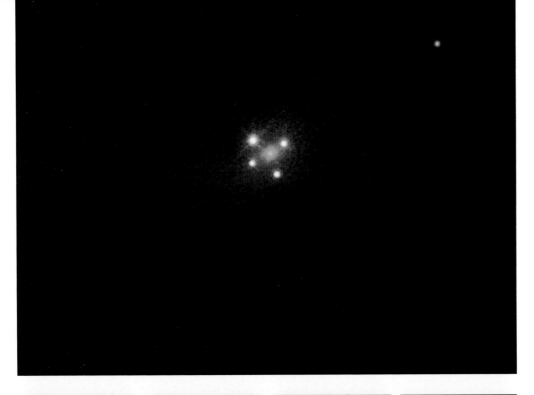

EINSTEIN CROSS

One of the eeriest and earliest pictures taken by Hubble was of a so-called Einstein cross. Seeing it for the first time, one astronomer joked that it looked like a giant LED flashlight being aimed in our direction by aliens. The photograph shows four images of a very distant quasar. The gravitational-lensing effect from a nearby normal galaxy has multi-imaged the quasar. In the center of the image is the gravitational lens. All that is seen of the nearer galaxy is its bright core.

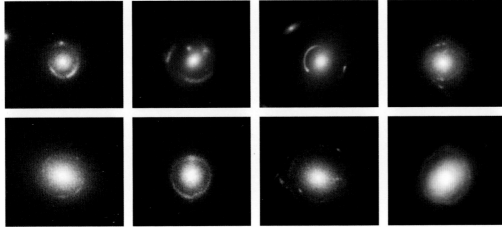

GALAXIES WITH AN EINSTEIN RING

This is a gallery of foreground galaxies that are remarkably aligned to much more remote background galaxies, creating the optical illusion of a ring around each galaxy. The ring is produced by the warping of space that smears the light from each background galaxy into an apparent circle. To find these rare alignments, the Sloan Digital Sky Survey examined a sample of 200,000 galaxies two to four billion light-years away. The spectral colors of the select galaxies suggested that light from two galaxies was superimposed. The spectral fingerprint was evidence of emissions from galaxies twice as far from Earth as the closer galaxies and directly behind them. Hubble's sharp vision was needed to actually see the ring around 16 galaxies identified for follow-up observations, some of which are shown here. By studying the amount of space warping that caused the arcs and rings, astronomers were able to precisely measure the mass of the foreground galaxies.

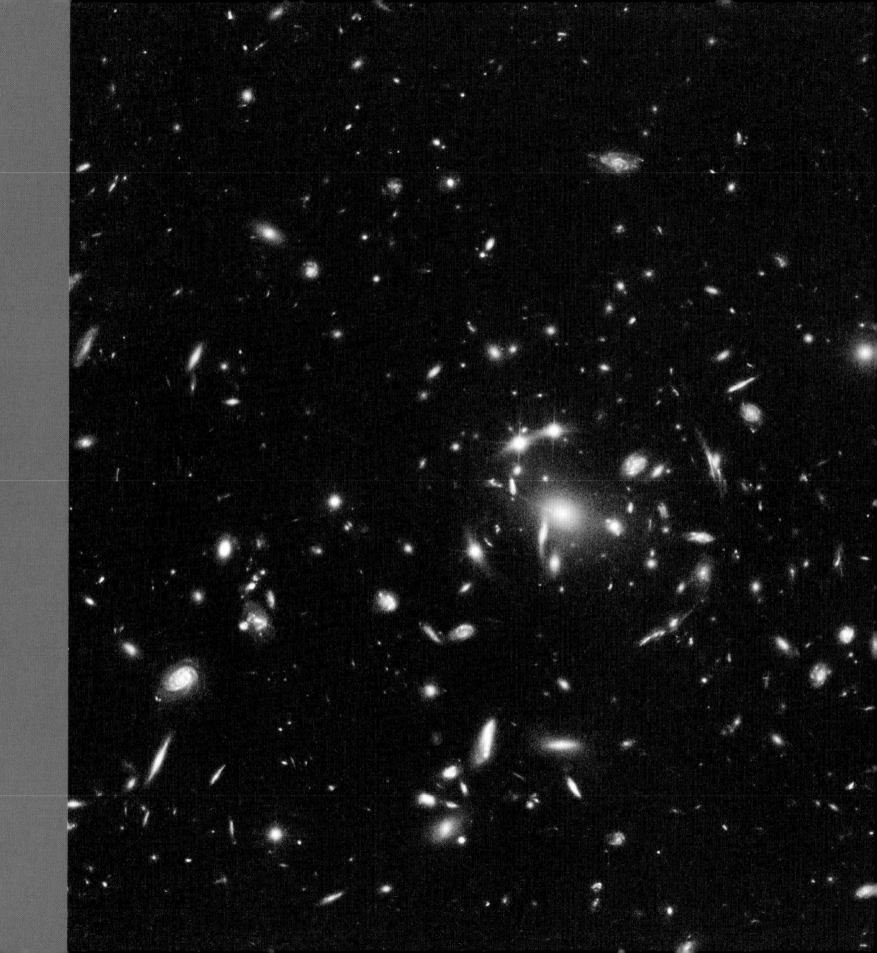

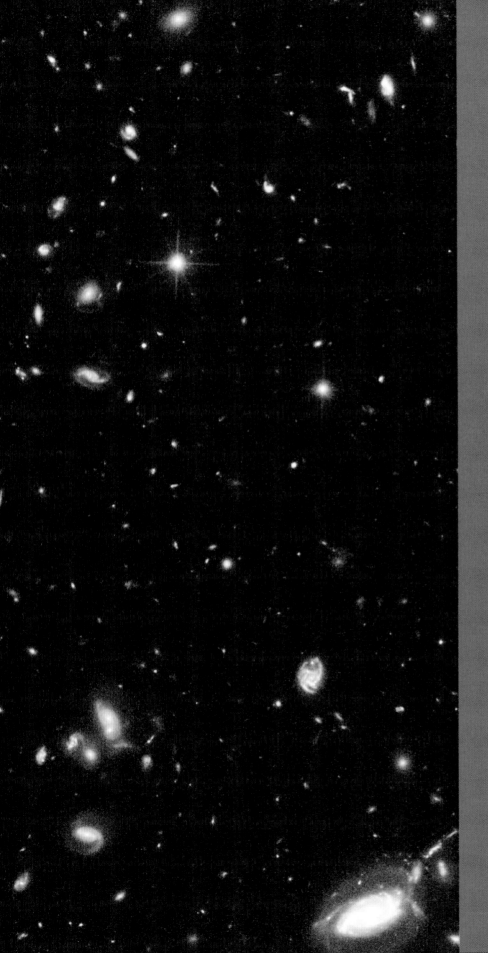

## FIVE-STAR GRAVITATIONAL LENS

Like a formation of airplanes seen at night, five starlike images of a single quasar are formed by the effects of a gravitational lens. The quasar is the brilliant core of a distant galaxy. It is powered by a black hole, which is devouring gas and dust and creating a gusher of light in the process. On its way toward us, the quasar's light passes through the gravity field of the big foreground galaxy cluster. As it does, the light is bent by the lens effect of the warped space to create five separate images of the quasar. Four of them are intensely white (two above, one left and one right of the galaxy cluster's core). The fifth quasar image is masquerading as the bright yellow core galaxy of the big galaxy cluster. The gravity-lens effect of the cluster also creates numerous images of other distant galaxies gravitationally lensed into arcs. At seven billion light-years (redshift z = 0.68), this galaxy cluster (catalog number SDSS J1004+4112) is one of the most distant clusters known. This is the first picture of a single quasar lensed into five images.

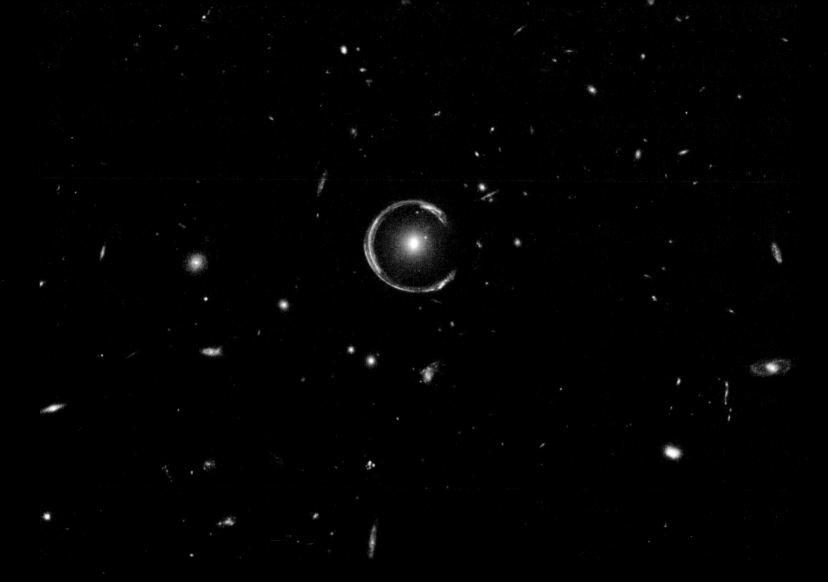

## HORSESHOE LENS GALAXY

A blue horseshoe shape encircles an old reddish galaxy. This is a near-perfect Einstein ring formed by a distant, young background galaxy, 10 billion light-years away, that is almost perfectly aligned with the foreground galaxy. But it is not precise enough to complete the ring. The gravitational lensing greatly brightens and magnifies detail that would normally be invisible to Hubble because of the great distance.

## SUPERCLUSTER

This massive cluster of galaxies is filled with the distorted shapes of distant background galaxies. If the cluster's gravity came only from the visible galaxies in the cluster and not dark matter, the distortions would be much weaker. The distorted galaxy images reveal that dark matter is more densely packed inside a galaxy cluster than previously thought. This would mean that all galaxy clusters formed perhaps 12 billion years ago, before the push of dark energy was strong enough to squelch large cluster formations.

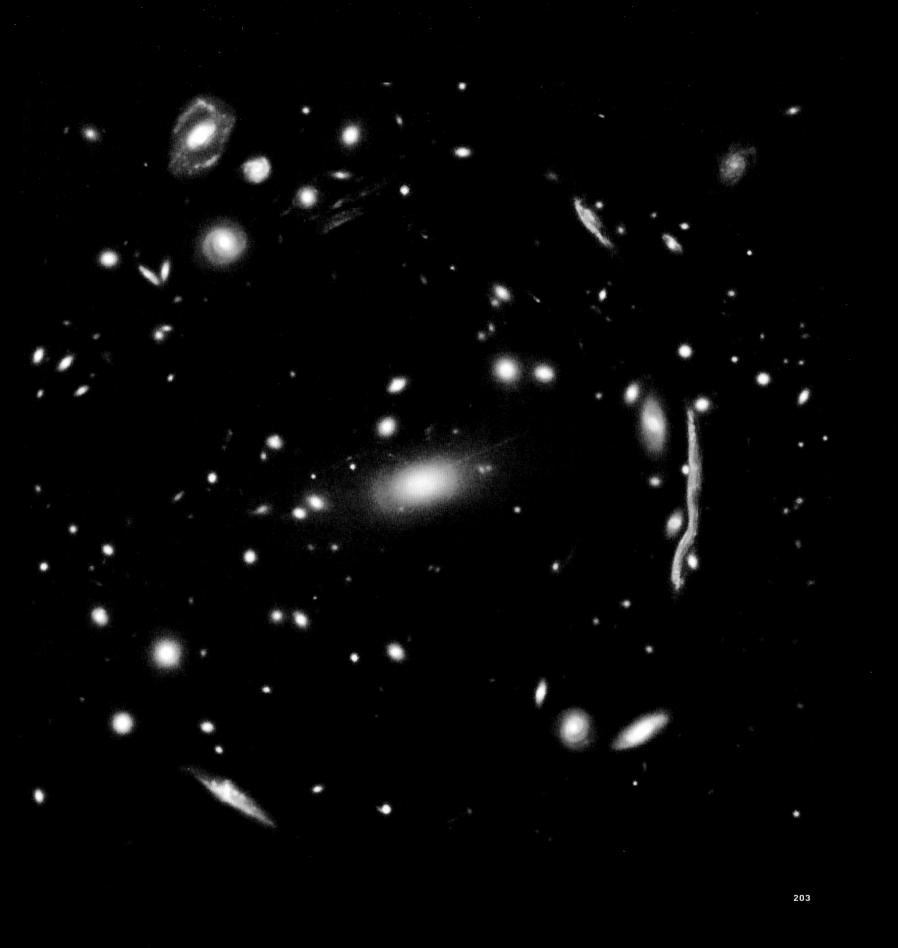

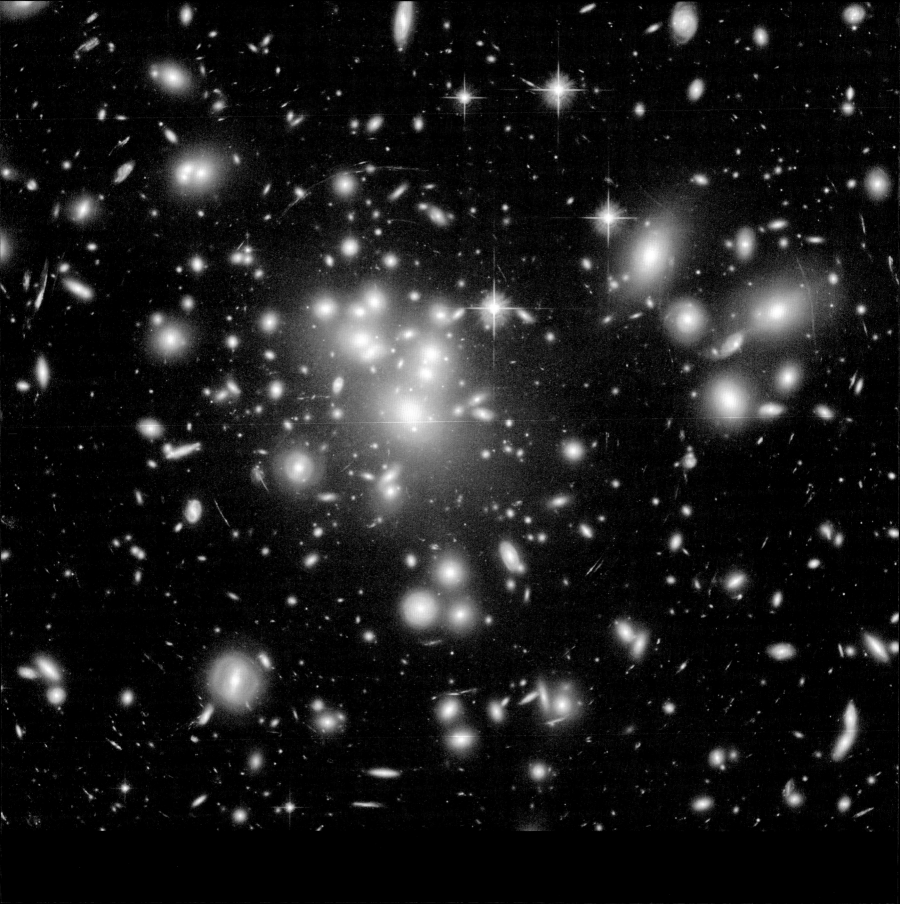

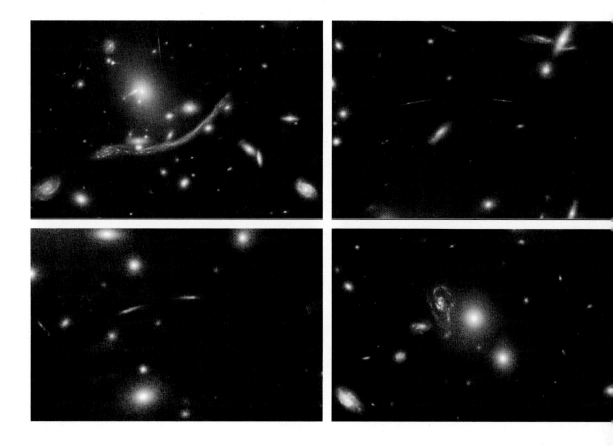

## GIANT GALAXY CLUSTER FILLED WITH ARCS

Hubble's vision is so sharp that it can uncover a cobweb of gravitational arcs embedded inside a massive foreground cluster of galaxies known as Abell 1689. Hubble sees 10 times as many arcs in this cluster as can be seen with ground-based telescopes. The light from hundreds of galaxies many billions of light-years away is smeared by the gravitational bending of light into delicate tracings of blue and red arcs. These are interspersed with the yellow-white glow of the galaxies in Abell 1689. The gravitational "lens" from the giant cluster is two million light-years across, the distance between the Milky Way and the Andromeda Galaxy.

## CLOSE-UP LOOK AT LENSING

These four images present a close-up look inside the galaxy cluster Abell 370. The lensing effect of the cluster's gravitational field is manifested as arcs and streaks and distortions of a sea of galaxies behind the cluster. The most striking image is a galaxy that looks like the head on a snake (upper left). This is not a real structure but, rather, multiple images of the galaxy smeared out by gravitational lensing. The spiral galaxy at lower right would typically look symmetrical, but its image has been pinched into a distorted shape by the warped space through which the galaxy's light has traveled. The arcs at upper right and lower left are typical lensing effects, in which the light from normal-looking galaxies has been stretched and multiplied.

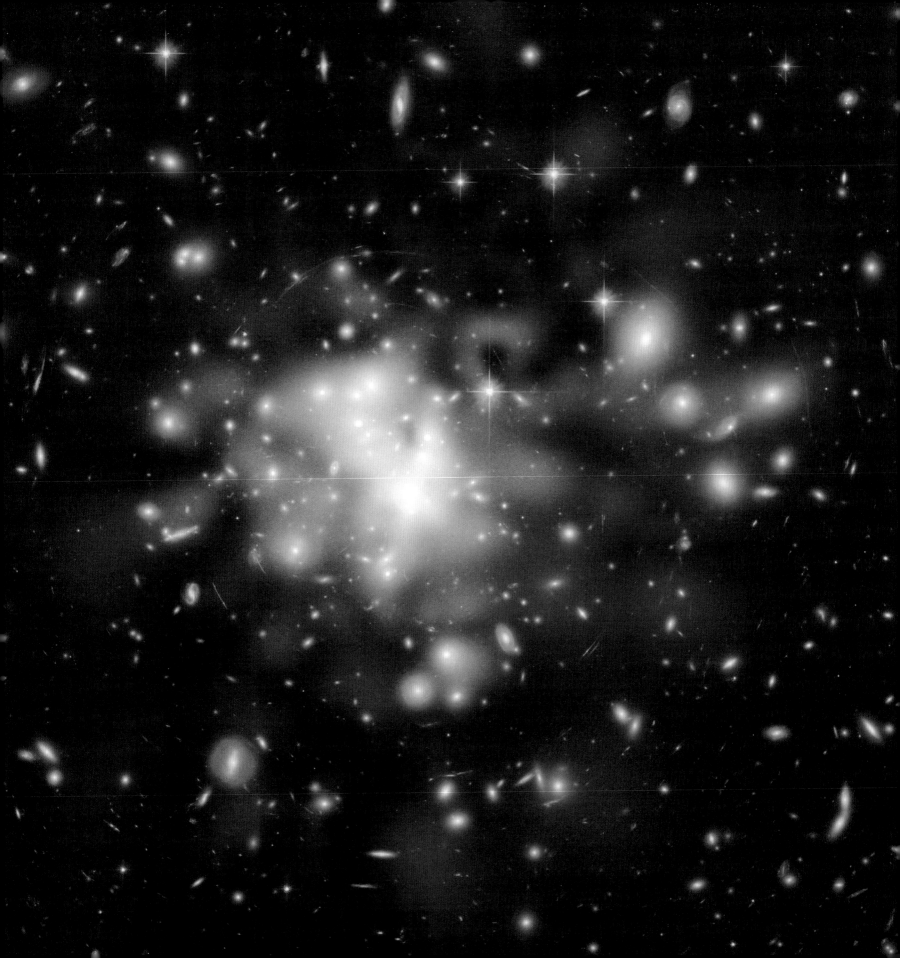

## DAZZLING DARK MATTER MAP

Facing page: This is the most detailed map of dark matter ever made. The lensing of dark matter in the foreground galaxy cluster creates a cobweblike set of concentric rings of stretched and distorted background galaxies. The blue is superimposed as a false color to plot the distribution of the dark matter in the cluster. Remarkably clumpy, the dark matter is crammed into the cluster's core, where the gravitational field of the cluster is strongest. The massive foreground galaxy cluster is 2.2 billion light-years away, while the background galaxies are billions of light-years more distant.

## 3-D DARK MATTER MAP

Hubble was used to create a three-dimensional map showing the distribution of dark matter in the universe. This map confirms that normal matter, largely in the form of galaxies, accumulates along the densest concentrations of dark matter. Stretching halfway back to the beginning of the universe, the map illustrates how dark matter has grown increasingly clumpy as it collapses under gravity. To assemble the map, Hubble surveyed a wide swath of sky nine times the sky area covered by the Earth's Moon. This allowed for the large-scale structure of dark matter to be seen by plotting the effects of gravitational lensing in a mosaic of 575 Hubble photographs. The dark matter map was constructed by measuring the shapes of half a million faraway galaxies. Their distances were provided by ground-based observations. Together, this data yielded the 3-D information.

## GALLERY OF LENSED GALAXIES

Hubble has provided a catalog of 67 gravitational-lens galaxies in the far-flung universe. That's five times more than all the known lensed galaxies before Hubble's launch, in 1990. The foreground galaxies doing the lensing are massive elliptical and featureless disk-shaped objects. Statistically, there could be nearly half a million similar gravitational lenses over the whole sky. These galaxies were found in a photographic catalog of two million galaxies and were identified by eye. The search process will be automated if computer software can be taught to be as discriminating as the human brain.

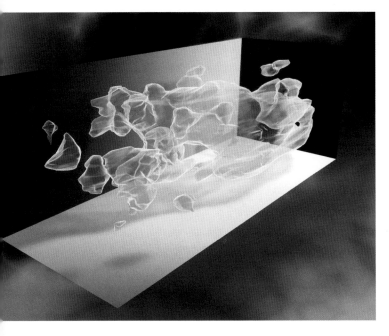

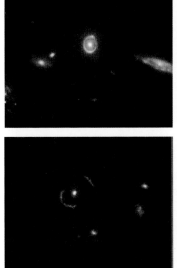

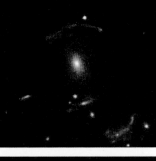

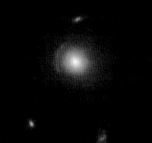

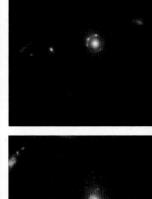

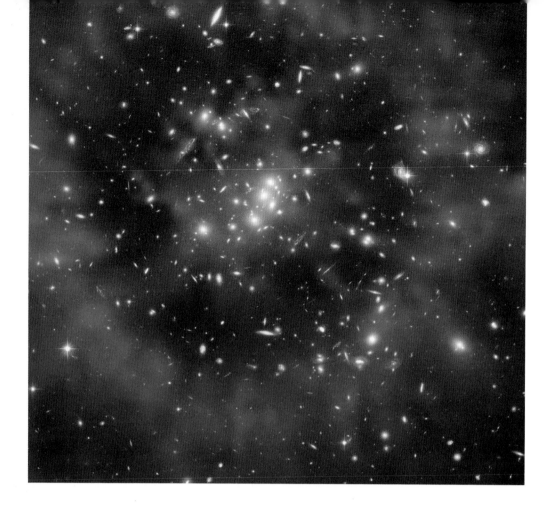

## DARK MATTER MAP IN SUPERCLUSTER

This photo is a composite of a visible-light image of the galaxy supercluster Abell 901/902 (taken at the ESO's La Silla Observatory in Chile) and a dark matter map derived from Hubble observations. The location and the density of the dark matter have been shaded red in this map. The composite image clearly shows that the clumps of dark matter coincide with the densest groupings of galaxies. Astronomers inferred the location of the dark matter by analyzing the effect of gravitational lensing, in which the light from more than 60,000 galaxies behind Abell 901/902 is distorted by intervening matter within the cluster. Using the observed, subtle distortion of the galaxies' shapes, researchers were able to plot the distribution of dark matter in the supercluster.

## DARK MATTER RING

One of the strongest pieces of observational evidence for dark matter is a ghostly ring that formed less than two billion years ago during a titanic collision between two massive galaxy clusters. The ring seen here is not an image of dark matter but a reconstruction of how dark matter is distributed in space based on mapping it through gravitational lensing.

The dark matter ring has a very unique structure that has never been seen in galaxy clusters. It is not simply caused by hot gas. The ring measures 2.6 million light-years across—more than the distance between the Milky Way and the neighboring Andromeda Galaxy. It is a stunning demonstration that dark matter behaves differently from normal matter. What's striking is the presence of a ripple in the mysterious substance, somewhat like the ripples created when wind blows across the surface of a pond. Pebbles on the bottom of the pond appear to change shape as the ripples pass over them. So, too, the background galaxies behind the ring show changes in their shapes due to the presence of the dense ring.

The dark matter structure looks like a ring because we are viewing the collision head-on. Computer simulations of galaxy cluster collisions explain the ripple. When two clusters smash together, the dark matter falls to the center of the combined cluster, then sloshes back out. As the dark matter moves outward, the pull of gravity causes it to slow down and begin to pile up.

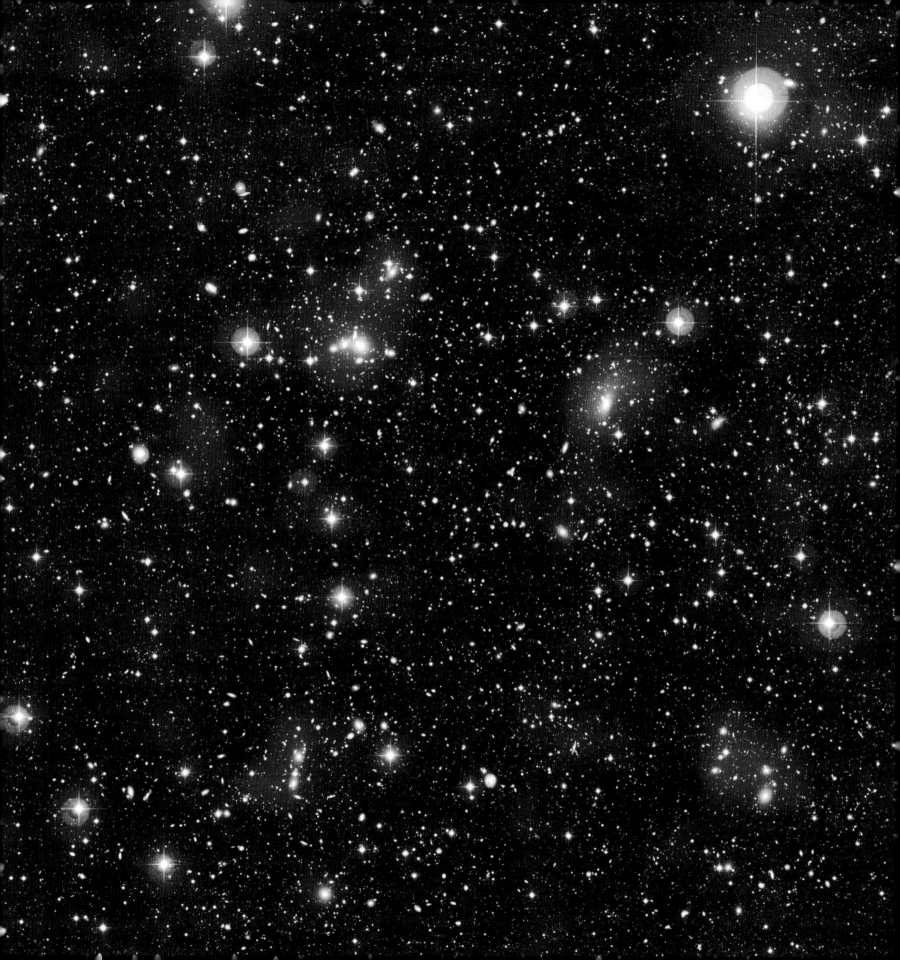

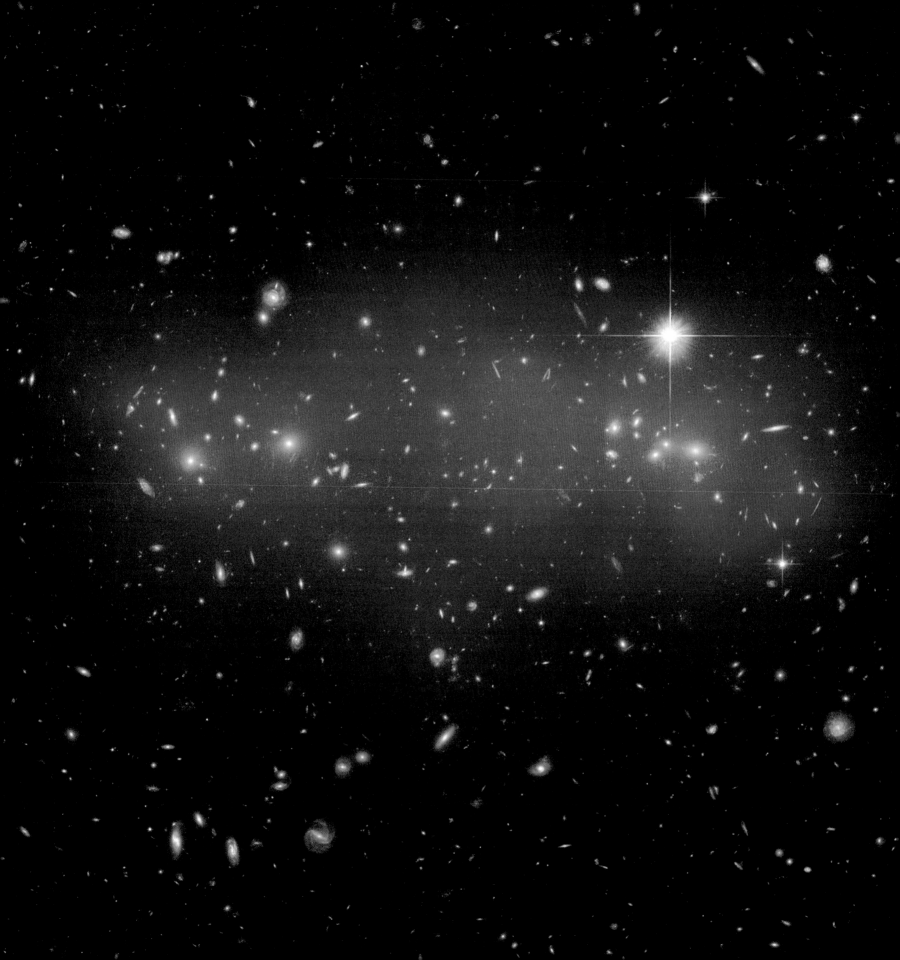

## CLUSTER COLLISION

This clash of enormous clusters of galaxies provides another piece of observational evidence that dark matter behaves differently from normal matter. The cluster, called MACS J0025.4-1222, formed when two large clusters tens of millions of light-years across collided. Such collisions are the most energetic events in the universe.

Hubble images were used to map the distribution of dark matter, while images from the Chandra X-ray Observatory show the glow of hot, normal matter that is in the form of plasma. The hot gas has been artificially tinted pink, and the dark matter is shaded blue. Few, if any, of the galaxies collided as the two clusters merged at speeds of millions of kilometers per hour, but the hot gas in the clusters (pink) did collide and slow down. The dark matter (blue), however, passed right through. The separation between the blobs of dark matter demonstrates that dark matter particles interact with each other only very weakly, if at all.

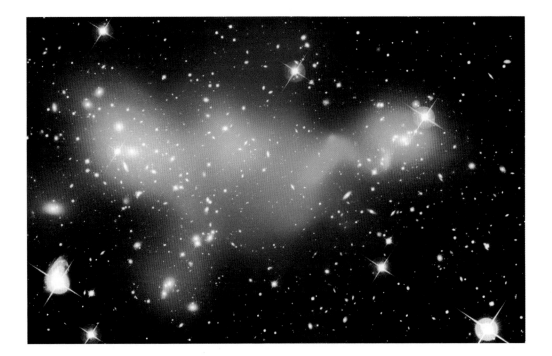

## DARK CORE

In an effort to untangle the flotsam and jetsam resulting from a collision between no less than three immense galaxy clusters, astronomers assembled this particularly garish-looking image. Super-imposed on the natural-color image of the galaxy smashup are false-color intensities showing the concentration of starlight (obtained from observations by the Canada-France-Hawaii Telescope), hot gas and dark matter in the resulting gigantic cluster. Starlight from galaxies is colored orange. The green-tinted regions show hot gas, as detected by NASA's Chandra X-ray Observatory. The gas is evidence of the collision itself. The blue-colored areas pinpoint the location of most of the mass in the cluster, which is dominated by dark matter. The dark matter map was derived from Hubble Wide Field and Planetary Camera 2 observations. The blend of blue and green in the center of the image reveals that a clump of dark matter resides near most of the hot gas, where very few galaxies are seen. This finding confirms previous observations of a dark matter core in the cluster. It's a perplexing "dark core," however, where the dark matter has somehow decoupled from the galaxies that shifted away from the center of the collision. This challenges basic theories of dark matter, which predict that galaxies should be anchored to the invisible substance, even during the shock of a collision.

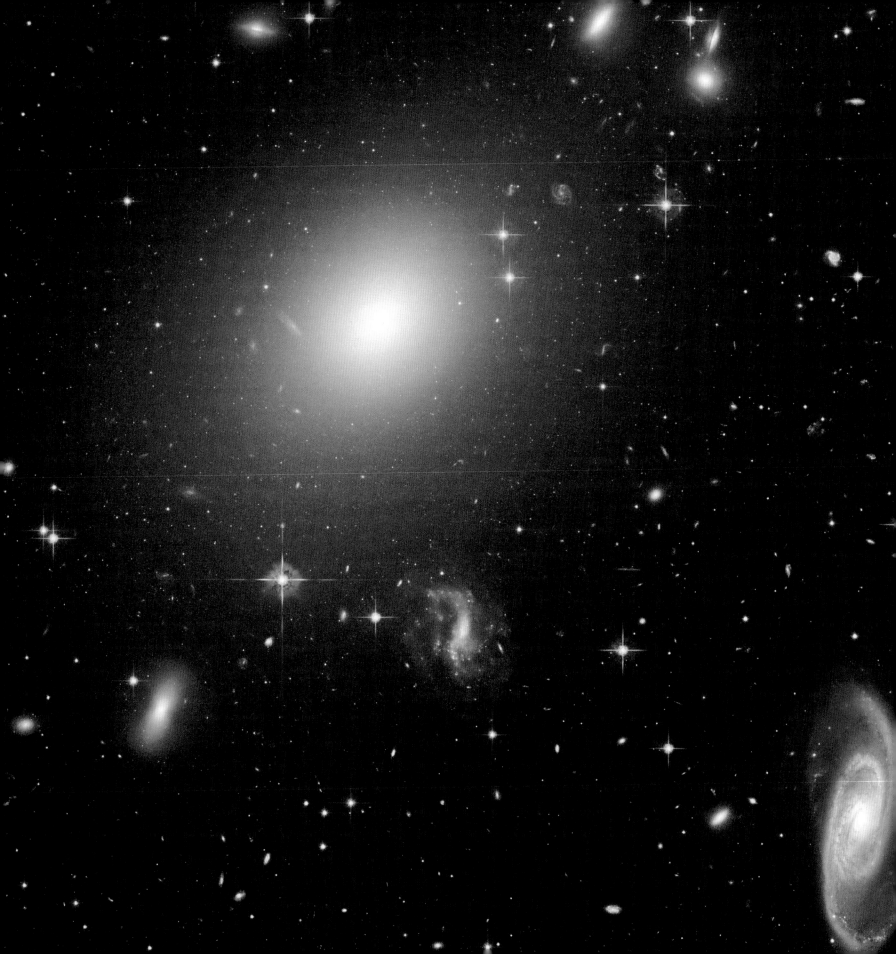

# EMPIRES OF STARS

## CHAPTER 8

Galaxies are the majestic city-states of the universe. Astronomers estimate there are at least 100 billion galaxies in the known universe. Yet less than a century ago, astronomers knew of just one galaxy: our Milky Way. At that time, it was thought that the Milky Way alone encompassed the entire observable universe.

If we could look down onto the disk of our galaxy, we would see a remarkable cosmic quilt. From our overhead view, we would notice lanes of bright stars tracing the galaxy's spiral arms, and the Sun would sit near the edge of the stellar lane called the Orion Arm.

All across the majestic disk, we would see cobwebs of dust, bright stars and glowing star-birth nebulas sprawling in all directions. It would be like flying over a great city at night, with avenues and boulevards stretching as far as the eye could see. Our attention would be drawn to the brilliant hub of the Milky Way pinwheel. Floating above the galaxy's hub are the globular star clusters. Each of these silent sentinels is a glittering jewel box containing at least 100,000 stars. They are mysterious, majestic and ancient. They are the first homesteaders of the galaxy and contain the universe's oldest stars.

Our tranquil bird's-eye view is misleading, however. Imagine that we could compress billions of years into just a few minutes, like a time-lapse movie. Behold a very different type of galaxy: Erupting and exploding stars, novas and supernovas go off like aerial bombs in a fireworks show. Even more frequently, the burnout of Sun-like stars casts off bubbly wisps of glowing gases. The effect resembles a stadium rock concert with camera flashes going off in a flurry of bursts.

Over millions of years, dark molecular clouds slam into the leading edges of the spiral arms. Compressed by such collisions, the clouds give birth to new stellar generations. Blisters form on the edges of the clouds, then explode with new stars and a pink glow of iridescent gases. Brilliant young star clusters form along the trailing edge of the spiral arms.

The globular clusters whirl around the galaxy's hub. Several barrel through the galaxy disk, where gravitational tidal stresses pull some of the globular clusters apart. Meanwhile, the monster black hole in the galaxy's hub occasionally flickers as it swallows wayward stars.

## MARKERS OF SPACE

The discovery of galaxies beyond ours by Edwin Hubble in the 1920s set astronomers on a path to understanding the origin and evolution of the universe. More than simply islands of stars floating in space, galaxies are markers of space. Their apparent movement away from us, also discovered by Edwin Hubble, is evidence that space is expanding and stretching like a rubber band.

The Hubble Space Telescope's views unveil a restless universe of galaxies interacting with one another over time. As with biological evolution, the visual evidence that galaxies have evolved

This family portrait of Abell S0740, a galaxy cluster approximately 450 million light-years distant, captures the diversity of galaxies in the universe. The dominant giant elliptical, ESO 325-G004, is a "cannibal" galaxy that has likely gravitationally attracted and swallowed dozens of smaller galaxies to attain its bloated dimensions. A variety of spiral galaxies dot the scene, with a "grand design" spiral at lower right. Myriad smaller, and hence more distant, galaxies fill the background.

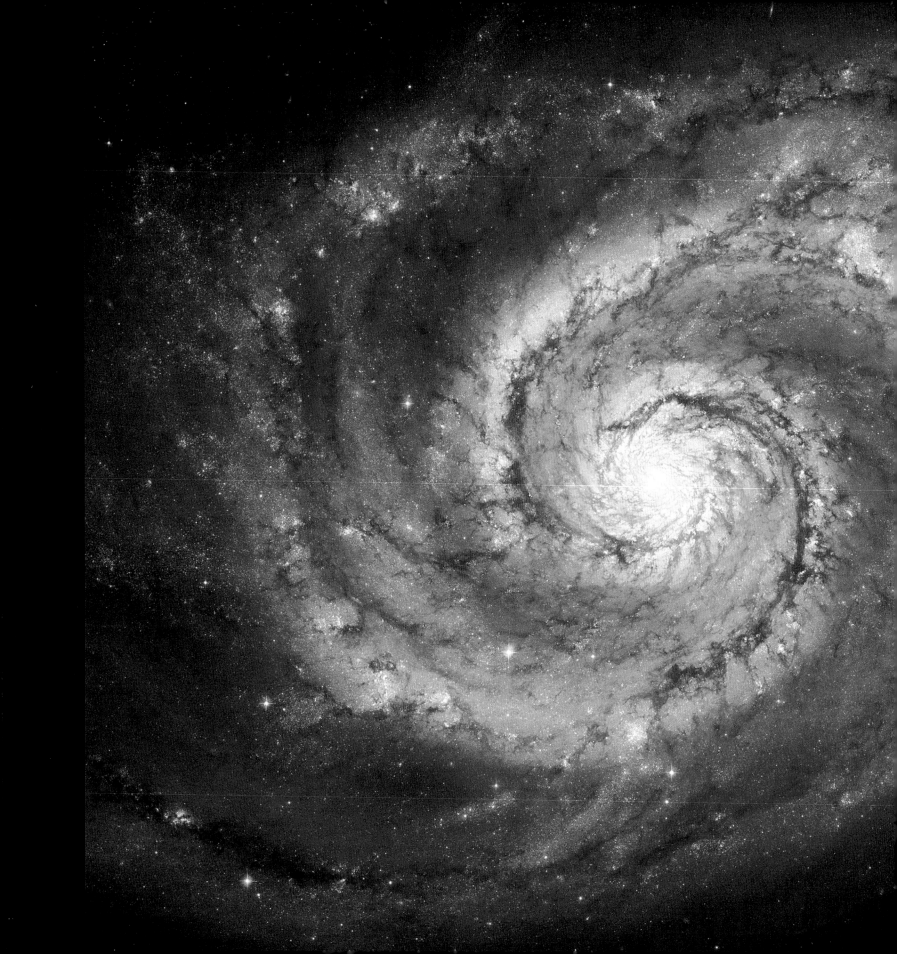

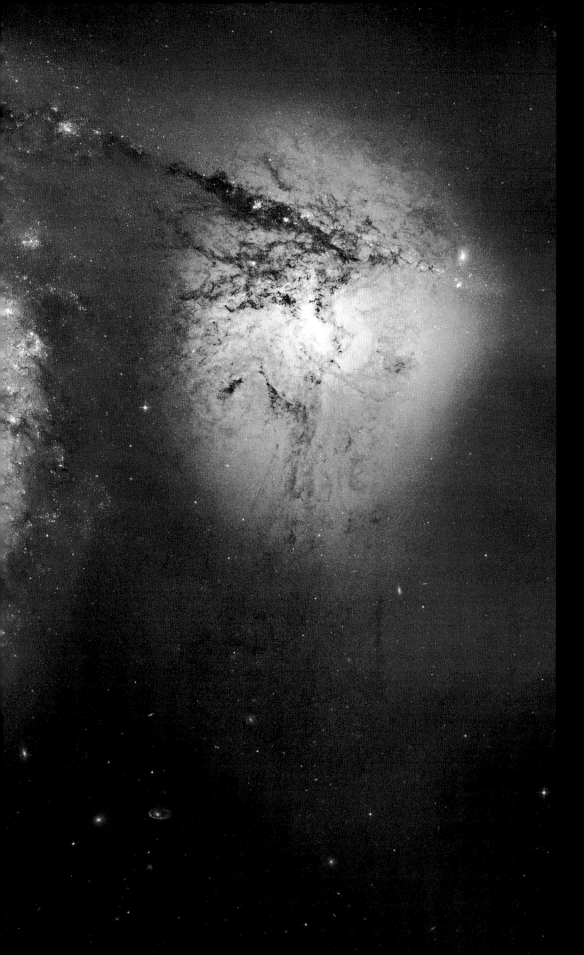

The stately Whirlpool Galaxy (M51) has enticed astronomers for nearly two centuries. Its two majestic spiral arms are star-formation factories that compress hydrogen gas to create clusters of new stars. In the Whirlpool, the assembly line begins with the dark clouds of gas on the inner edge of each spiral arm, then moves to bright pink star-forming regions and ends with the brilliant blue star clusters along the outer edge. The galaxy's arresting whirlpool shape was first sketched in 1845 by Irish nobleman and astronomer Lord Rosse, using the world's largest telescope. He correctly hypothesized that spiral nebulas, as they were then called, contained so many fine stars that most telescopes could not resolve them individually.

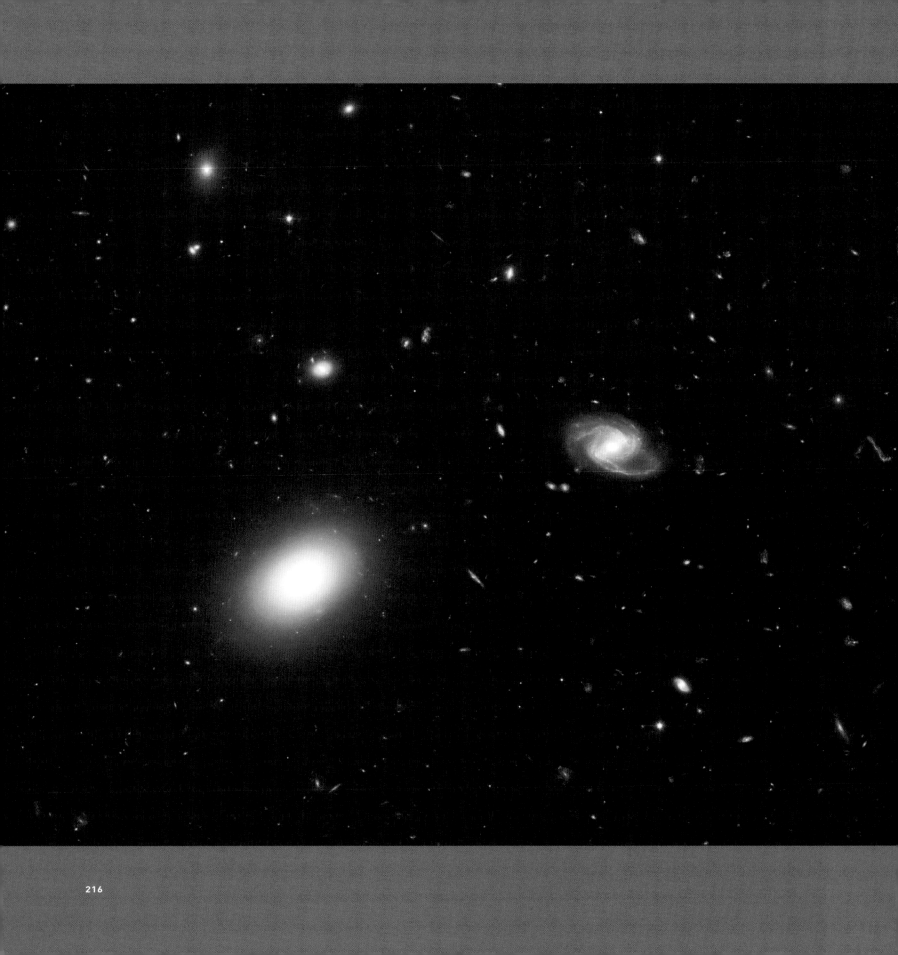

is overwhelming. Hubble's deep-space pictures reveal a "Wild West" universe where galaxies, which were smaller and did not have the classic spiral or elliptical shapes, collided and merged frequently. Our own Milky Way formed over a construction period spanning billions of years and has undergone numerous encounters in which it has assimilated much smaller galaxies caught in its gravitational field.

Like snowflakes, no two galaxies are exactly alike, though the fundamental physics is the same. Hubble's views of majestic, fully formed galaxies present a tapestry of wonder.

In a more visceral way, galaxies impart a staggering sense of infinity to the universe. Assuming our Milky Way is a typical galaxy, then there are 100 trillion trillion planets scattered throughout the universe. If none of them beyond Earth are inhabited, it would be a terrible "waste of space," lamented Scottish essayist Thomas Carlyle in the mid-1800s. "If they be inhabited," he continued, "what a scope for misery and folly."

GALAXY PANORAMA
(following 10 pages)

Perhaps better than anything else in this book, the image on the next spread opens a window on the universe of galaxies—arguably, the Hubble Space Telescope's greatest gift so far. The image reveals a rich tapestry of thousands of galaxies stretching back through most of the universe's history. The closest galaxies in the foreground emitted their observed light about a billion years ago. The most distant galaxies, a few of the very faint red specks, are seen as they appeared more than 13 billion years ago. The image combines a broad range of colors, from the ultraviolet, through visible light and into the near infrared. Such a detailed view of the deep universe in this combination of color, clarity, accuracy and depth has never before been assembled. The panorama shows galaxy shapes that, at each earlier epoch, appear increasingly chaotic as galaxies grew through accretion, collisions and mergers. The galaxies range from the mature spirals and ellipticals in the foreground to smaller, fainter, irregularly shaped galaxies, most of which are farther away and, therefore, existed further back in time. The smaller galaxies are considered the building blocks of the large galaxies we see today.

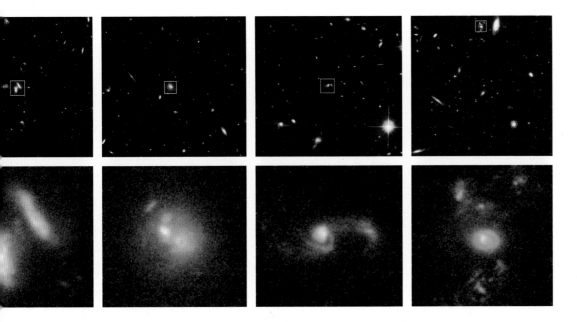

Galaxies in the early universe were far more unruly than they are today, as these magnified views of remote galaxies reveal. The samples of "pathological" galaxies from Hubble deep surveys show that collisions were much more common eight to nine billion years ago.

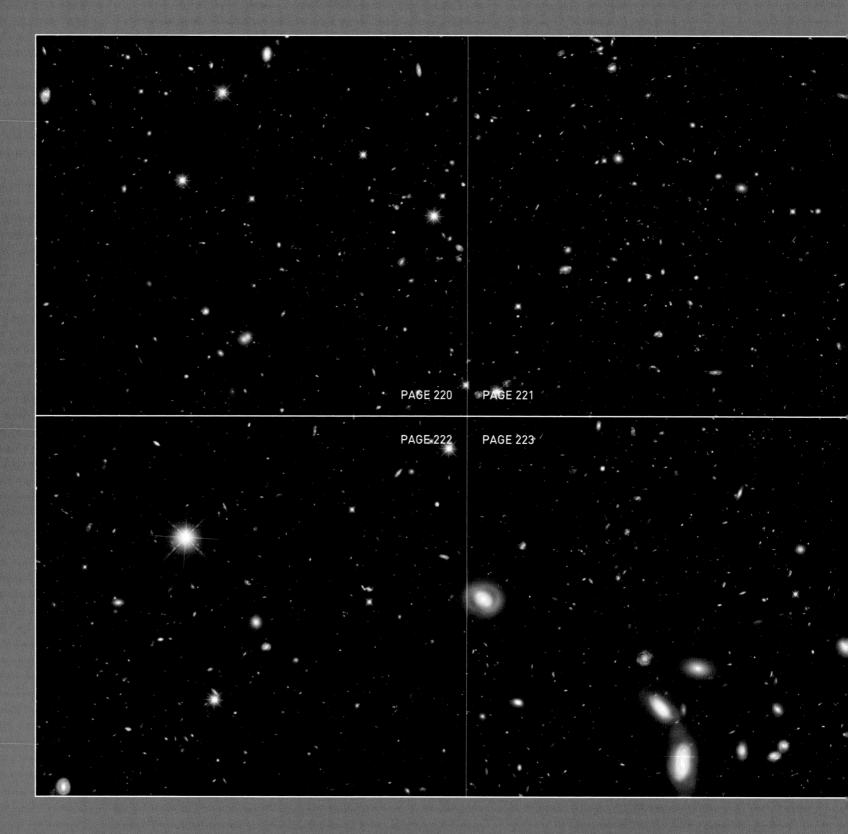

PAGE 220    PAGE 221

PAGE 222    PAGE 223

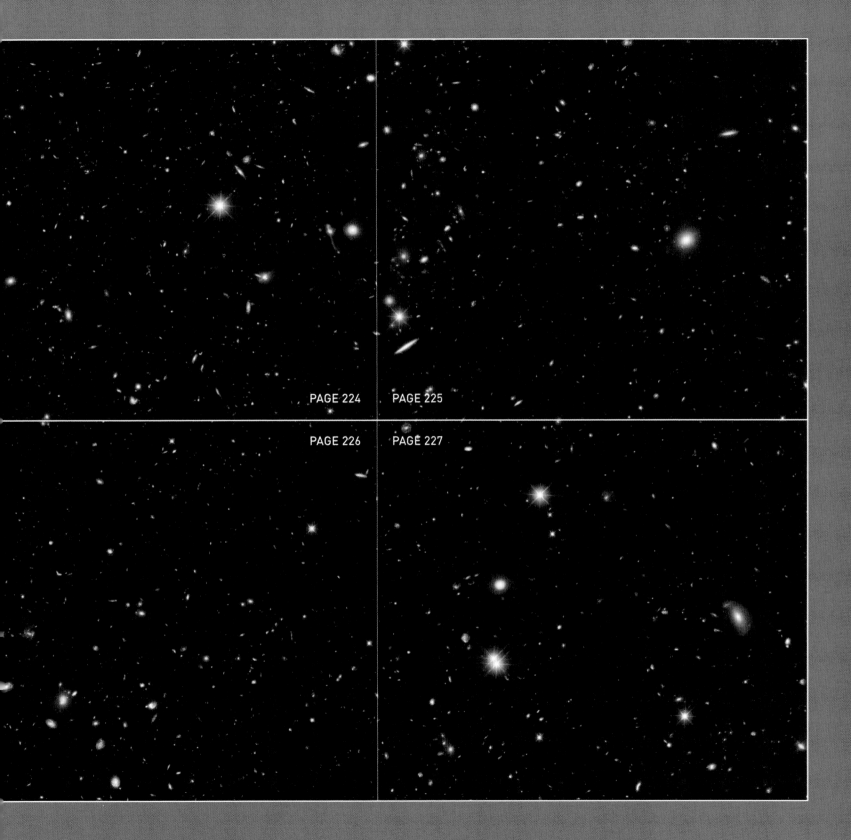

PAGE 224    PAGE 225

PAGE 226    PAGE 227

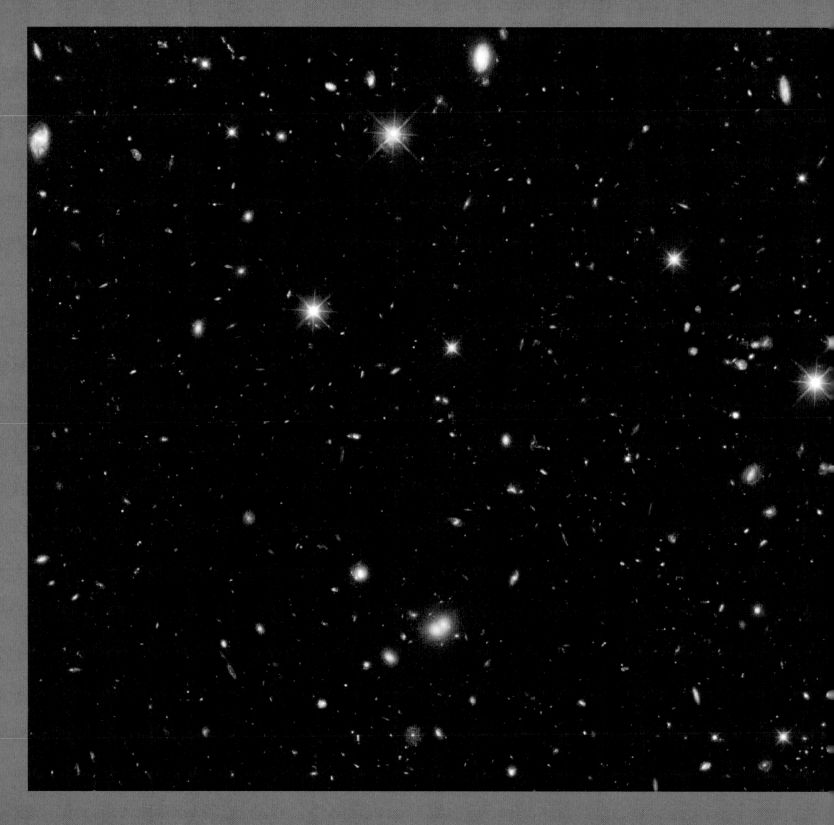

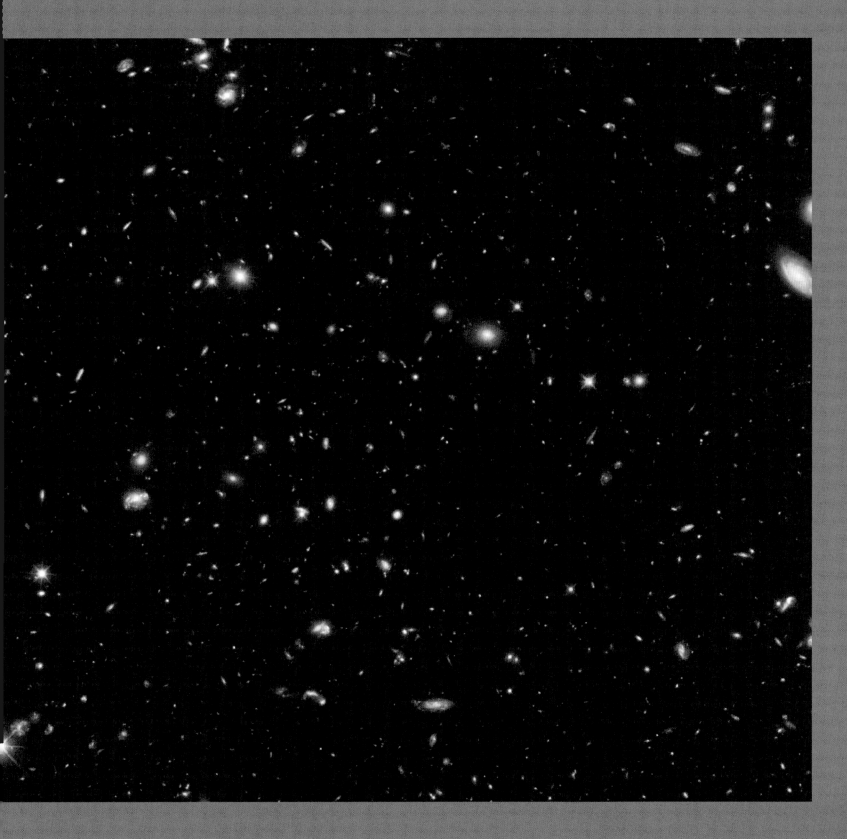

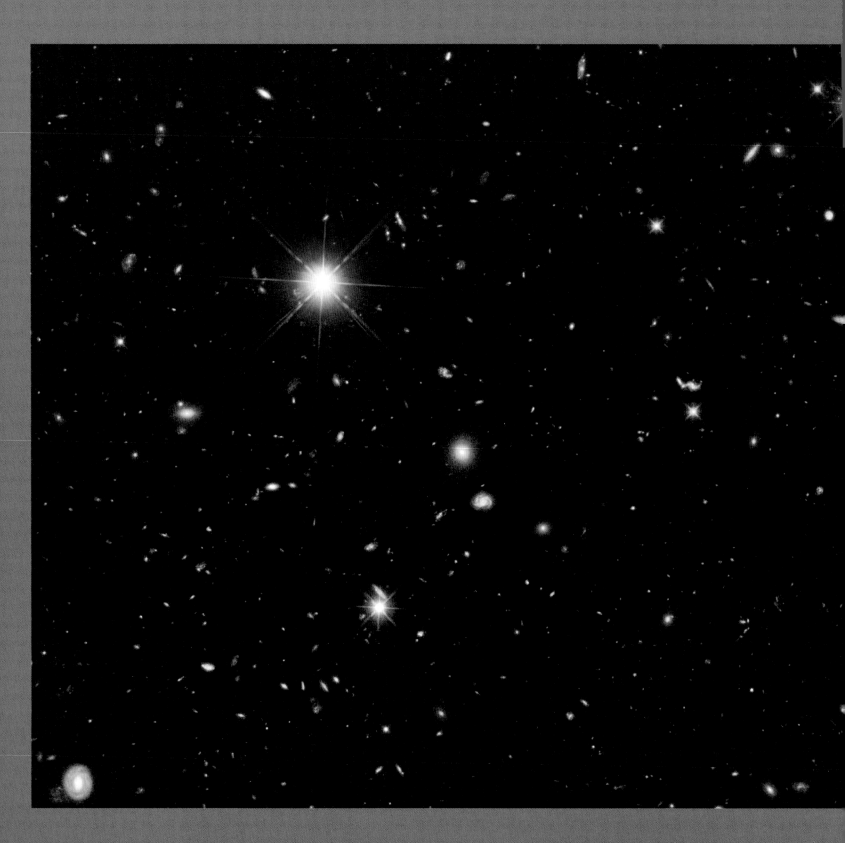

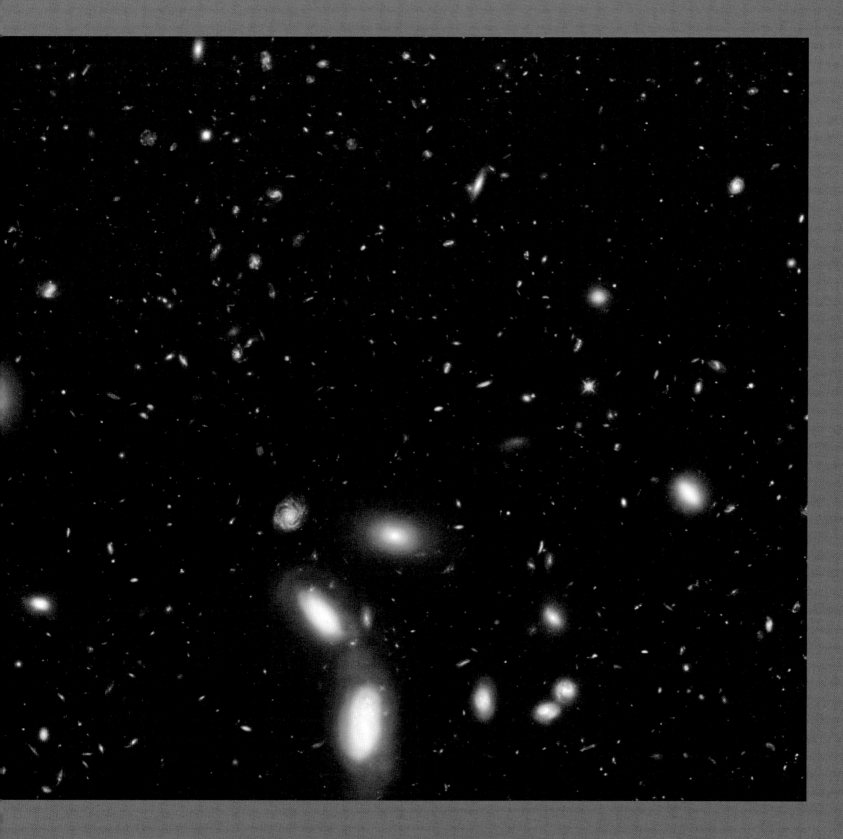

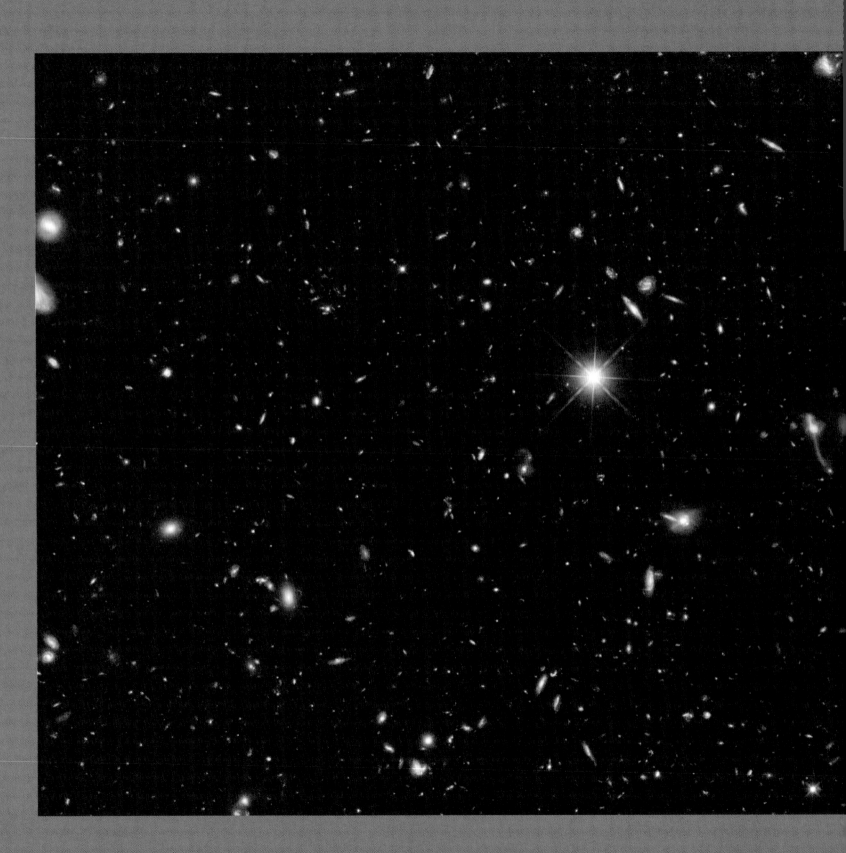

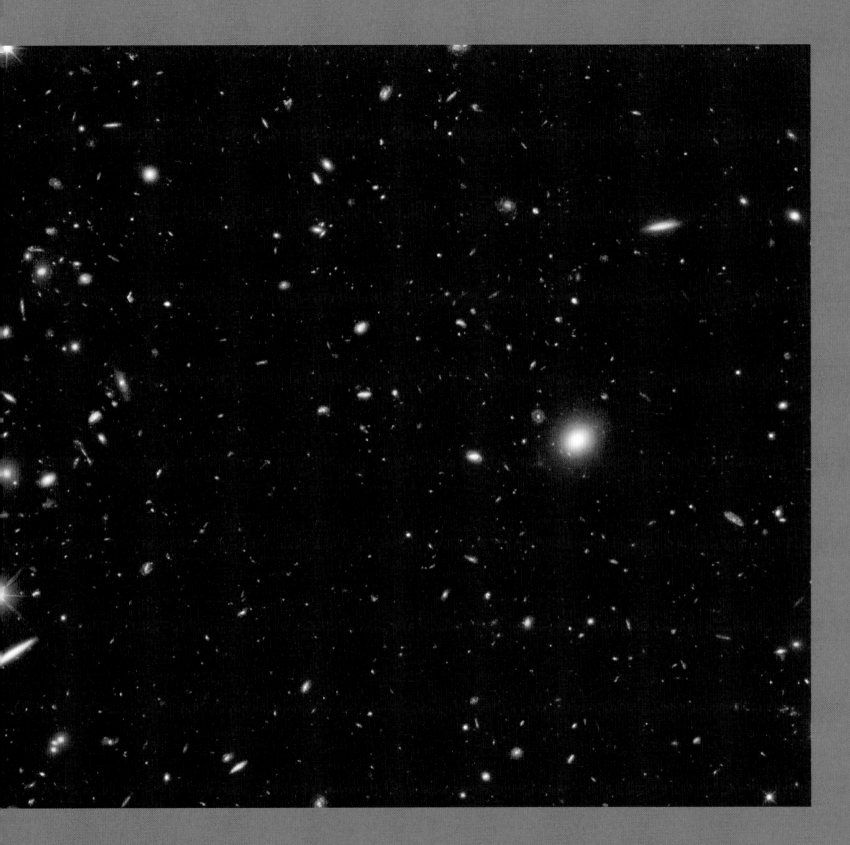

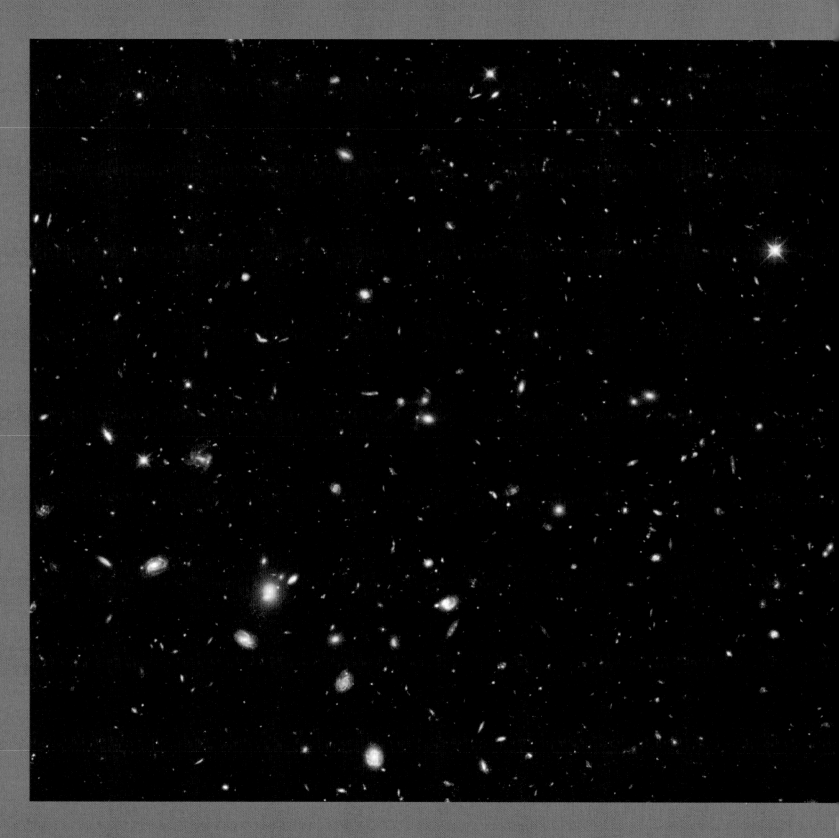

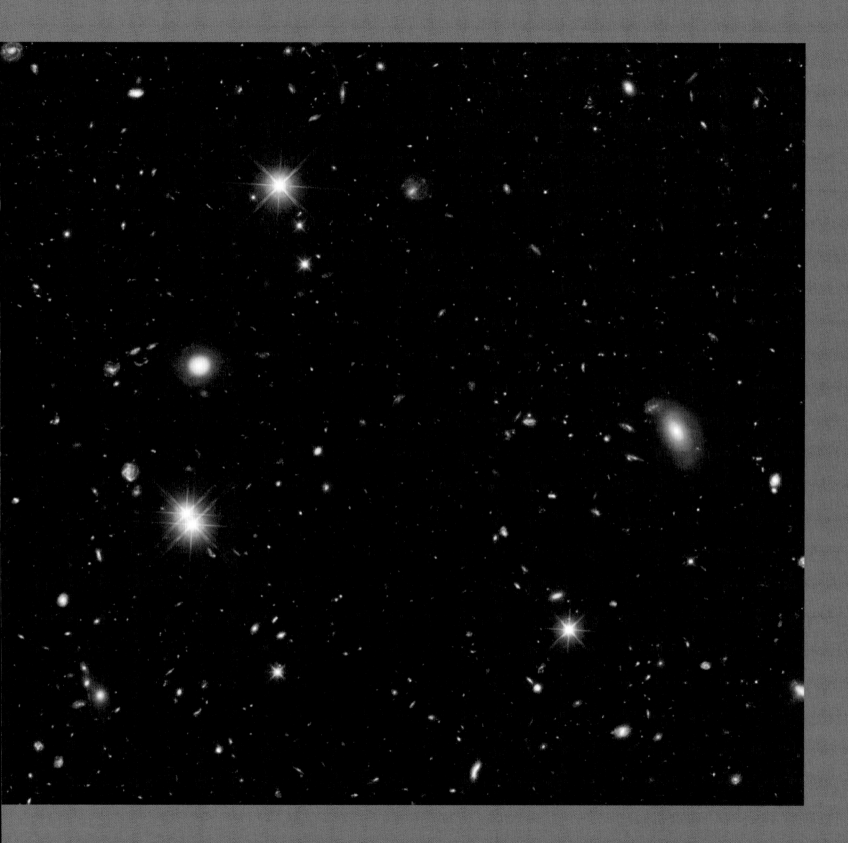

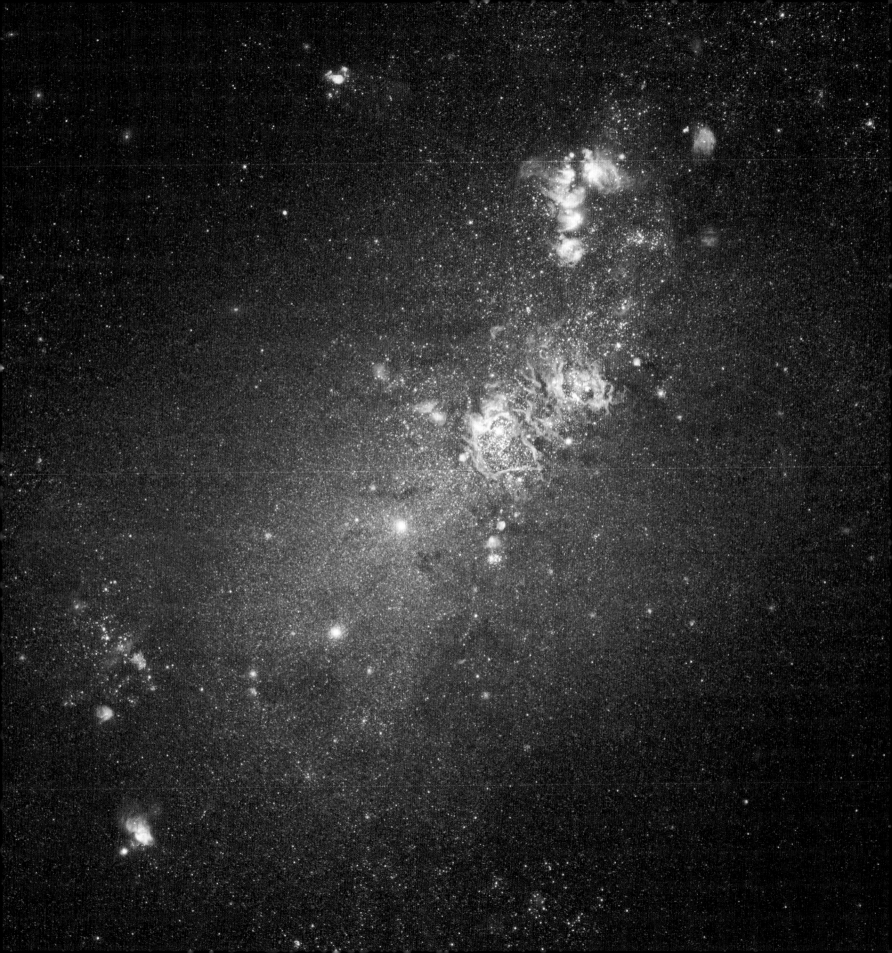

## DWARF GALAXY NGC4214

This blobby collection of stars is ablaze with gas clouds and new-born suns. Hubble's sharp vision and the dwarf galaxy's close prox-imity—only 10 million light-years away, in the constellation Canes Venatici—make NGC4214 an ideal laboratory to research the triggers of star formation and evolution. This object is prototypical of the myriad young galaxies in the early universe that started the galaxy-construction process. But NGC4214 is a late bloomer. After lingering in intergalactic space for billions of years, it is only now beginning to ignite its production of stars. Explod-ing stars and stellar winds have carved out great bubble structures in the galaxy and have sculpted filaments of glowing primordial hydrogen. Bright, aging red super-giant stars pop out of the image, as well as hot, young blue stars, giving the picture an unusual granularity.

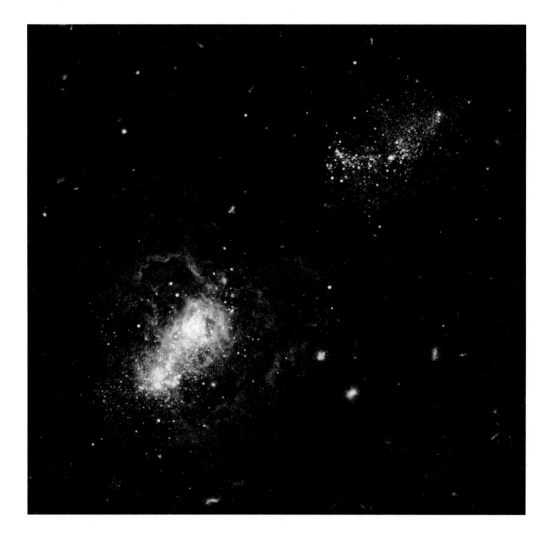

## BORN-AGAIN GALAXY

The concentrated bluish white knots embedded in the heart of I Zwicky 18, a dwarf irregular galaxy, are two major starburst regions where stars are being produced at a furious rate. This gives I Zwicky 18 a youthful appearance, resembling that of galaxies typically found only in the early universe. However, Hubble also observed faint, older stars within this galaxy, suggesting that it formed at the same time as most other galaxies but has undergone a firestorm of star birth over the past 500 million years.

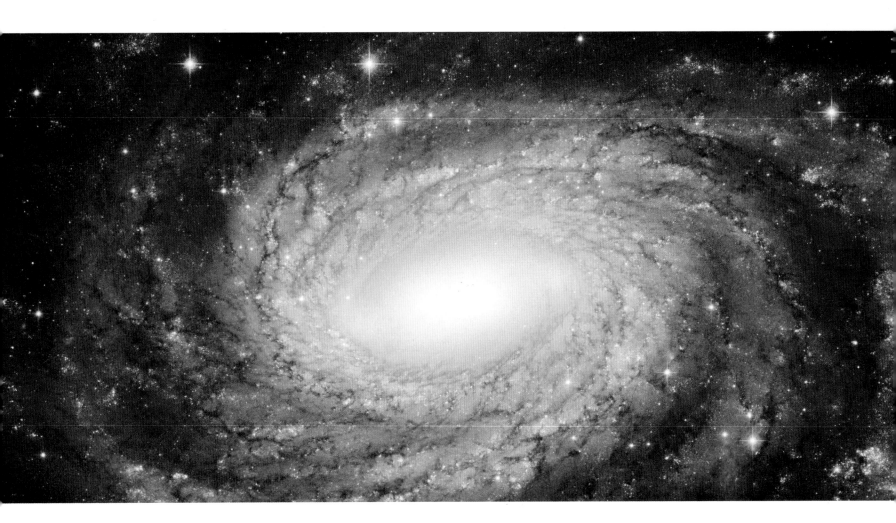

NGC6384

Star birth in this relatively quiescent middle-aged galaxy has declined. Noticeably missing are pinkish nebulas that are the sites of new star formation. Radiation and stellar winds from superhot, young blue stars have cleared out the remaining gas, shutting down any further production of stars. A bright concentration of starlight marks the galaxy's center. Spiraling outward, dust lanes are silhouetted against the population of whitish middle-aged stars. Much younger blue stars trace the spiral arms.

M74

The "grand design" spiral galaxy M74 displays perfectly symmetrical spiral arms dotted with clusters of young blue stars and pinkish star-birth regions. Are the spiral arms winding up or unwinding? Neither. The arms mark areas where density waves trigger star birth. Spiral galaxies, therefore, tend to appear stable for long intervals. M74 is roughly 32 million light-years away.

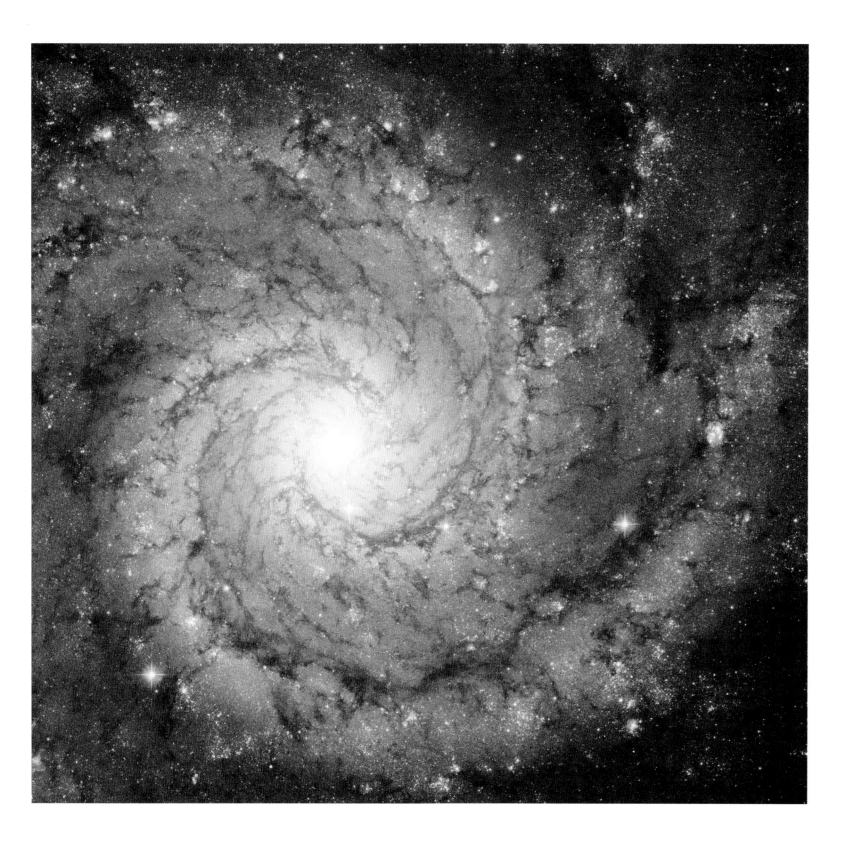

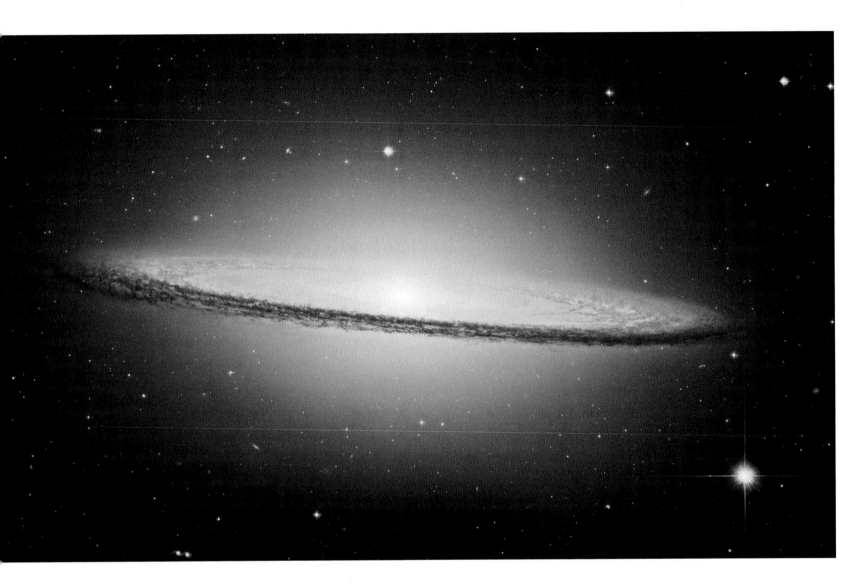

SOMBRERO GALAXY

The classic, nearly perfectly edge-on Sombrero Galaxy (M104) has a bulbous core of older stars that is encircled by thick dust lanes. Slightly more than half the diameter of our Milky Way Galaxy, M104 is 50,000 light-years across and lies 28 million light-years away.

NGC5866

This striking view leaves little doubt that spiral galaxies are actually flattened pancakelike disks of stars. Tilted almost precisely edge-on to our line of sight, NGC5866 has a crisp, razor-sharp dust lane dividing it into two halves. This Hubble image highlights the galaxy's architecture: a subtle, reddish bulge surrounding a bright nucleus, a blue disk of stars running parallel to the dust lane and a transparent outer halo. NGC5866 is roughly two-thirds the diameter of our Milky Way.

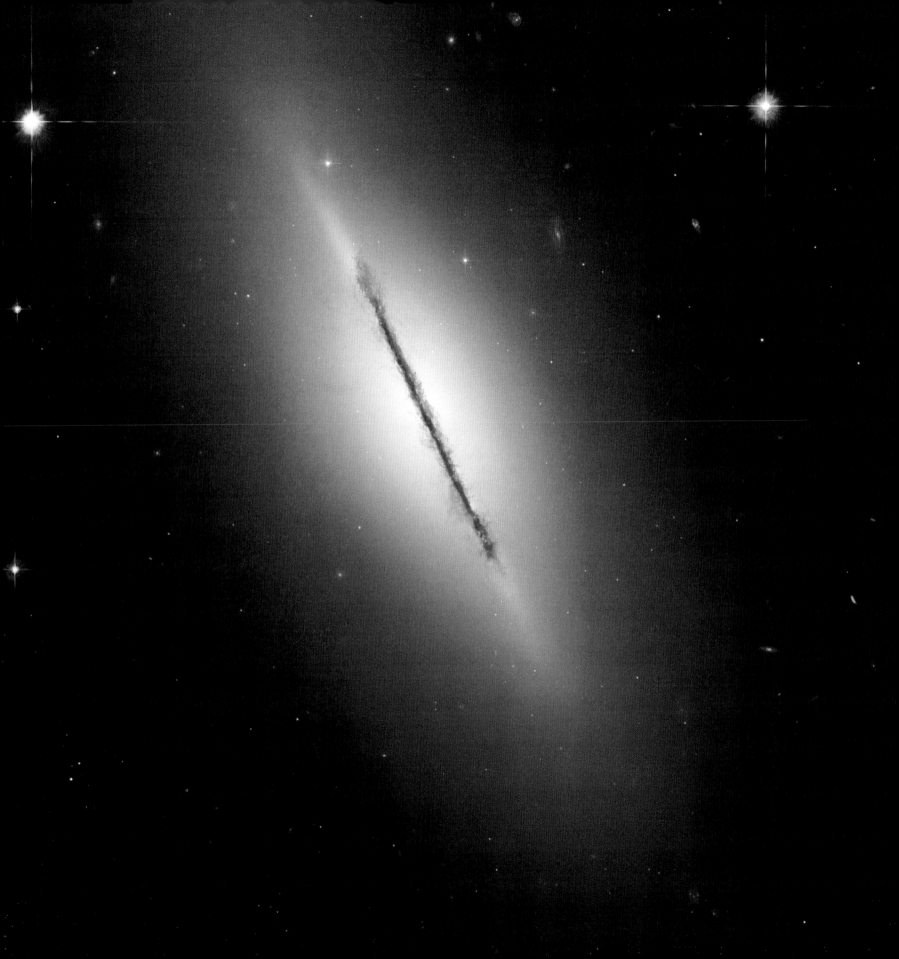

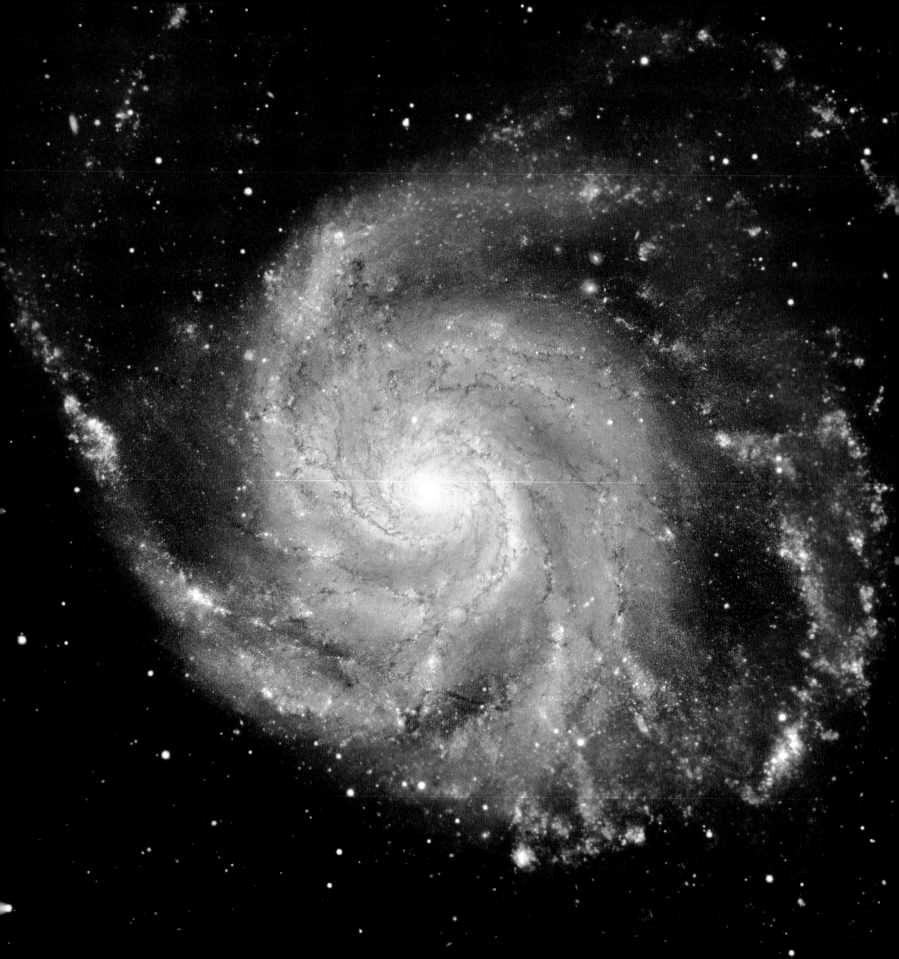

## M101

Nicknamed the Pinwheel Galaxy, this gargantuan galaxy is nearly twice the diameter of our Milky Way and contains one trillion stars. It could have ten trillion planets or more. At least 100 billion of its stars are similar to the Sun in terms of temperature and longevity. M101's disk is so thin and transparent that Hubble easily sees many more distant galaxies lying behind it. Since the Pinwheel Galaxy is 25 million light-years away, the light we're viewing was emitted at the beginning of the Miocene epoch, when mammals flourished and the mastodon first appeared on Earth.

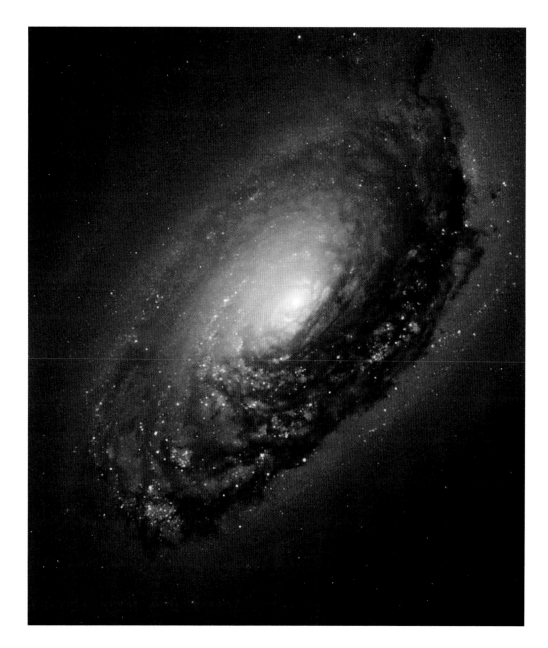

## M64

Nicknamed the Black Eye Galaxy for obvious reasons, M64 formed as a result of the collision of two galaxies approximately one billion years ago. The unique dark band of absorbing dust is the product of two counterrotating disks of gas and dust left over from the collision. Vigorous star formation is occurring in the shear region where the rotating gases are colliding and compressing.

## NGC1672

Dual dust lanes extend straight outward from the nucleus of this barred spiral galaxy, then follow the inner edges of the galaxy's spiral arms. Delicate curtains of dust partially obscure and redden the light of the stars and galaxies behind them by scattering blue light. A striking illusion in this view is that galaxies far behind NGC1672 appear to be embedded in the big galaxy's disk. In reality, they are vastly more distant. A few bright foreground stars inside our own Milky Way appear as dazzling diamondlike objects adorning the big galaxy, but these stars are only a tiny fraction of the galaxy's distance of 65 million light-years.

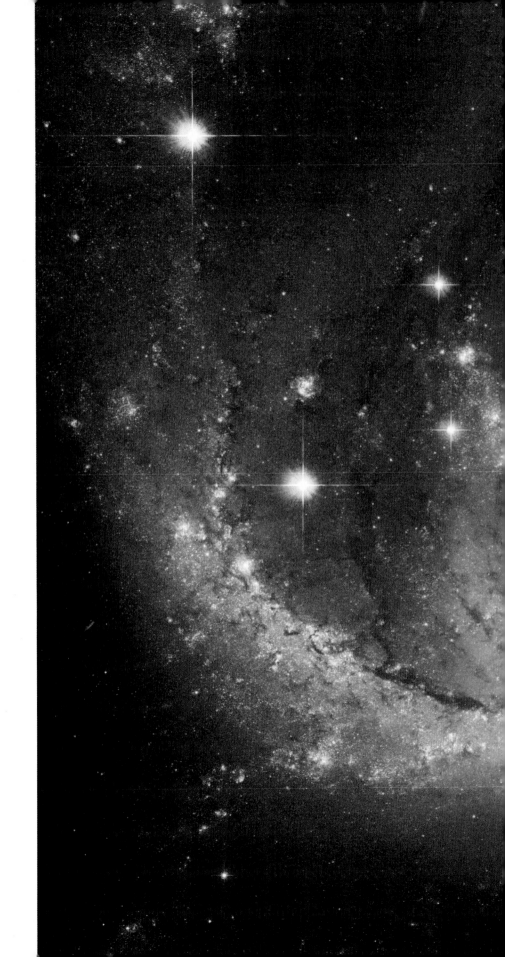

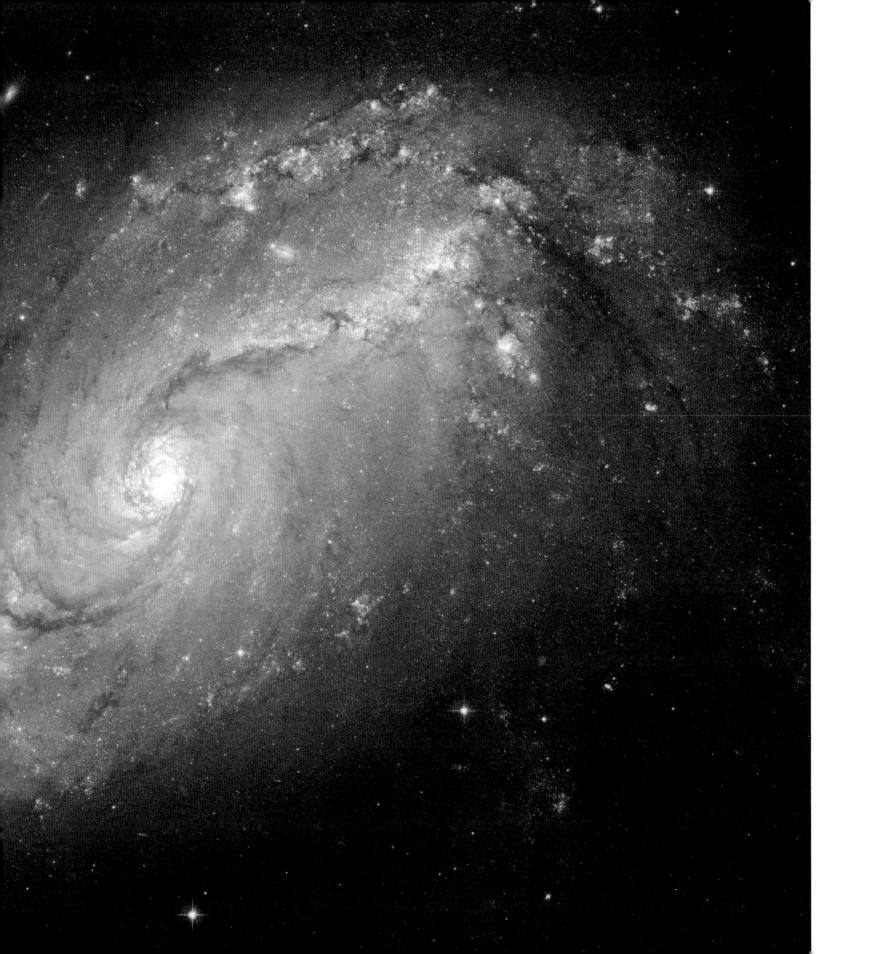

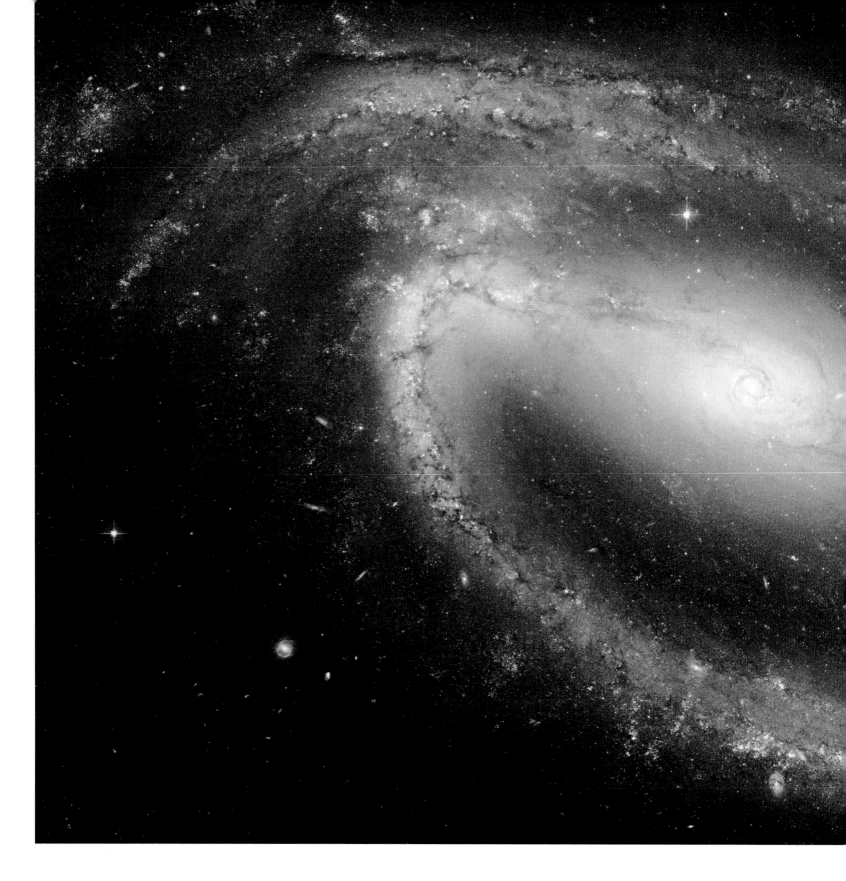

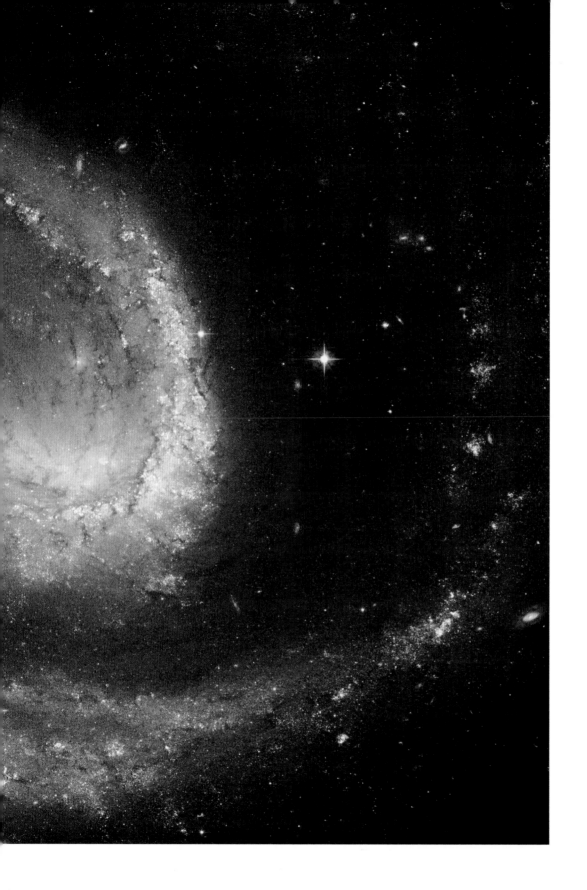

## NGC1300

This striking galaxy is the poster
child for all barred spiral galaxies,
which differ from normal spiral
galaxies in that the arms of the
galaxy do not spiral all the way in
to the center. Instead, they are
connected to the ends of a straight
bar of stars that contains the
nucleus at its center. Blue and red
supergiant stars, star clusters and
star-forming regions can be seen
across the spiral arms. Dust lanes
trace out fine structures in the disk
and bar. Numerous more distant
galaxies are visible in the background. About 115,000 light-years
across, NGC1300 is 70 million
light-years from Earth.

NGC1132

The elliptical galaxy NGC1132 displays the smooth distribution of stars typical in such galaxies. Conspicuously missing are blue stars, glowing nebulas and dust. Giant elliptical galaxies are probably the result of a merger of two or more spiral galaxies. Following the collision, an eruption of star birth ensues, which ultimately blasts away any remaining gas and dust needed for making future generations of stars.

ESO 510-G13

This strangely warped edge-on
galaxy looks like a pizza crust
tossed into the air. The warping
is telltale evidence of a close
encounter with a smaller galaxy.
Gravitational forces twist the disk
structure into a potato-chip shape.
Eventually, the disturbances will
dampen out, and ESO 510-G13
will once again take on the appearance of a normal galaxy.

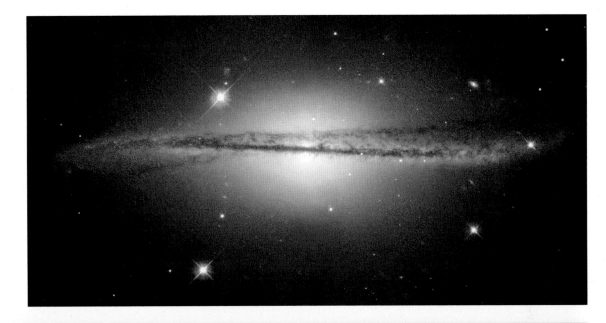

NGC634

Sweeping lanes of stars are evident
in this oblique view of a majestic
spiral galaxy. They have a striking
ripple appearance resembling the
classic wave structure seen when
a stone is tossed into a pond.

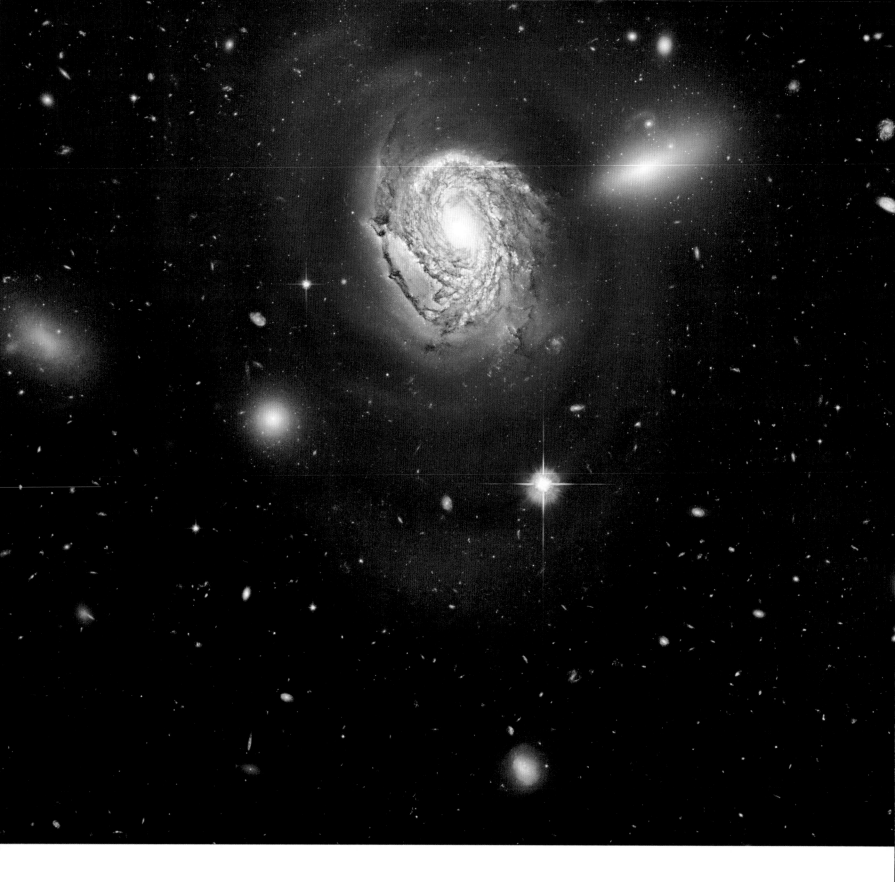

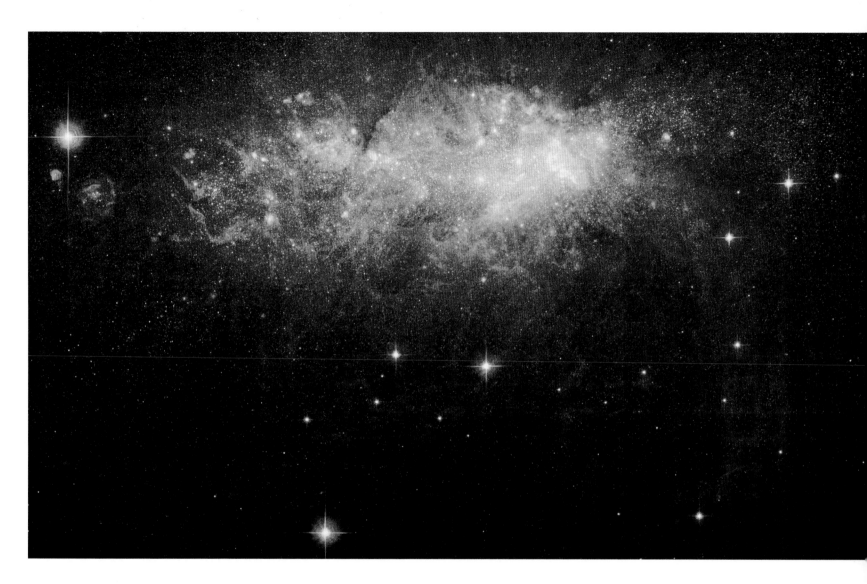

NGC1569

Size alone does not indicate a galaxy's level of activity. Sparkling with the light from millions of newly formed young suns, this tiny nearby dwarf galaxy is building stars at a prodigious rate—100 times greater than that observed in our home galaxy, the Milky Way. The stellar fireworks started 100 million years ago and are sweeping across the galaxy as supernova explosions compress nearby clouds of gas. The centerpiece is a grouping of three giant star clusters, each containing more than a million stars.

NGC4911

This giant pinwheel galaxy is close to the center of the Coma Galaxy Cluster, one of the densest nearby collections of galaxies in the universe. Tugged by the gravitational pull of neighboring galaxies, wispy outer spiral lanes extend far beyond the gas- and star-rich inner spirals.

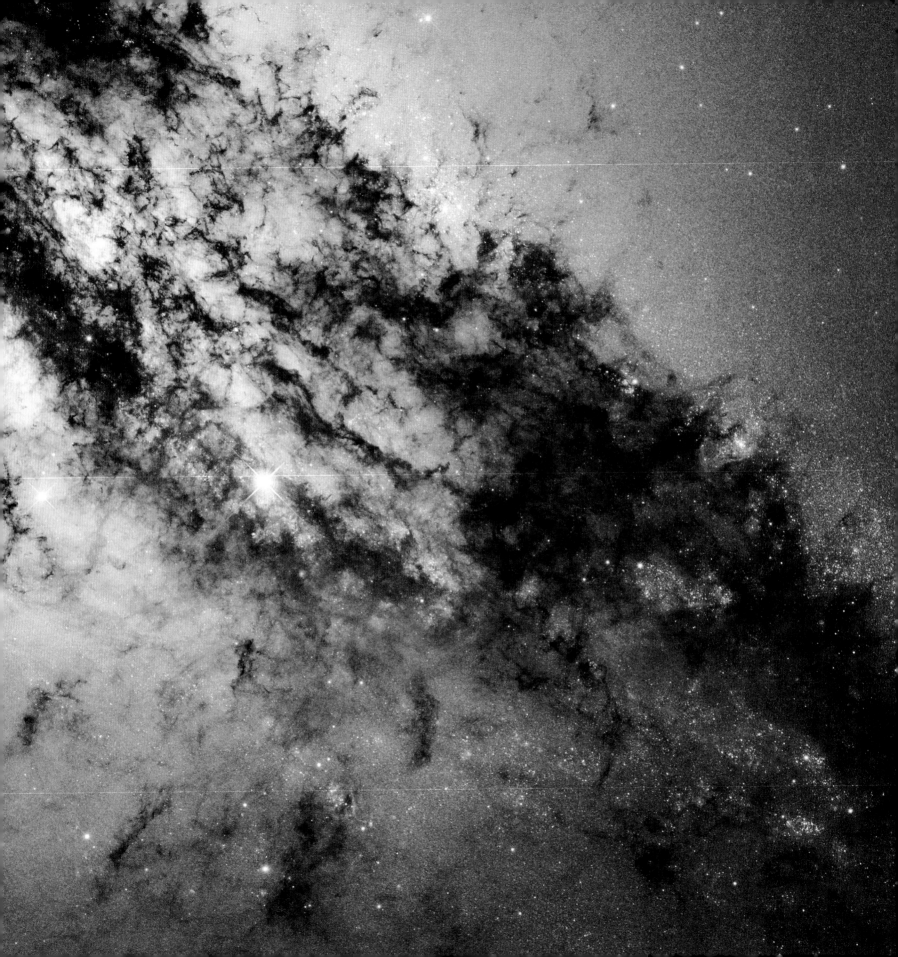

# COLLIDING GALAXIES

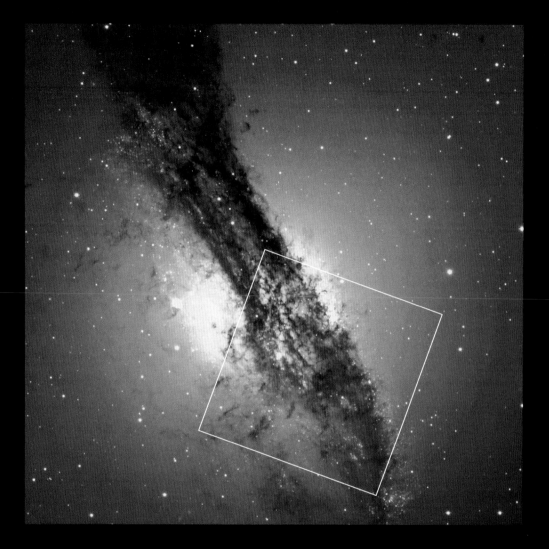

CENTAURUS A

Facing page: Menacing-looking dark lanes of dust crisscross Centaurus A, telltale evidence that this giant elliptical galaxy swallowed a smaller companion. Hubble captures the vibrant glow of young, blue star clusters, and its infrared vision penetrates into regions normally obscured by the dust. The warped shape of Centaurus A's disk of gas and dust is a testament of the past collision and merger with another galaxy. The resulting shock waves cause hydrogen gas clouds to compress, triggering a gusher of new star formation. At a distance of just over 11 million light-years, Centaurus A contains the closest active galactic nucleus to Earth. The galaxy's hub is home to a supermassive black hole that ejects jets of high-speed gas into space. In the large-telescope image above, we see a wider view of the entire galaxy taken from the Earth's surface.

## HICKSON COMPACT GROUP 31

This collection of warped, ancient galaxies is on its way to becoming a large elliptical galaxy. The entire system is aglow with star birth, triggered when close encounters between the galaxies cause hydrogen gas to compress and collapse to form stars. The four small galaxies are extremely close together, within 75,000 light-years of each other. It is a rare local example of what astronomers think was a common event in the distant universe.

## RING GALAXY

This unusual blue ring of stars is 150,000 light-years across, larger than our Milky Way. Called AM 0644-741, the galaxy is a member of the class of so-called ring galaxies, which are unique examples of how collisions between galaxies can alter their structure in striking ways. A star ring forms when an intruder galaxy plunges through the disk of another galaxy. The resulting gravitational shock of the collision drastically changes the orbits of stars and gas, causing them to rush outward. The effect is somewhat like the ripples in a pond when a stone is tossed into it. As the ring "tsunami" snowplows outward, gas clouds collide and are compressed. The clouds then contract under their own gravity to form an abundance of new stars. The Ring Galaxy is about 300 million light-years away.

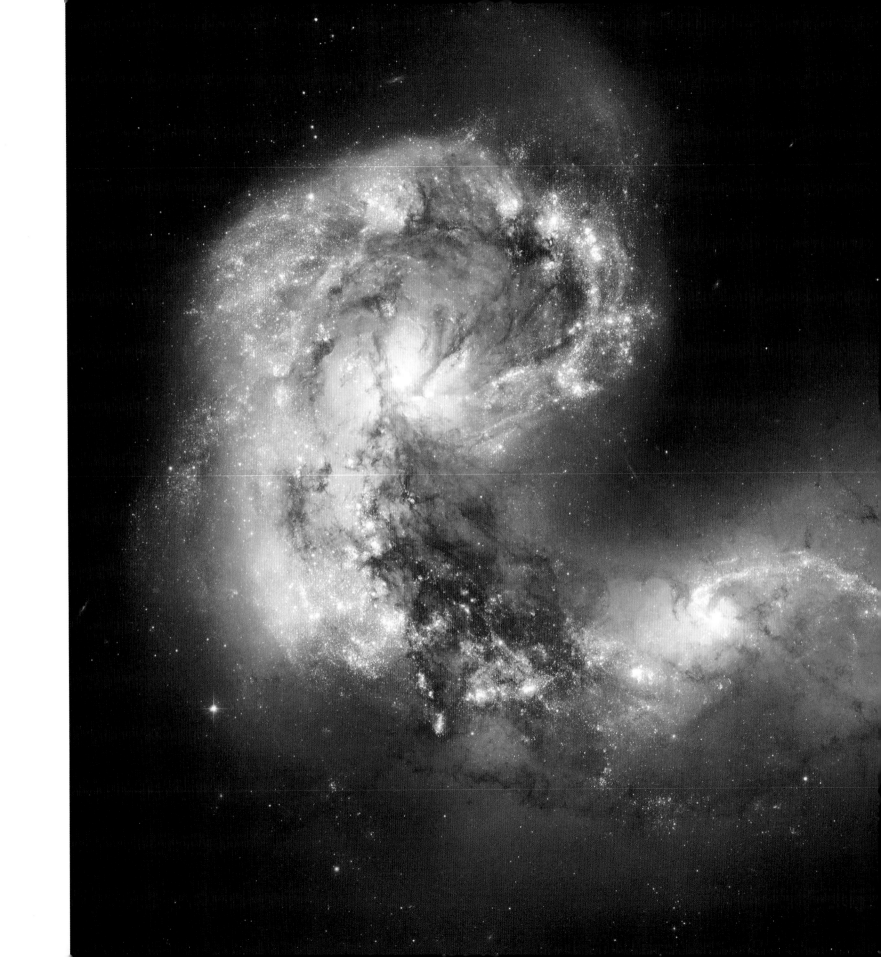

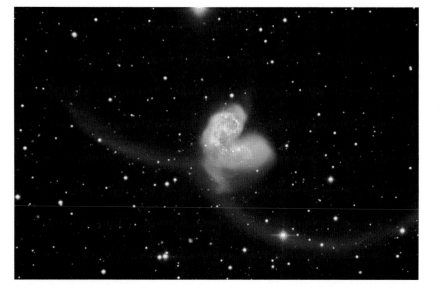

ANTENNAE GALAXIES

The Antennae is a prototypical example of what happens when two galaxies that are nearly the same size collide. Billions of stars will be formed during this major smashup. The two spiral galaxies began interacting with each other a few hundred million years ago. Nearly half of the faint objects in the Hubble Space Telescope image at left are young star clusters born in the collision. The two orange blobs to the left and right of center are the cores of the original galaxies and consist mainly of old stars crossed by filaments of dust. The brightest and most compact of the star-birth regions are called "super star clusters" and are unlike anything seen in the Milky Way Galaxy. About a hundred of the most massive clusters will survive to form regular globular clusters, similar to those found in our own Milky Way. The "tails" seen in the wide view above, taken from a ground-based observatory, are gas and dust flung off as the collision evolves.

**TADPOLE GALAXY**
Seen against a seemingly
infinite backdrop of galaxies,
this odd-looking galaxy with
the long streamer of stars and
gas appears to be racing through
space, like a runaway pinwheel
firework. Dubbed the Tadpole
Galaxy, UGC 10214 is distorted
by the tidal tug of a very blue,
compact galaxy visible just above
the more massive galaxy. Strong
gravitational forces from the interaction created the long tail of
debris that stretches out more
than 280,000 light-years, twice
the diameter of the Milky Way.

## MICE GALAXIES

These colliding galaxies are aptly nicknamed The Mice because of their long, streaming tails of stars and gas. The pair will eventually merge into a single giant elliptical galaxy. Myriad clusters of hot, young blue stars can be seen in the galaxy at left. Their formation was triggered by the tidal forces of the gravitational interaction. Material can also be seen flowing between the two galaxies. The Mice are 300 million light-years away.

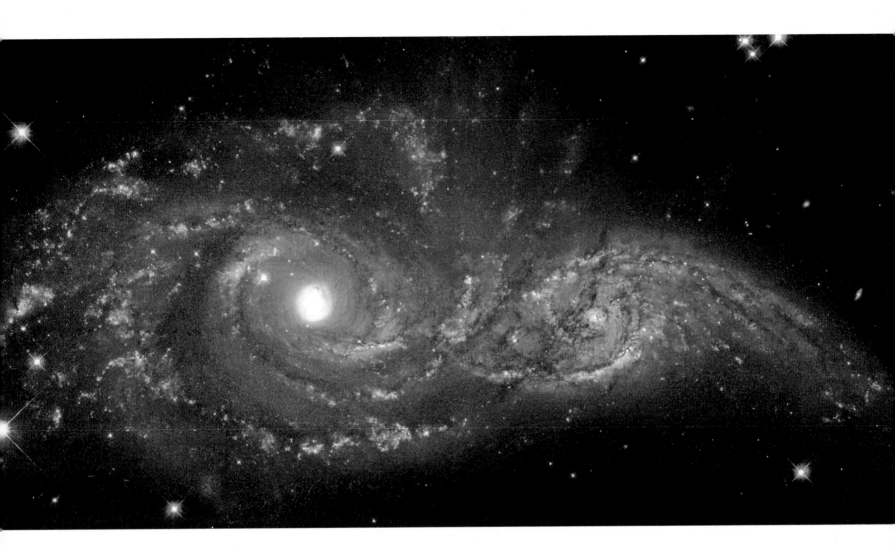

NGC2207 and IC2163

Two passing spiral galaxies create the uncanny appearance of a pair of owl eyes. Strong tidal forces from NGC2207, the larger galaxy at left, have stretched its smaller companion (IC2163). As is typical of interacting galaxies, stars and gas have been flung into long streamers extending 100,000 light-years into space. The pair made its closest approach about 40 million years ago. IC2163 is destined to be pulled back and will swing closer to the larger galaxy in the future. Ultimately, the two will probably merge to form an elliptical galaxy.

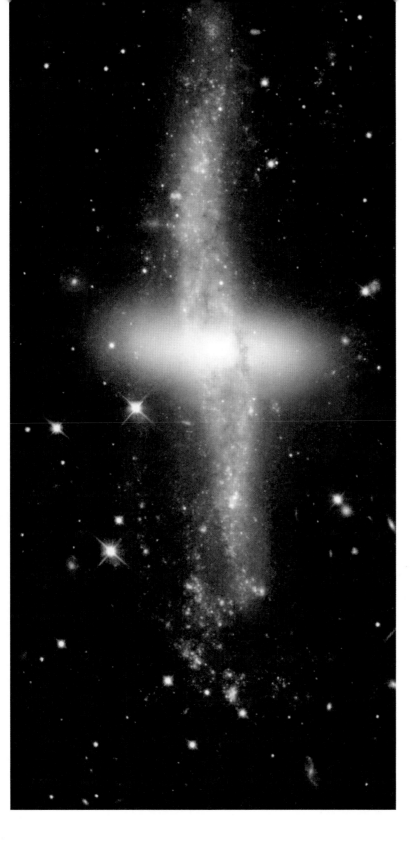

POLAR RING GALAXY

In this polar ring galaxy (one of only 100 known in the universe), what is left of one galaxy following a collision has become the rotating inner yellow disk of old reddish stars in the center. When the two galaxies collided, the gas from the smaller galaxy was stripped off and captured by the larger (vertical) galaxy, forming a new ring of dust, gas and stars around the inner galaxy, almost at right angles to the old disk. Although "collision" is the word astronomers use to describe the interaction of galaxies like these and the pair on the facing page, the stars in each galaxy rarely come into contact with one another. The average distance between stars in a typical galaxy is comparable to that of a golf ball on a golf course in Boston and one on a course in New York City.

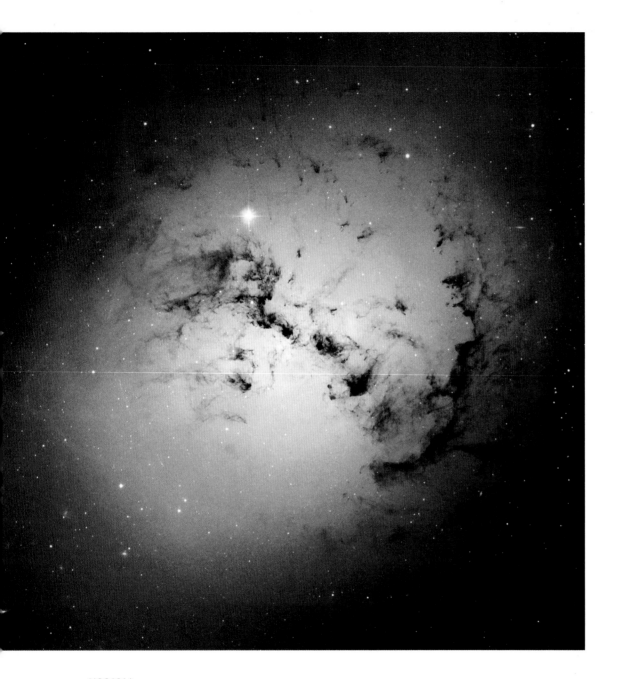

ARP 273
A cosmic waltz between two galaxies is the result of gravitational tidal distortion from their close proximity to each other. Despite the fact that they are separated by tens of thousands of light-years, a tenuous tidal bridge of material stretches between the pair. The swath of blue across the top is the combined light from clusters of bright, hot, young blue stars. The smaller, nearly edge-on companion galaxy shows intense star formation at its nucleus, which was probably triggered by the interaction. More close encounters and an eventual merger are the likely future of this galaxy duo.

NGC1316
These complex loops and blobs of dust are evidence of a gigantic collision between an elliptical galaxy and one or more spiral galaxies billions of years ago. The inner regions of the resulting galaxy have an intricate system of dust lanes and patches, which are probably the shreds of a smaller galaxy.

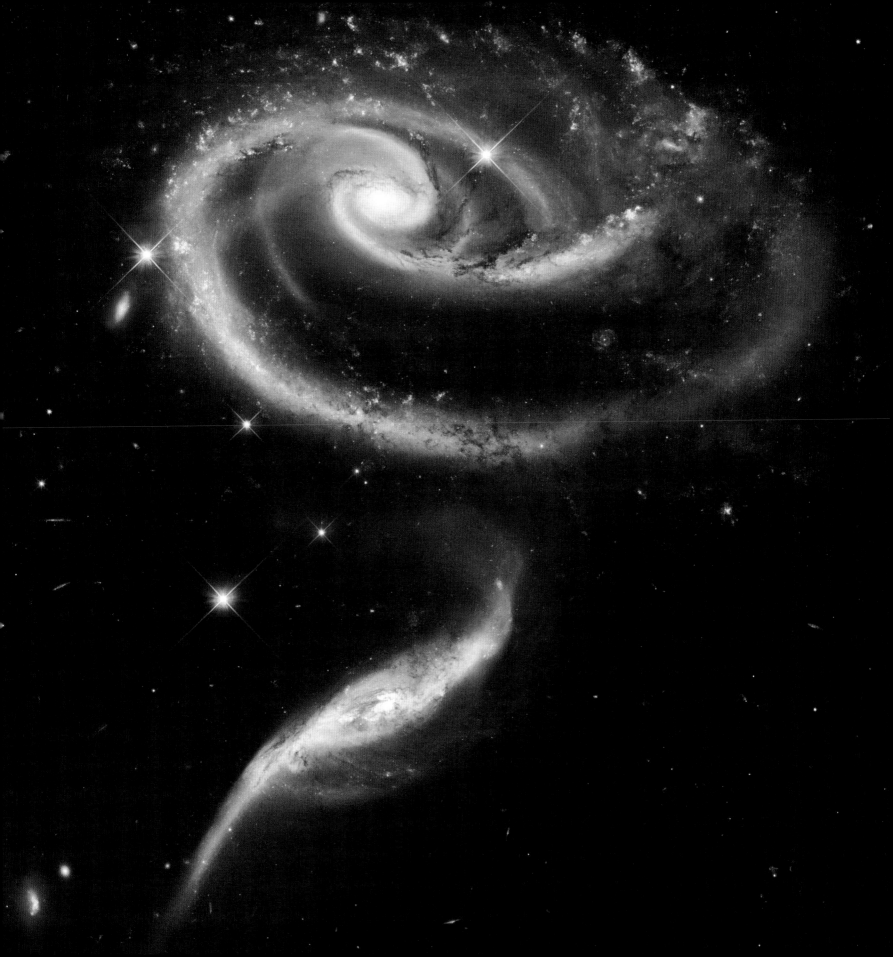

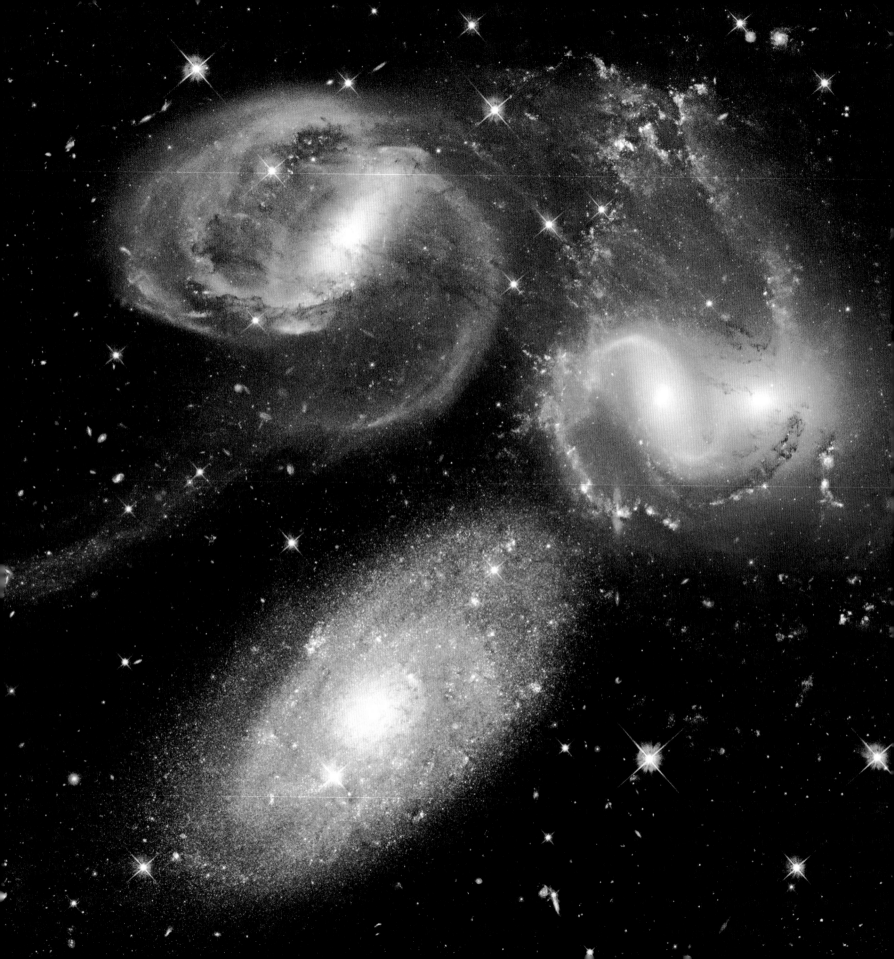

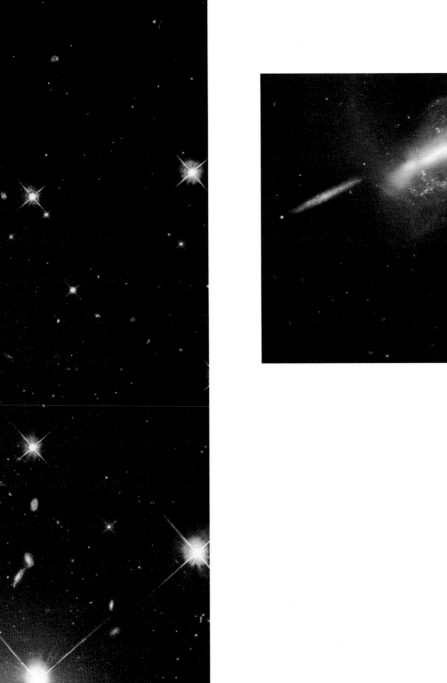

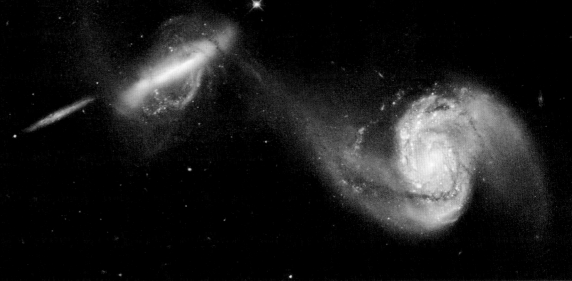

## ARP 87

Drawn by the tidal tug of gravity,
a diaphanous stream of stars,
gas and dust flows from the galaxy
at right and forms an enveloping
arm around its companion, an
edge-on spiral galaxy. Some of
the stars and gas from the large
galaxy have been trapped by the
gravitational pull of the smaller
one, as can be seen in the
corkscrew-shaped material.

## STEPHAN'S QUINTET

One of the most famous examples of interacting galaxies is Stephan's Quintet. Three of the galaxies
have distorted shapes, elongated spiral arms and long, gaseous tidal tails containing myriad star clusters. The interactions among the galaxies have sparked a frenzy of star birth in the pair of intertwined
galaxies just above center. This drama is being played out against a rich backdrop of far more distant
galaxies. The galaxy at lower left is in the foreground and not part of the grouping. It is 40 million light-
years from Earth, while the remaining members of the quintet reside 290 million light-years away.

257

NGC5775

Star-forming regions in this edge-on galaxy glow fiercely in shades of red from fluorescing hydrogen. NGC5775 is heading toward a collision with another galaxy just outside the field of view, although it won't happen for hundreds of millions of years. Such turbulent galaxy histories are common here in the richly populated Virgo cluster of over 2,000 galaxies.

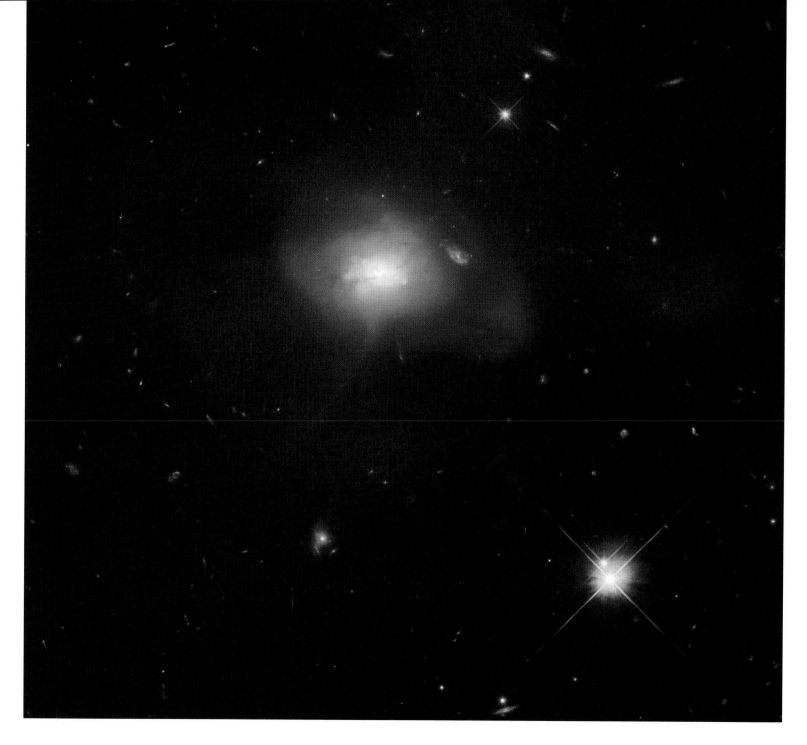

AFTERMATH OF A POWERFUL COLLISION

This is a rare detailed look at the result of a billion-year-long collision of two spiral galaxies each the size of our Milky Way. Under gravity's tidal pull, the galaxies were stretched and torn apart as their matter became wrapped around a common hub of gravity. The starry population was scattered into random orbits, then settled down into a new elliptical galaxy. Some ribbons of the remnant dust lanes obscure parts of the galaxy's central bluish core, but they will disappear in time.

## M82

The core of this nearly edge-on spiral galaxy looks as if it is erupting. In reality, it is undergoing an explosion of new star birth that is propelling hydrogen far outside the galactic plane. This effervescent activity started about 100 million years ago, probably as the result of a close brush with M81, its neighboring spiral galaxy.

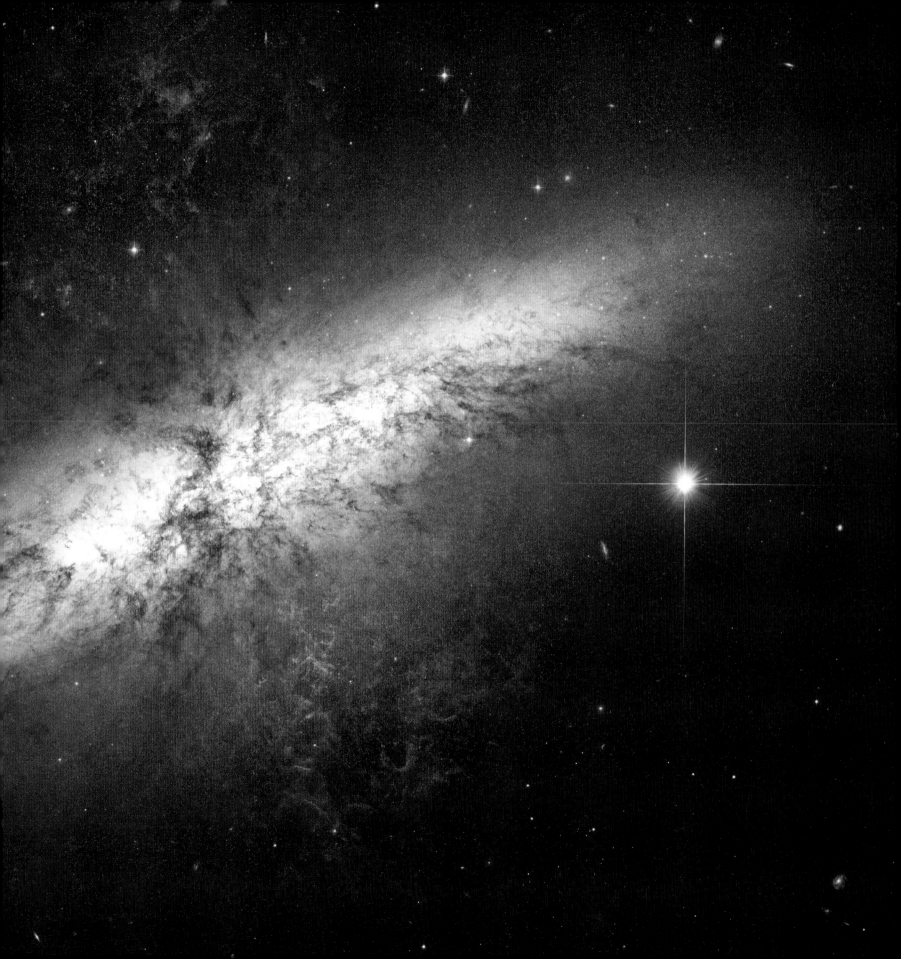

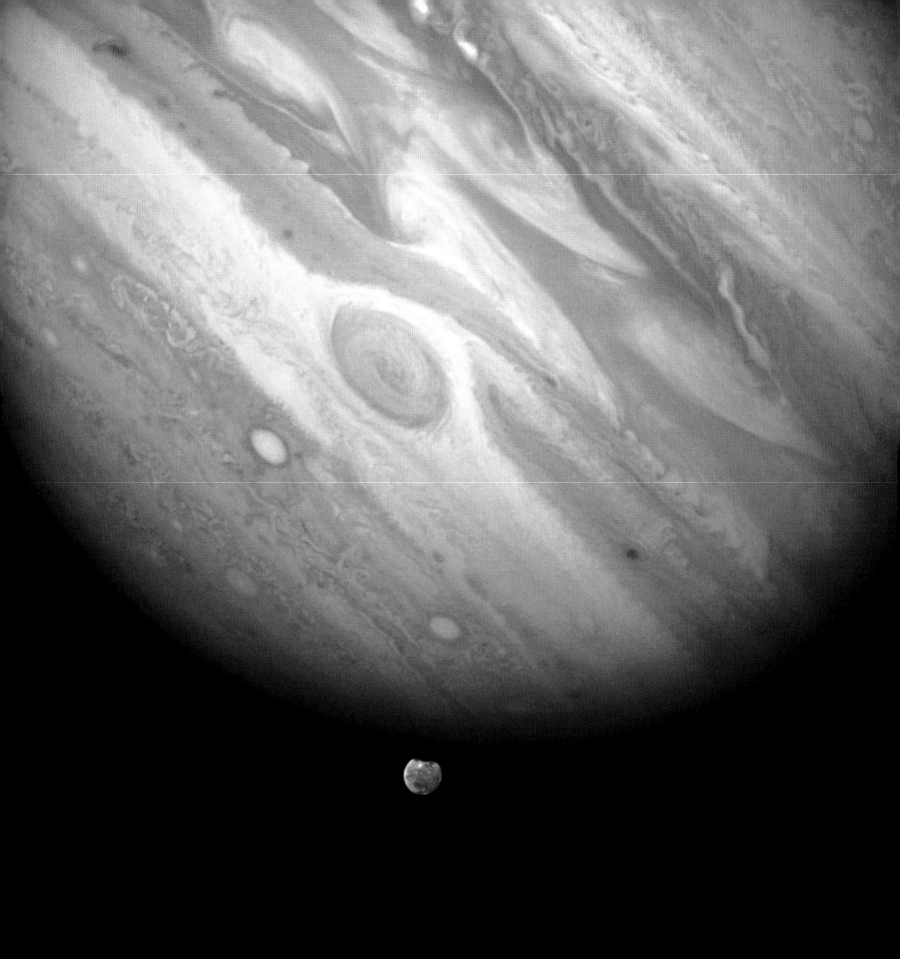

# NEIGHBOR WORLDS: THE PLANETS

## CHAPTER 9

Although the Hubble Space Telescope was designed primarily to probe the most distant reaches of the universe, it also provides exquisitely sharp views of the Earth's companion worlds in the solar system.

On January 7, 1610, Galileo pointed his newly constructed telescope toward the planet Jupiter. He immediately saw the planet as a small disk, not a point of light like the stars. Over the next few nights, Galileo noticed tiny starlike objects beside the planet changing position from night to night. He soon realized these were moons orbiting Jupiter in the same way the Moon orbits Earth—but there were four of them.

These and other revelations by Galileo—phases of the planet Venus, mountains and plains on the Moon, myriad stars in the Milky Way—changed humanity's perception of the heavens from a mysterious, magical realm to a place where real worlds accompany Earth. These profound first discoveries were the initial links in a chain of telescopic finds that continues to this day, with Hubble now regarded as the most important telescope since Galileo's, four centuries earlier.

In 1946, astronomer Lyman Spitzer was the first scientist to lay out in detail the advantages of having a large telescope in orbit far above the interference and distortion of the Earth's atmo - sphere. Spitzer pointed out that advances in rocketry during the Second World War had suddenly turned dreams of space exploration—and the possibility of orbiting a large telescope in space— from science fiction to reality.

It happened fast. From Sputnik 1, the first satellite to orbit Earth, in October 1957, to the first astronauts landing on the Moon, in July 1969, was a brief 12 years. Then, during the 1970s, designs for the space telescope were refined, first for a telescope with a 120-inch-diameter primary mirror, then the final 94-inch size. It was built in the 1980s and launched in 1990 by the space shuttle Discovery. This same period marked the first exploration of the planets of our solar system by robotic spacecraft. By 1986, all seven of the major planets—Mercury, Venus, Mars, Jupiter, Saturn, Uranus and Neptune—had been examined close-up by flyby probes.

As elucidating as they were, however, the flybys provided only snapshots of the planets. The outer planets, with their turbulent atmospheres and season variations, are ever changing. Over more than two decades, Hubble has chronicled numerous and often unexpected changes on planets with atmospheres.

In 1994, for instance, not long after the telescope's vision was repaired, Hubble gave astronomers a ringside seat to a once-in-a-millennium spectacle: the collision of multiple comet fragments with the planet Jupiter. Without Hubble's grand vision, scientists would have missed an extraordinary space drama play out on a daily basis over the period of a week.

Today, when vast dust storms swirl on Mars or a comet smashes into Jupiter or cloud turbulence disrupts the normally tranquil atmosphere of Saturn, researchers command Hubble to take a look. This has beautifully complemented the sometimes fleeting views by spacecraft.

On April 9, 2007, the Hubble Space Telescope caught Jupiter's largest moon Ganymede playing peekaboo. About one-third the Earth's diameter, Ganymede is seen here just before it ducks behind the giant planet. Three images taken by the Wide Field and Planetary Camera 2 using red, green and blue filters were combined to produce this unique color portrait.

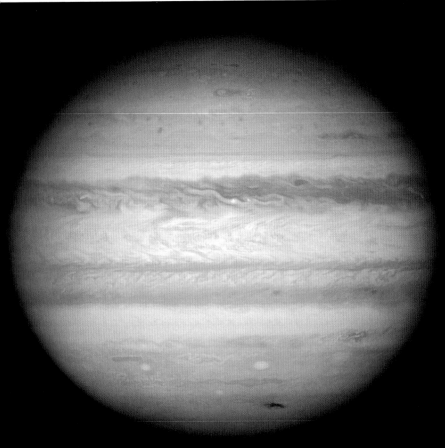

July 23, 2009

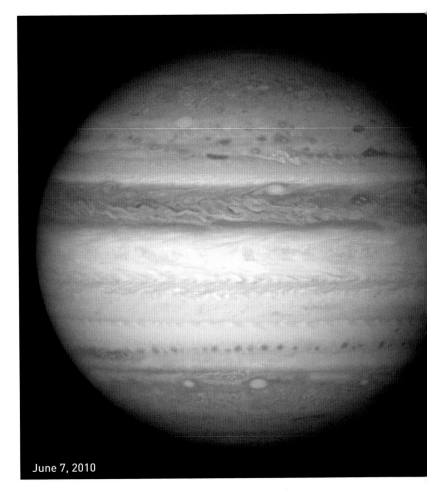

June 7, 2010

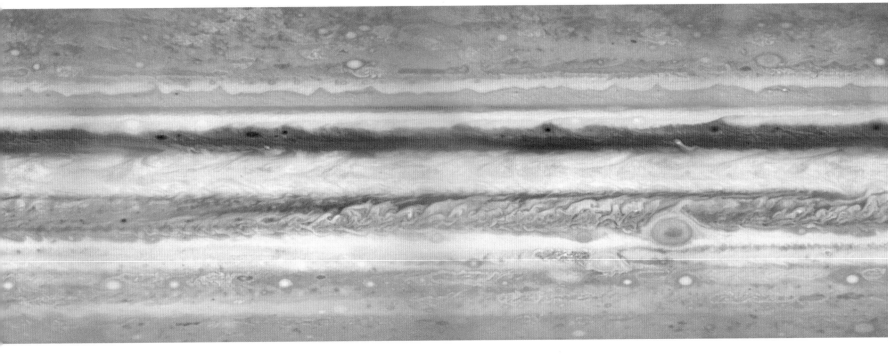

## JUPITER'S STRIPES

Since the mid-1600s, astronomers' sketches of the face of Jupiter have shown two prominent cloud bands similar in appearance to the Hubble image at far left. But a few times each century, the south belt disappears for one or two years, as happened during 2010 (left). The entire visible surface of the giant planet is clouds and haze. The different hues are from the varying amounts of sulfur and ammonia compounds in the mostly hydrogen atmosphere. Jupiter's rapid rotation spins these clouds into belts defined by altitude and composition. The belt that disappeared in 2010, then reappeared in 2011, is twice as wide as Earth and 20 times the Earth's diameter in length. But the mechanism that controls its occasional disappearance remains unknown.

## MAPPING JOVIAN CLOUDS

Imagine peeling Jupiter's atmospheric "skin" like an orange and flattening it. The result would be something like the Mercator projection at left. It was made from digitally flattening several Hubble images of Jupiter to show the entire planet in one image.

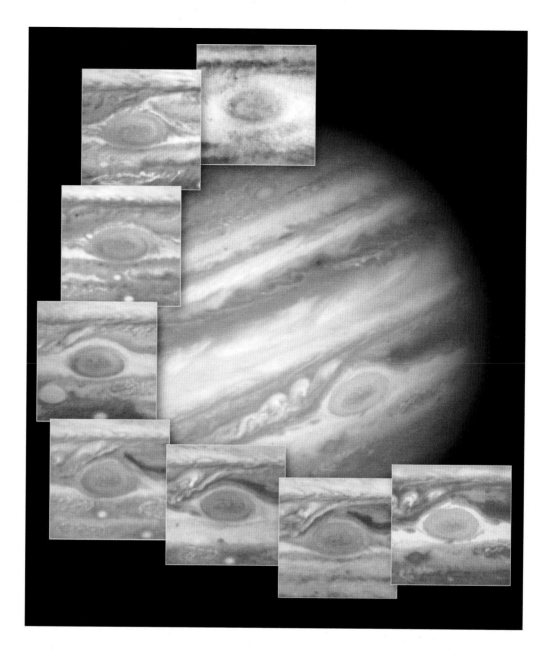

## JUPITER'S GREAT RED SPOT

When 17th-century astronomers first turned their telescopes to Jupiter, they noted a conspicuous reddish oval on the planet. The Great Red Spot is still present in Jupiter's atmosphere, more than 300 years later. This feature is now known to be a vast storm, spinning like a cyclone, with winds reaching speeds of about 400 kilometers per hour. The largest-known storm in the solar system, the Red Spot has a diameter of 25,000 kilometers, almost twice the size of the entire Earth. The shape and intensity of the Great Red Spot change over the years, as these Hubble images show.

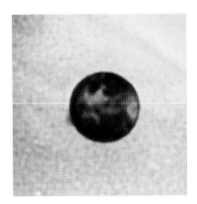

**VOLCANO ON JUPITER MOON**
A 400-kilometer-high plume of gas and dust (fuzzy orange patch just off moon's left edge) from a volcanic eruption on Io, Jupiter's innermost moon, was captured by the Hubble Telescope as the satellite passed in front of Jupiter (blue background) in July 1996. Io is the most volcanically active body known in the solar system. To form an eruptive cloud this size, the material must have been blown out of the volcano at more than 3,000 kilometers per hour. Io's volcanic eruptions propel gas and dust hundreds of kilometers into space. The ejection seen here is from Pele, one of Io's most powerful volcanoes.

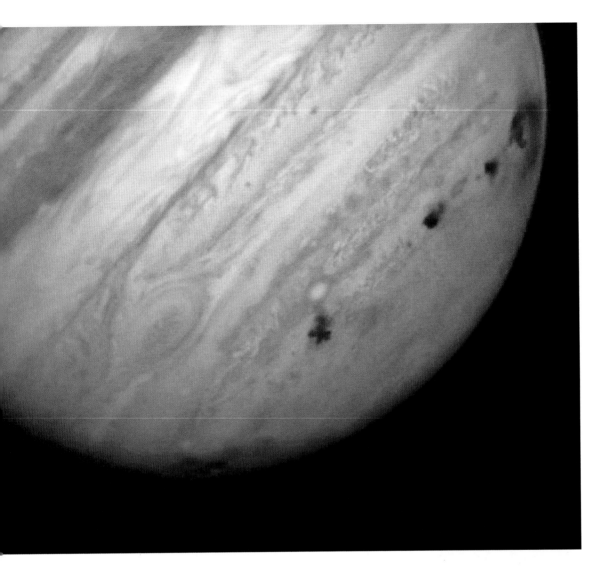

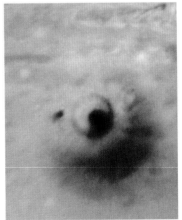

**IMPACTS ON JUPITER FROM COMET SHOEMAKER-LEVY 9**
From July 16 to 22, 1994, twenty fragments of Comet Shoemaker-Levy 9 smashed into Jupiter. The impacting pieces caused enormous explosions and left dark bruises on Jupiter that were visible in the telescopes of astronomy hobbyists for weeks. The Hubble Space Telescope got the best views (above and right) of this once-in-a-lifetime celestial event.

IO AND ITS SHADOW
As Jupiter's volcanic moon Io passes above the turbulent clouds of the giant planet on July 24, 1996, it casts a conspicuous black shadow that sweeps across the face of Jupiter at 17 kilometers per second. Io is about the size of the Earth's Moon.

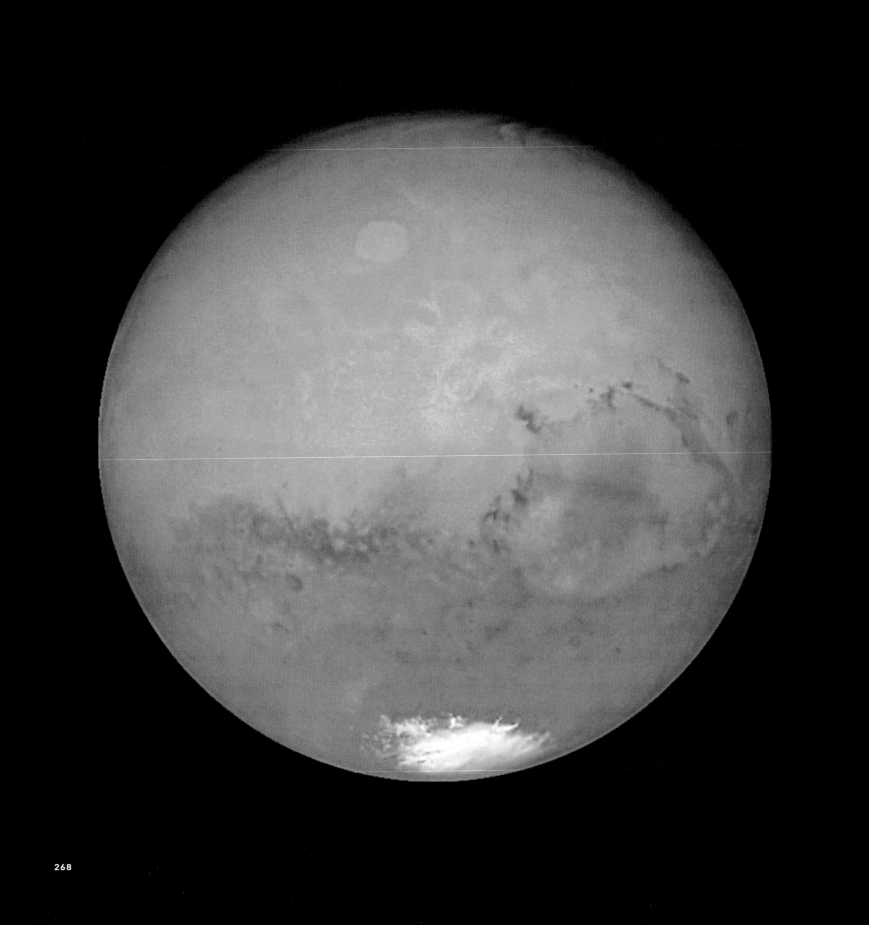

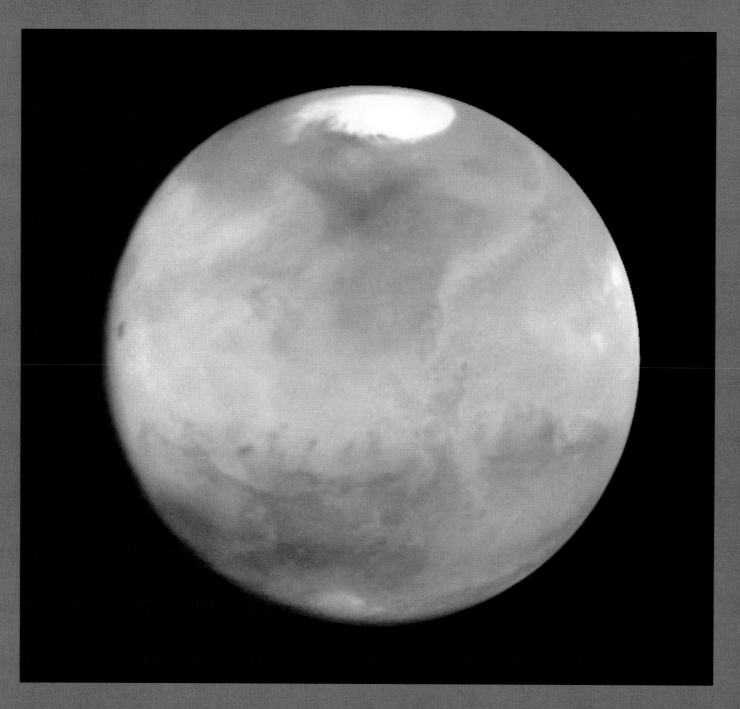

## HUBBLE VIEWS MARS

The south polar cap of Mars is prominent in the Hubble portrait on the facing page, taken in August 2003, when the planet made its closest approach to Earth in more than 60,000 years. Although the Martian polar caps are water ice and frozen carbon dioxide, water does not exist in liquid form on the surface of Mars, making the red planet an extreme desert world. In the Hubble view above, the north polar cap is prominent, along with thin, water-ice-crystal clouds. Reddish areas are fine dust deserts; darker regions are more gravelly.

## THE MOST EARTH-LIKE PLANET

In the late 1800s, telescopes became powerful enough to begin to reveal features on Mars that reminded astronomers of Earth: deserts, dark regions that change with the Martian seasons and polar caps that expand in Martian winter and contract in summer. Occasionally, clouds appear and dust storms kick up. This fueled more than half a century of lively debate about the possibility of higher forms of life on the red planet. But when the first few spacecraft to fly by Mars revealed a landscape more like the Moon than Earth, this idea was finally put to rest.

## SHRINKING MARS POLAR CAP

These images were created by assembling mosaics of pictures taken by Hubble in October 1996 and in January and March 1997 and projecting them to appear as they would if seen from above the north pole of Mars. The first view is late winter in the northern hemisphere, showing a huge deposit of frozen carbon dioxide—essentially, some of the Martian atmosphere frozen to the ground. The last image shows the first day of Martian summer, when maximum sunlight reaches the polar region. The frozen carbon dioxide is gone, and just a residual cap of water ice remains.

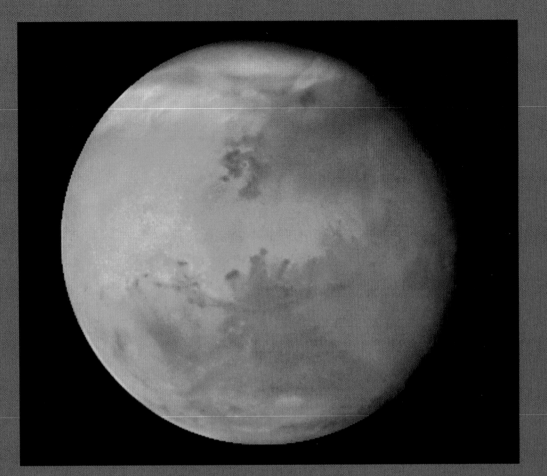

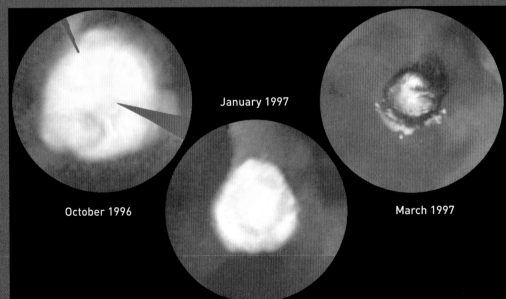

January 1997

October 1996

March 1997

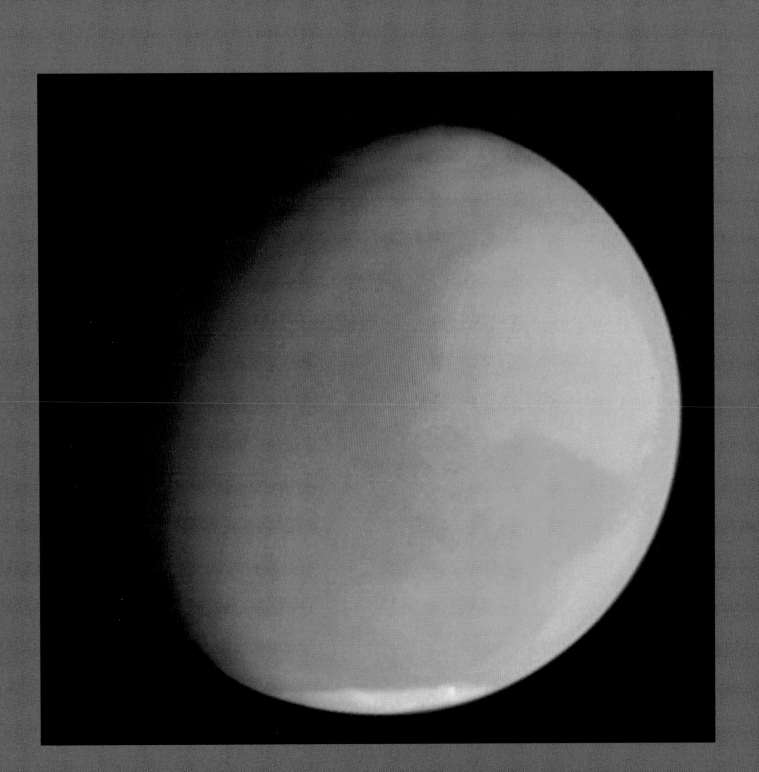

**MARS GLOBAL DUST STORM**
In October 2001, Hubble gave astronomers a ringside seat to the biggest global dust storm seen on Mars in several decades. Larger by far than any ever seen on Earth, the Martian dust storm raised a gargantuan cloud of dust that engulfed the entire planet for several months. Mars' carbon dioxide atmosphere is less than 1 percent as dense as the Earth's atmosphere, although winds can reach 250 kilometers per hour.

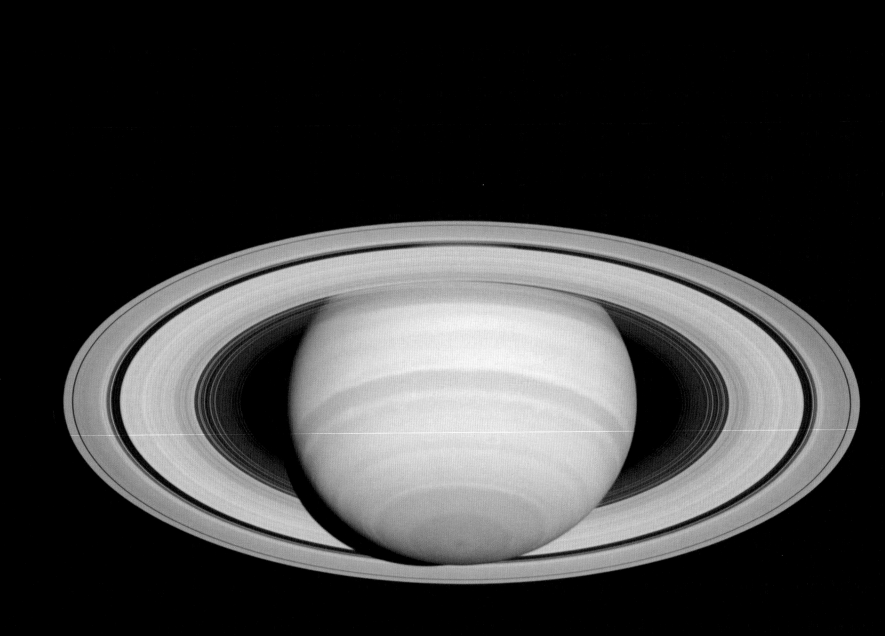

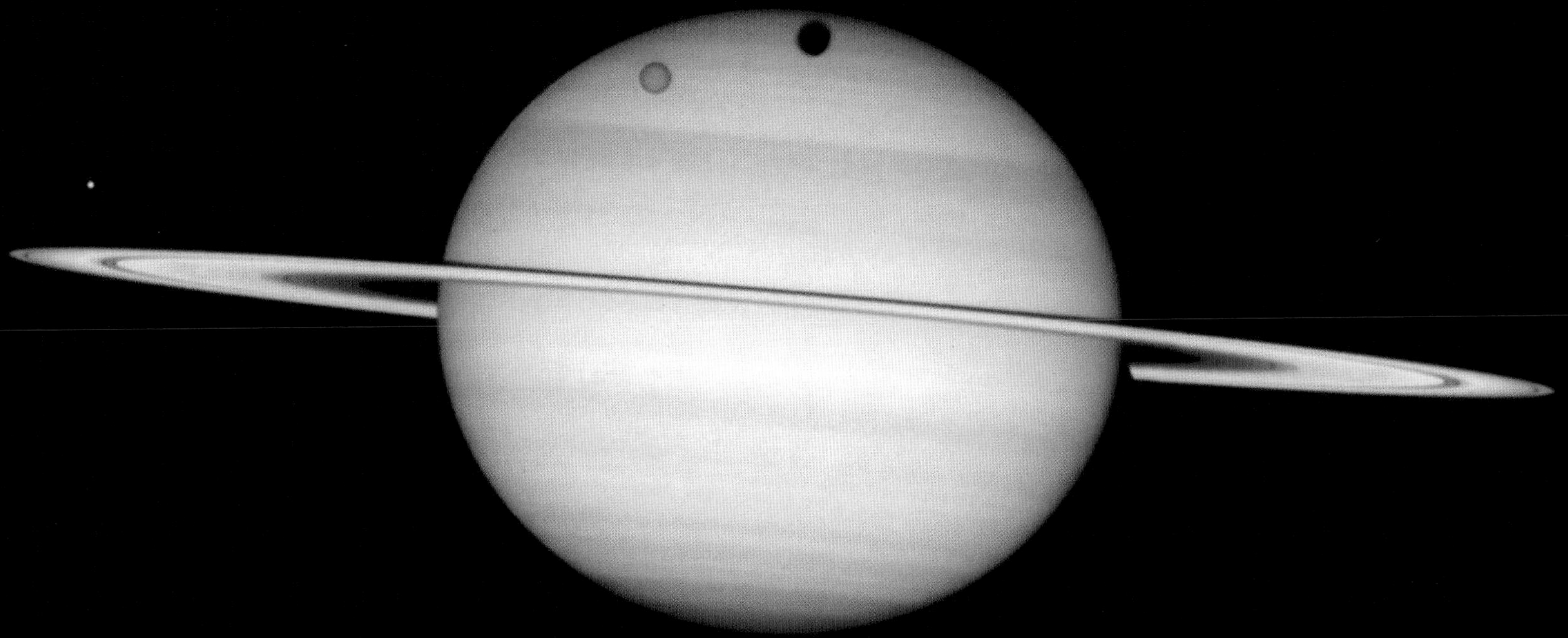

**GLORIOUS SATURN**

Taken eight years apart, these exquisite Hubble portraits of Saturn show the famous rings nearly edge-on, above, and open at maximum, facing page, to our line of sight from Earth. Some of the larger of Saturn's more than five dozen moons are seen above, including most prominently Titan, the largest, casting its inky shadow on the planet. The rings are composed of trillions of icy particles that probably originated with the collision of large moons aeons ago. The delicate structure of the rings seen in the tilted view is maintained by the gravitational influence of Titan and Saturn's other large moons.

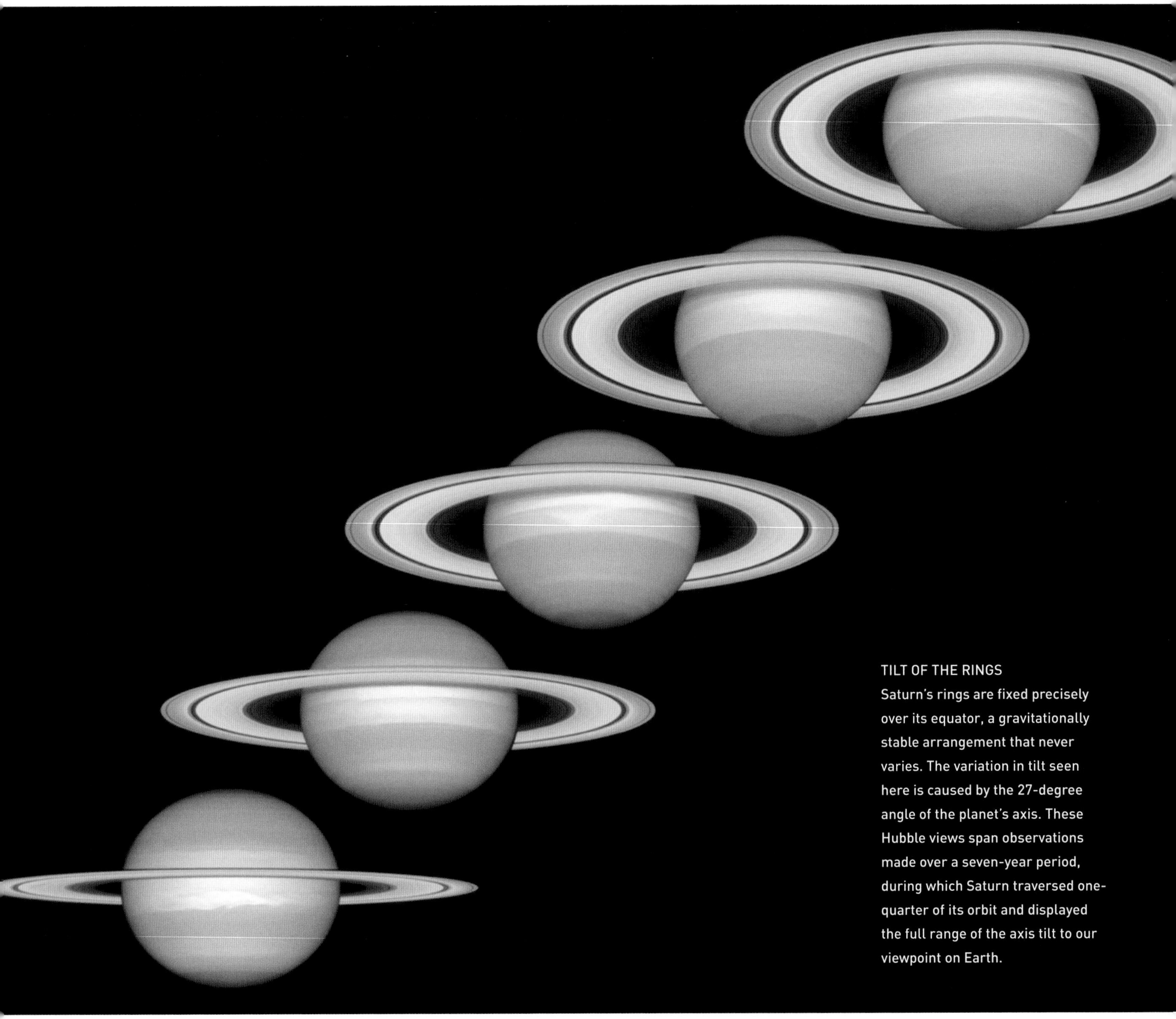

## TILT OF THE RINGS

Saturn's rings are fixed precisely over its equator, a gravitationally stable arrangement that never varies. The variation in tilt seen here is caused by the 27-degree angle of the planet's axis. These Hubble views span observations made over a seven-year period, during which Saturn traversed one-quarter of its orbit and displayed the full range of the axis tilt to our viewpoint on Earth.

## SATURN AURORAS

Auroras have been observed on Saturn by Hubble and also by the Saturn-orbiting Cassini spacecraft using ultraviolet detectors. The ultraviolet results were then superimposed on Hubble images of Saturn taken at about the same time. Planets in our solar system known to have significant auroras are Earth, Jupiter and Saturn. Hubble has detected less intense auroral activity on Uranus and Neptune.

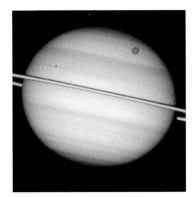

## MOONS ON SATURN

On February 24, 2009, an extraordinary quadruple moon transit occurred on Saturn. In addition to Saturn's largest moon, Titan, three much smaller moons are in front of the planet: Dione and Enceladus (white dots at left center; Dione is the larger of the two), along with their black shadows, and Mimas (on right at edge of Saturn).

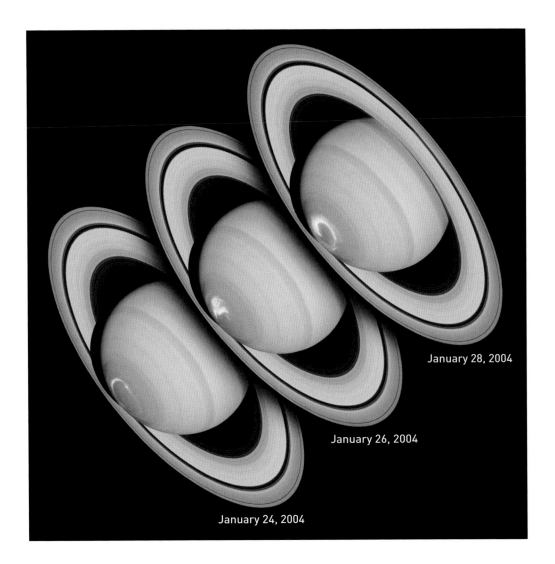

January 28, 2004

January 26, 2004

January 24, 2004

SATURN'S RING SHADOW

This Cassini spacecraft close-up of Saturn is one of the few images in this book not obtained by Hubble. It was included here because it complements the Hubble Saturn pictures so well. The shadow cast on the planet by the incredibly thin ring system is one of the most striking sights in nature.

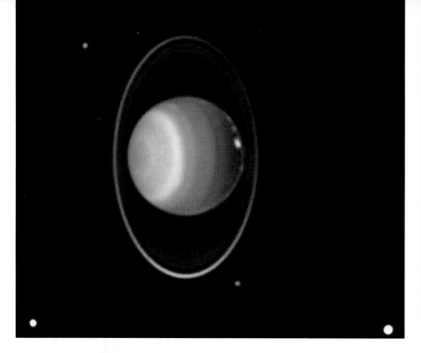

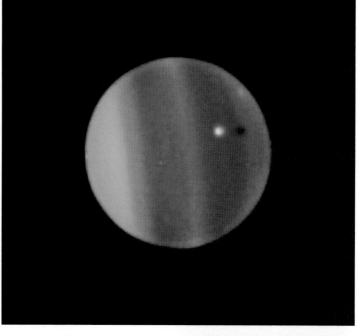

## URANUS PORTRAITS

The Hubble Space Telescope's infrared view of Uranus, above, reveals four major rings surrounding the planet and the four largest of its more than two dozen known moons. Infrared shows clouds in Uranus's methane atmosphere more readily than visible-light cameras. Hubble captured a moon traversing the face of Uranus, above right, along with its shadow. The white dot is the icy 1,150-kilometer-diameter moon Ariel, which is casting its shadow onto the cloudtops of Uranus. The moons of Uranus orbit the planet in such a way that they rarely cast shadows on the planet's surface.

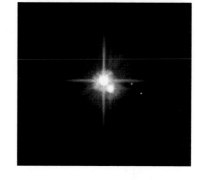

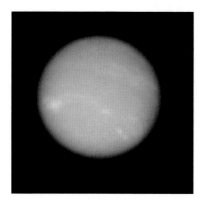

## NEPTUNE

Only one-third the size of Jupiter and more than seven times farther from Earth than the giant planet, Neptune is a small target for telescopes at the Earth's distance. Nevertheless, with the infrared capabilities of the Wide Field Camera 3, Hubble has detected clouds on the remote world.

## PLUTO

Pluto could almost be regarded as a double planet, since its largest moon, Charon, is half its diameter. Two much smaller moons, Nix and Hydra (farthest out), are visible in this Hubble image. Vague light and dark patches have been detected on Pluto by Hubble, as seen in the illustration on page 14. In 2006, the International Astronomical Union reclassified Pluto as a dwarf planet, largely because of its size (smaller than the Earth's Moon) and its similarity to other bodies that orbit the Sun beyond Neptune.

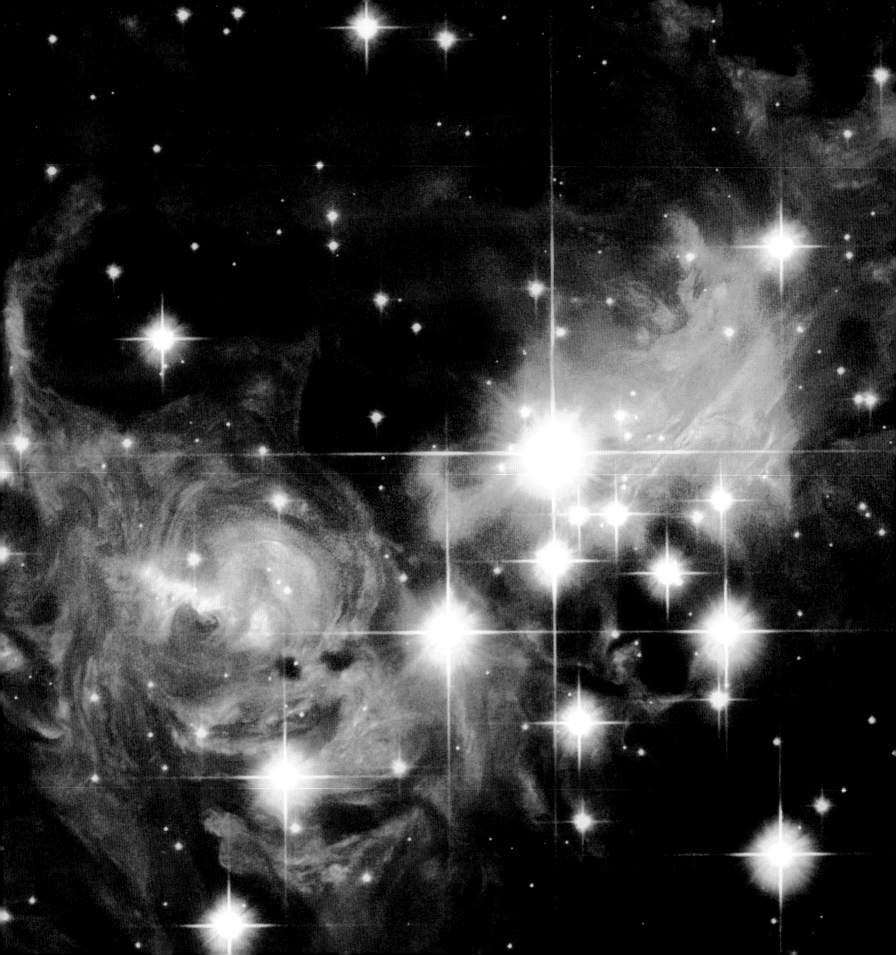

# HUBBLE'S STRANGE UNIVERSE

## CHAPTER 10

Hubble's cosmic portfolio is full of grand views of familiar celestial objects: galaxies, nebulas, planets and myriad stars. But some Hubble images are downright bizarre. They display eerie objects, mysteriously forged in space and time. Among the strangest pictures are those of events that come and go unexpectedly. Many of these we've never seen, or not seen clearly, until Hubble's sharp vision was turned on them.

The universe is molded and sculpted by the interplay of a few simple forces: gravitational and magnetic fields; electromagnetic radiation; winds of charged particles; and kinetic forces from bodies colliding. Some objects, shaped by random and chaotic events of nature, can look astonishingly balanced and symmetrical, offering our senses the surprising mystery and beauty of the universe. Without a three-dimensional view of a target—which is impossible given the vast distances between the stars—it can be difficult deducing its true shape. Perhaps the most bizarre example of this is the so-called light echo. Radiation from an erupting or exploding star can rebound through space, like an earthly echo across a canyon, creating the illusion of an expanding ring. The Hubble image on the facing page is an example of a light echo.

Besides going bump in the night, some space phenomena go flash in the night. Hubble can detect such outbursts and does so as part of a robotic team—a fleet of high-energy orbiting observatories. Hubble acts as a forensics detective, conducting follow-up work to pinpoint the nature of the object from which the flash of light emanated.

The solar system is a lively and unpredictable celestial "mosh pit." Beginning in the late 1950s, scientists took a step-by step approach to reconnoitering the solar system's planets, moons, asteroids and comets with unmanned space probes. Every single planet that has been visited by spacecraft has held its share of surprises and offered up unpredicted details. The strategy of exploration was first to swiftly fly by planets, then to send planetary orbiters that would loop around these worlds to carry out global photographic surveys. The landers followed, poking, probing and chemically sniffing interesting locations. Hubble continues to monitor the planets long after the probes have left or died out. Some of the solar system's planets and moons look deceptively stable in spacecraft snapshots. But the gas-giant outer planets, in particular, have roiling atmospheres that are a dramatic example of chaotic processes gone wild. Understanding the shifting spots, whorls and jet streams in planetary atmospheres is as challenging as trying to predict the pattern cream will make when swirled into a cup of hot coffee.

Hubble has served as a security camera for scrutinizing disruptive events in the solar system. During Hubble's operation, there have been a series of unexpected, if not unimagined, transitory events. Especially interesting to astronomers are collisions between solar system objects. Earth is a victim of interplanetary smashups too, and seeing what happens when these occur elsewhere is like watching a series of lab experiments from a safe distance.

The bizarre phenomenon of a light echo painted this cosmic display of shadow, light and color captured by Hubble. Details on page 283.

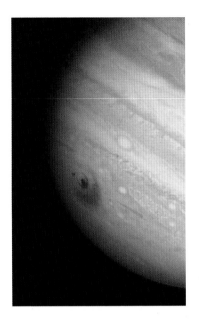

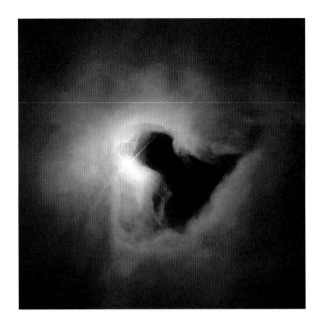

COMET IMPACTS ON JUPITER
This once-in-a-millennium view
shows what happens when a comet
strikes a planet. In fact, the dark
debris fallout seen here is from
two of 23 fragments of Comet
Shoemaker-Levy 9 that slammed
into Jupiter in July 1994. The
large bruise, about the diameter
of Earth, was created by the impact
of fragment G on July 18. The
oblique entry path caused the
crescent debris splash.

KEYHOLE
This eerie dark shape resembles a chess pawn or a ghostly "shadow man" rearing his head out of the
mist. The object lies inside the reflection nebula NGC1999. Like fog around a street lamp, a reflection
nebula shines because the light from an embedded source illuminates its dust; the nebula does not
emit any visible light of its own. NGC1999 lies close to the Orion Nebula, about 1,500 light-years away.
Hubble astronomers thought the black patch was a dense, doorknob-shaped cloud of dust and gas that
was blocking background starlight. The European Space Agency's Herschel Space Observatory, the
most powerful infrared telescope ever flown in space, was aimed at the dark patch with the intent of
seeing through the dust in infrared light. But it saw nothing. Astronomers finally concluded that the
patch looks black because it is empty space. Narrow jets of gas from some of the young stars in the
region may have punched a hole in the nebulous walls of dust and gas.

MAGNETIC MONSTER
Facing page: The core of galaxy NGC1275 looks like a multi-tentacled octopus. The delicate arms are
shaped by strong magnetic fields around the core, which is home to an erupting supermassive black
hole. The fields entrap gas, making long filaments that stretch out beyond the galaxy. Around one
million times the mass of the Sun, a typical thread is only 200 light-years wide and extends up to
20,000 light-years into space. The filaments form when cold gas from the core of the galaxy is dragged
out into the wake of rising bubbles of hot gas expelled by the black hole. How these filamentary structures have been able to withstand the hostile environment of the galaxy cluster in which NGC1275
is embedded remains a mystery. They should have heated up, dispersed and evaporated by now or
collapsed under their own gravity to form stars.

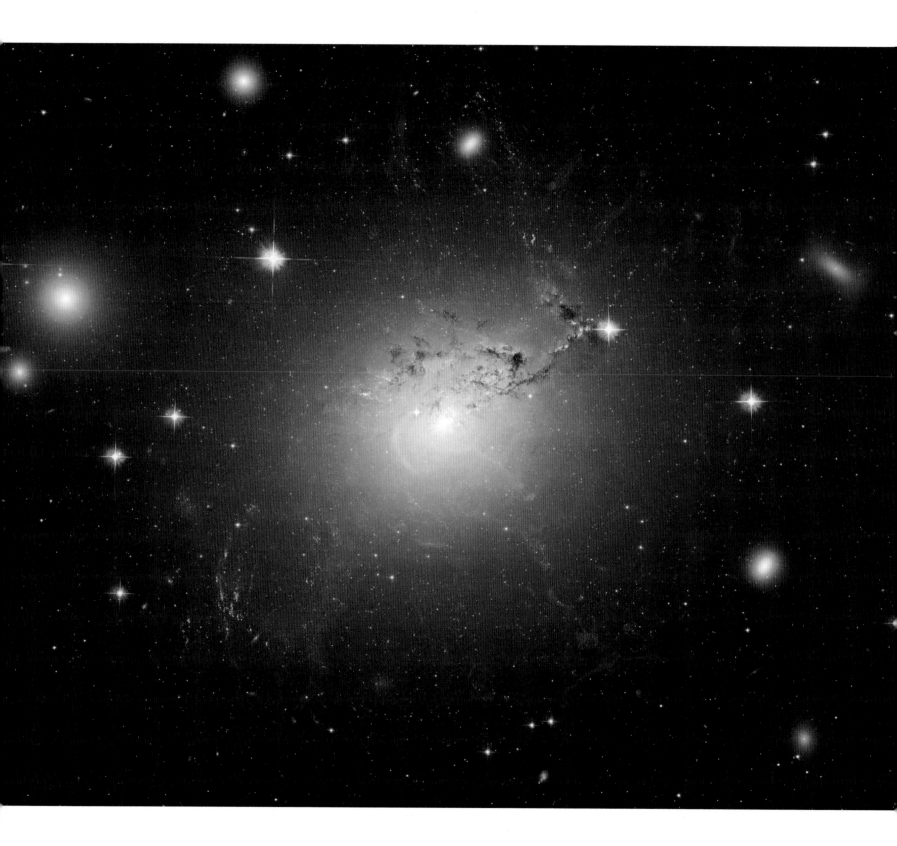

## HANNY'S OBJECT

This strange "green goblin" in the black abyss of space doesn't resemble anything yet seen among the universe of galaxies. It was found in the Sloan Digital Sky Survey, a long-term mapping of large areas of the night sky with a 2.5-meter telescope. It superficially resembles the bubble of a supernova remnant inside our galaxy. But follow-up observations have shown that the enigmatic green blob is 700 million light-years away, at the same distance as its neighbor galaxy IC2497 (seen at top of image). This was a shocking discovery, because it means that the green blob is gigantic—the size of the entire Milky Way Galaxy.

A Dutch schoolteacher, Hanny van Arkel, discovered the object as part of the Galaxy Zoo program, which enlists the help of amateur astronomers in classifying galaxies. She dubbed it Hanny's Voorwerp (Dutch for "object"). Astronomers turned Hubble's gaze on the mystery object. Hubble images revealed delicate filaments of gas and a pocket of young star clusters that are only a couple of million years old. The theory is that IC2497's rambunctious core produced a quasar, an energetic light beacon powered by a black hole. The quasar shot a broad beam of light in the direction of Hanny's Voorwerp, illuminating the cloud. The green color is from glowing oxygen.

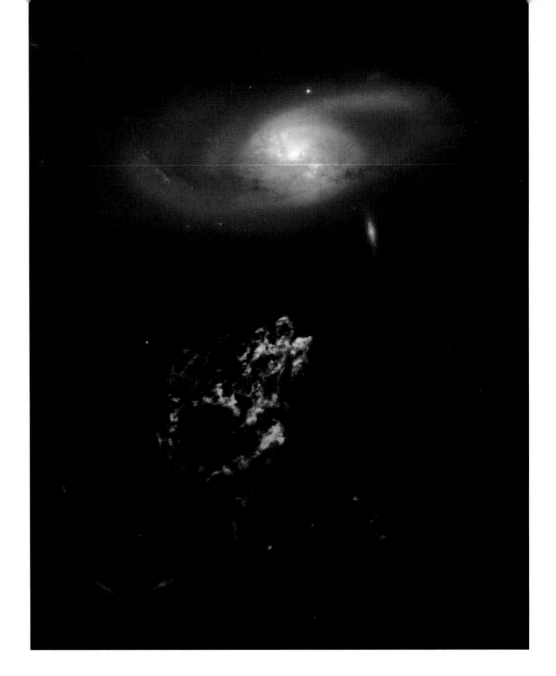

## QUASAR SHADOW PLAY

Illuminated by a quasar blast from the nearby galaxy IC2497 (seen at top of image), the unusual green blob known as Hanny's Voorwerp is thought to be part of a huge dim ring of gas that encircles the galaxy. The quasar may have cast a shadow on the blob, causing an apparent gaping hole about 20,000 light-years wide in the bizarre object. Hubble reveals sharp edges around the opening, suggesting that an object close to the quasar may have blocked some of the light and projected a shadow onto Hanny's Voorwerp, something analogous to a fly on the lens of a movie projector casting a shadow on the movie screen. The quasar outburst may have also triggered star formation.

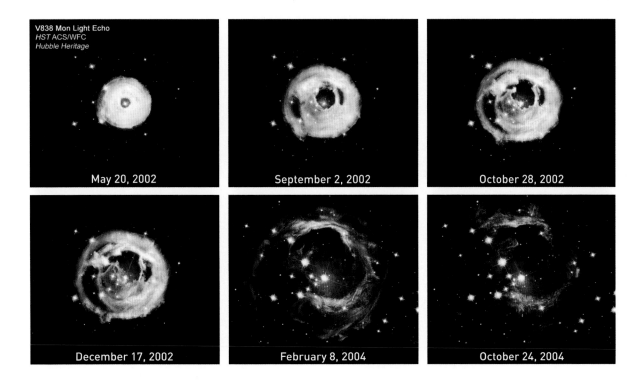

V838 Mon Light Echo
*HST* ACS/WFC
*Hubble Heritage*

May 20, 2002     September 2, 2002     October 28, 2002

December 17, 2002     February 8, 2004     October 24, 2004

## V838 MONOCEROTIS LIGHT ECHO

Very few objects significantly change in apparent shape and size over the course of an astronomer's lifetime or even generations of astronomers. This particular target is unique in many respects. It presents to Hubble astronomers the grandest illusion in the universe: a light echo. Because light travels at a finite speed across space, all the light from a transient event, such as the outburst of a star, does not arrive at Earth at the same time. It can rebound off nearby clouds of dust and arrive later as an "echo." This is identical in principle to sound echoing off a canyon wall. And even though light travels much faster than sound, space is so, well, spacious that the reverberation can be followed for years.

On January 6, 2002, V838 Monocerotis, a previously unknown variable star on the edge of our galaxy, brightened to 600,000 times the luminosity of the Sun, making it briefly one of the brightest stars in the galaxy. Astronomers determined that the brightening was caused by a rapid expansion of the star's outer layers. This caused the star to cool and turn noticeably red. The star may have swallowed a smaller companion star, stirring up its nuclear furnace to unleash a blast of radiation.

At first glance, Hubble's images—taken over the course of several years—appear to show an expanding spherical shell of debris. But what we're seeing is light from the outburst reflecting off surrounding dust clouds. This light arrives later than the initial flash, producing an illusion of expanding rings.

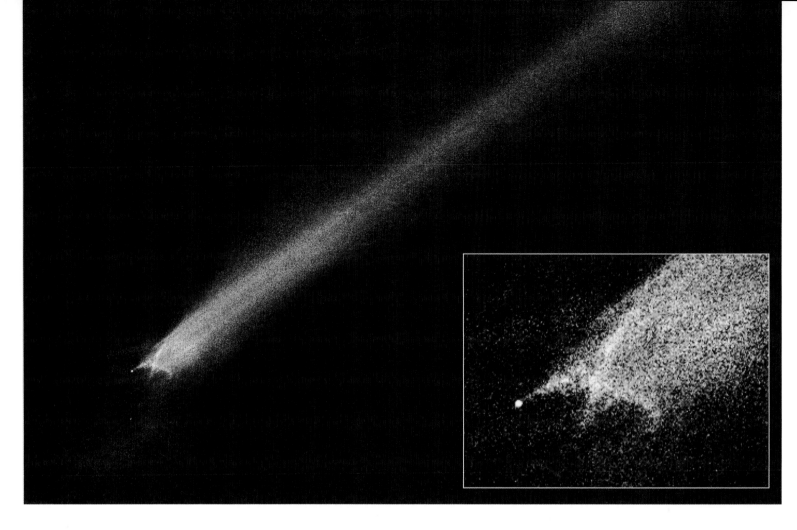

## ASTEROID COLLISION

In early 2009, astronomers photographed a curious object with a long tail. Normally, this would have been cataloged as a comet, except that it was in the wrong place—the asteroid belt. Comets originate in the far fringes of the solar system. Hubble was called upon to take a close look at the 100-meter-wide object at the head of the tail-like structure. An uncanny, never-before-seen X-shaped pattern was resolved by Hubble. This was no simple comet cloud.

The most straightforward explanation is that two asteroids collided, the smaller body blasting a crater into the larger asteroid and shooting out streamers of dust. The violent encounter was as powerful as the detonation of a small atomic bomb. The thickest streamers formed an X-pattern. Hubble astronomers expected the debris field to expand dramatically, like shrapnel flying from an exploding hand grenade. But, instead, they found that the object was expanding very slowly. By measuring the expansion rate, they calculated that they had missed the explosive event by a year.

Based in part on this observation, astronomers estimate that flying mountain-sized asteroids collide, on average, once a year. This collisional grinding of large bodies into smaller objects may be one of the main processes by which asteroids are destroyed and keep dusting interplanetary space with debris. The two asteroids themselves were probably relics from impacts between larger asteroids tens or hundreds of millions of years ago.

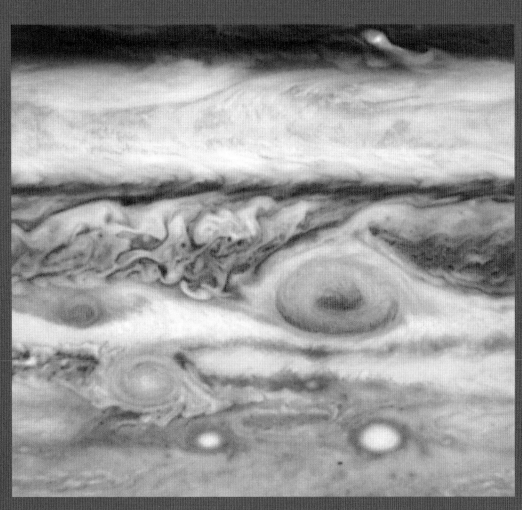

## RED SPOT PAC-MAN

In 2008, Jupiter began to look as if it had come down with measles. A small version of Jupiter's Red Spot, dubbed Red Spot Jr., had appeared in 2006; then a smaller third spot, the Baby Red Spot, materialized alongside the classic Great Red Spot. Both of the new red spots were previously white. The abrupt change of color indicates that the swirling storm clouds rose hundreds of kilometers above Jupiter's top cloud deck to heights comparable to that of the Great Red Spot. One explanation is that the storms may be powerful enough to dredge material from deep within Jupiter's atmosphere and lift it to higher altitudes, where solar ultraviolet radiation, via some unknown chemical reaction, produces the familiar brick color.

Baby Red Spot and the Great Red Spot were on a collision course in 2008. In the Hubble picture sequence below, Baby Red Spot moves ever closer to the Great Red Spot until it is caught up in the Earth-sized storm's anticyclonic spin. In the final image, the Baby Red Spot is deformed and pale in color and has been spun to the right of the Great Red Spot (arrow). Meanwhile, Red Spot Jr., at the bottom edge, has skirted past its big brother apparently unscathed—at least on this Jovian lap.

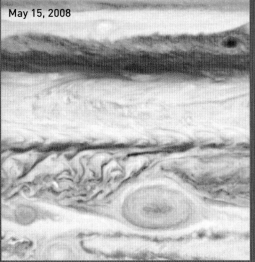

May 15, 2008

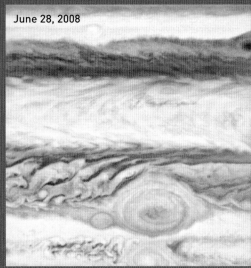

June 28, 2008

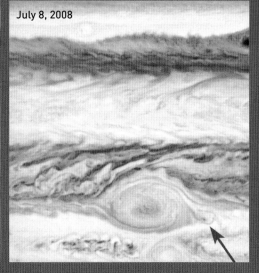

July 8, 2008

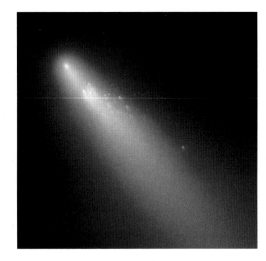

## DEATH OF A COMET

In 2006, Comet Schwassmann-Wachmann 3 disintegrated right in front of Hubble's cameras, giving astronomers an unprecedented opportunity to study the demise of a comet nucleus. The comet broke apart just after passing near the Sun on a roller-coaster orbit. Comets are thought to be fragile agglomerations of ice and dust forged in the outer recesses of the solar system when Earth and the other planets were forming 4.6 billion years ago. The heat of the Sun, plus its gravitational tug, caused this cometary bundle of fragments to become unglued. Hubble images showed house-sized pieces breaking into smaller chunks. The rapid rotation of the comet's weakened nucleus might also have caused it to explosively pop apart due to the outburst of trapped volatile gases, like corks bursting from a case of champagne bottles. Several dozen mini-fragments can be seen trailing behind each main fragment, probably associated with the ejection of chunks of surface material.

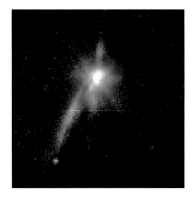

## RUNAWAY PLANET?

In 1997, Hubble took an image that was compelling and unbelievable. A very red—and therefore cool—pinpoint object was perched at the end of a ghostly finger of illuminated dust stretching 200 billion kilometers from a young binary star system (brightest object). Formally designated TMR-1C, the object lies about 450 light-years away, in the constellation Taurus. The enigmatic finger was interpreted as having formed when a planet was gravitationally ejected from the binary system. But associations can be deceiving, and follow-on observations from ground-based telescopes in 2009 revealed that the phantom object had gotten brighter and bluer, suggesting it may be a young protoplanet surrounded by a thick, spinning disk of dust, which might explain the variability of the planet's light. For now, this object is still in the unsolved-mystery file.

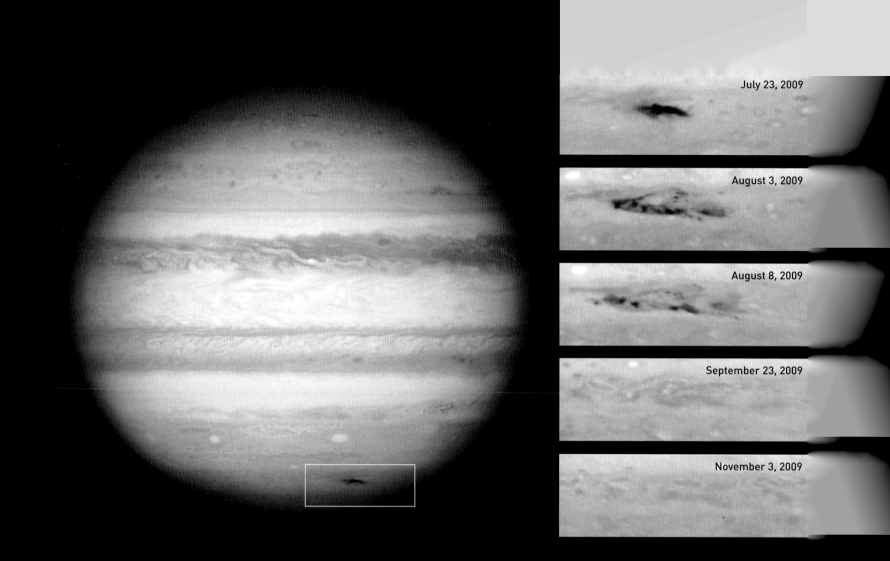

July 23, 2009

August 3, 2009

August 8, 2009

September 23, 2009

November 3, 2009

## ROGUE ASTEROID

An unexpected unknown object struck Jupiter on July 19, 2009, leaving a dark bruise the size of the Pacific Ocean. (Yes, that "small" black scar is that big!) No one saw the collision, but it was captured on an amateur astronomer's "surveillance video" of the planet. Soon, observatories around the world, including Hubble, were zeroing in on the Jovian bruise. Astronomers had an uncanny sense of déjà vu about the feature—and for good reason. In mid-July 1994, scientists had watched pieces of Comet Shoemaker-Levy 9 carpet bomb Jupiter.

By comparing crisp Hubble images of the two events that happened 15 years apart, astronomers concluded that the culprit may have been an asteroid roughly 500 meters wide. The oval shape of the "footprint" on Jupiter indicates that the intruder came in at a shallow angle, unleashing the energy of a few thousand Hiroshima atomic bombs in one blast. The impact tossed dark dirtlike material from the asteroid into Jupiter's atmosphere, which was then smeared out over several weeks by the planet's unrelenting winds.

This represents the first time in history that the immediate aftermath of an asteroid striking a planet has been observed. The image underscores the fact that the solar system is a rowdy neighborhood. Jupiter impacts were thought to occur every few hundred to few thousand years, but unpredictable smashups may happen more frequently. There may be so many uncataloged smaller bodies in the solar system that they might show up anytime to wreak havoc. In 1834, for example, British astronomer George Airy reported a dark feature in Jupiter's southern belts that appeared nearly four times as large as shadows

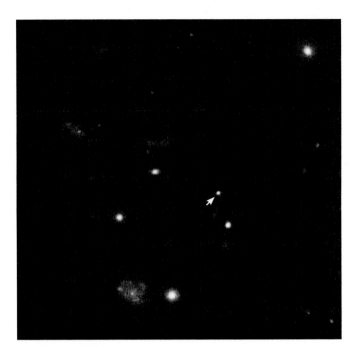

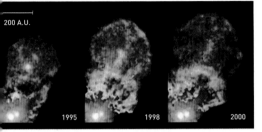

200 A.U.

1995    1998    2000

## STAR BLOWING BUBBLES

Expanding bubbles of glowing gas that are being blown out from the young binary star system XZ Tauri can be seen in these Hubble images. Gas from an unseen disk around one or both stars is being channeled through magnetic fields surrounding the binary system, then forced out into space at nearly 480,000 kilometers per hour. There was a dramatic change in XZ Tauri's appearance between 1995 and 1998. In 1995, the bubble's edge was the same brightness as its interior. When Hubble took another look in 1998, however, the edge was suddenly brighter. This is probably the result of the hot gas at the leading edge of the bubble cooling off and is the first time astronomers have seen such a cooling zone "turn on" around a young star.

## BRIGHTEST GAMMA-RAY BURST

On March 19, 2008, for nearly one full minute, a single star in the universe was as bright as 10 million galaxies. If you had been standing outside under a clear, dark sky anywhere in North America at 2:12 a.m., EDT, and happened to be looking at the constellation Boötes, you would have seen a dim fifth-magnitude star suddenly appear out of the darkness. At a distance of seven billion light-years, this is the most remote unaided-eye object ever visible from Earth. But nobody saw it. The star, called GRB 080319B, also emitted a brilliant flash of gamma rays, and NASA's Swift satellite, a gamma-ray-burst watchdog, detected the X-ray flash. This Hubble picture, taken on April 7, shows the fading optical counterpart of the titanic blast (arrow). Hubble astronomers had hoped to see the host galaxy where the burst originated. They were surprised to see that even three weeks after the event, the light from the explosion was still drowning out the galaxy's light. Called a long-duration gamma-ray burst, such an event may be caused by the death of a very massive star, perhaps 50 times the mass of the Sun. Such an explosion, sometimes dubbed a "hypernova," is more powerful than an ordinary super-nova explosion and far brighter. This is partly due to the fact that its energy is concentrated into a nar-row blowtorchlike beam. When such a beam happens to be aimed directly at Earth, it appears incred-ibly bright for its distance. At least one gamma-ray burst is detected daily somewhere in the universe.

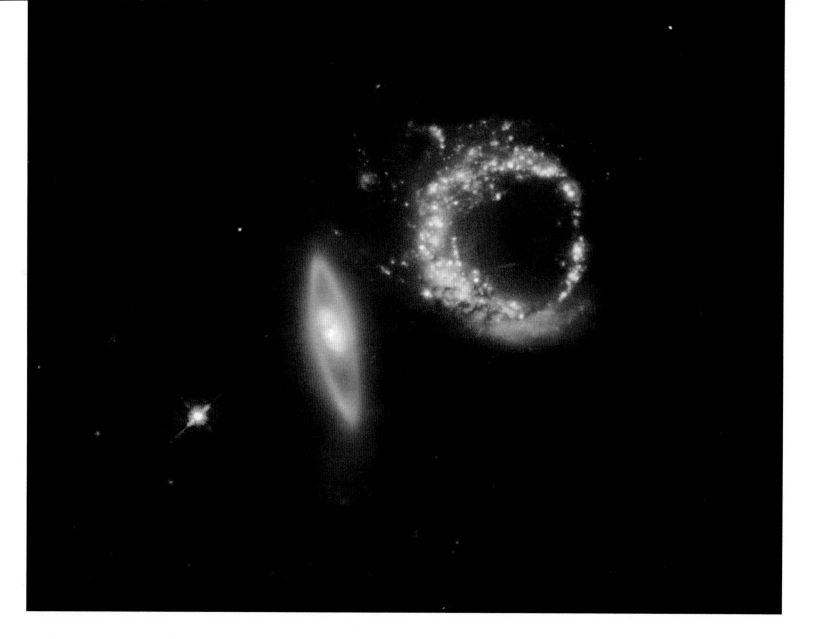

## THE 'PERFECT 10'

This strange-looking pair of galaxies called Arp 147 happen to be oriented so that they appear to be skywriting the numeral 10. The galaxy on the left, the number "1," is relatively undisturbed, but it does have a striking smooth ring of starlight not commonly seen in galaxies. It is tilted nearly edge-on to our line of sight. The galaxy on the right, resembling a "0," has a clumpy blue ring of intense star formation.

The blue ring likely formed after the galaxy now on the left passed through its companion galaxy. At the point of impact, a propagating density wave was generated and spread outward to form the dough-nut shape. The density wave collided with material that was moving inward due to the gravitational pull of the two galaxies, producing shock waves and dense gas and stimulating star formation. The dusty reddish knot at the lower left of the blue ring probably marks the location of the nucleus of the galaxy before it was clobbered.

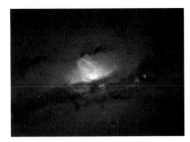

### BLACK HOLE BUBBLE

A glowing apparition appears to rise from a black cauldron in this eerie photo. Hot bubbles emerge from a dark band of dust at the core of galaxy NGC4428 (another bubble, below the dust band, is barely visible). The bubbles are caused by material being blown into space from a black hole that is engorged with infalling dust and gas. Twin jets of matter eject material as they are blasted outward by the intense radiation and magnetic fields in the black hole's immediate vicinity. The jets eventually slam into a wall of dense, slow-moving gas, producing the glowing material.

### RUNAWAY STARS

Bizarre tadpole and boomerang shapes are produced as runaway stars plow through regions of dense interstellar gas, creating brilliant arrowhead structures and trailing tails of glowing gas. An arrowhead, or bow shock, forms when the star's powerful stellar winds (streams of matter flowing from the star) slam into surrounding dense gas. The phenomenon is similar to that seen when a speeding boat pushes through water on a lake. Depending on the star's distance from Earth, the bullet-nosed bow shock can be up to one-quarter of a light-year long. The bow shock indicates that the star is traveling fast—more than 200,000 kilometers per hour—roughly five times faster than a typical young star.

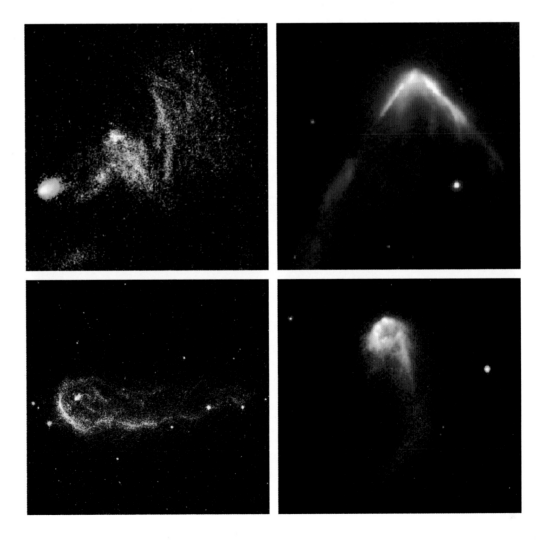

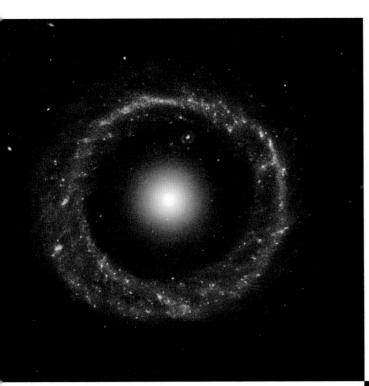

GALAXY SILHOUETTE

In an extraordinary chance align-
ment, a face-on spiral galaxy lies
precisely in front of a larger spiral.
This lineup provides us with a rare
opportunity to see in silhouette
what would typically be dark material inside the foreground galaxy.
The skeletal lanes of interstellar
dust are the raw material for building new generations of stars. A
small red patch near the center
of the image is the bright nucleus
of the background galaxy.

BULL'S-EYE GALAXY

A ring of hot blue stars pinwheels
about the yellow nucleus of this
nearly perfectly tilted face-on
galaxy. Dominated by clusters of
young massive stars, the blue ring
contrasts sharply with the yellow
nucleus of mostly older stars. This
wheel-inside-wheel formation can
be caused when a galaxy speeds
through the middle of a companion
galaxy, producing a "splash" of star
formation. The blue ring of stars
could also be the shredded remains
of a galaxy that passed nearby. This
close encounter happened about
two to three billion years ago.

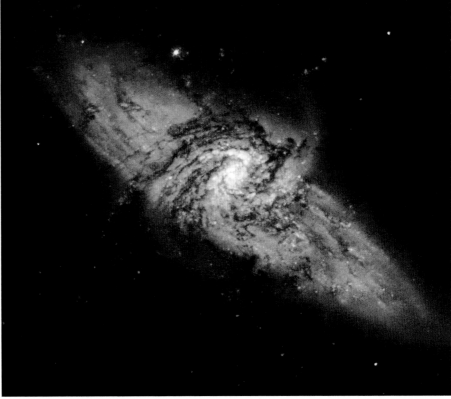

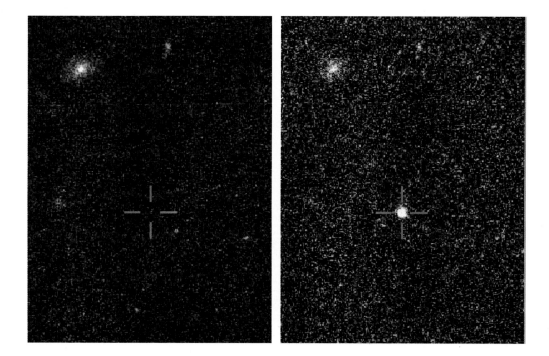

## MYSTERIOUS BEACON

While flashes of light from somewhere near or far in the universe are common events for astronomers, an object that appears out of nowhere has all the trappings of a celestial ghost, because it defies conventional explanations. Hubble discovered this cosmic beacon (right photo) on February 21, 2006, while astronomers were searching for the glow of distant supernovas. But the behavior of this phenomenon did not match that of a supernova. It steadily rose in brightness for 100 days, then dimmed back to oblivion over the next 100 days. This brightness behavior has never been recorded for any other type of celestial event. Supernovas rapidly peak after no more than 70 days and slowly fade away, and they are found in galaxies. Yet searches through various astronomical survey catalogs for the source of this light have not uncovered any evidence of a star or a galaxy at the location of the mystery flash. Follow-up observations in 2009 at Caltech's Palomar Observatory, in California, indicated that the object is halfway across the known universe.

## DEATH STAR

Like a cosmic weapon run amok, a powerful beam of energy is blasting a nearby galaxy. An immense jet of gas traveling at nearly the speed of light is emanating from the vicinity of the supermassive black hole in the larger galaxy at lower left. Its companion galaxy has apparently swung into the path of the jet. The jet impacts the galaxy at its edge and is then disrupted and deflected, much like a stream of water from a hose fans out after hitting a wall at an angle. Jets produced by a supermassive black hole transport enormous amounts of energy far from the black hole, enabling them to affect matter on a vastly larger scale than the size of the black hole itself. The effect of the jet on the smaller galaxy is likely to be substantial, because the pair of galaxies, known as 3C321, are extremely close to each other, just 20,000 light-years apart—almost touching, in galactic terms.

A composite of images taken at several different wavelengths—X-rays recorded by the Chandra X-ray Observatory (purple); visible and ultraviolet wavelengths from the stars seen by Hubble (red and orange); and radio emissions detected by the Very Large Telescope array (blue)—shows how the jet from the main galaxy is striking its companion galaxy and projecting into deep space.

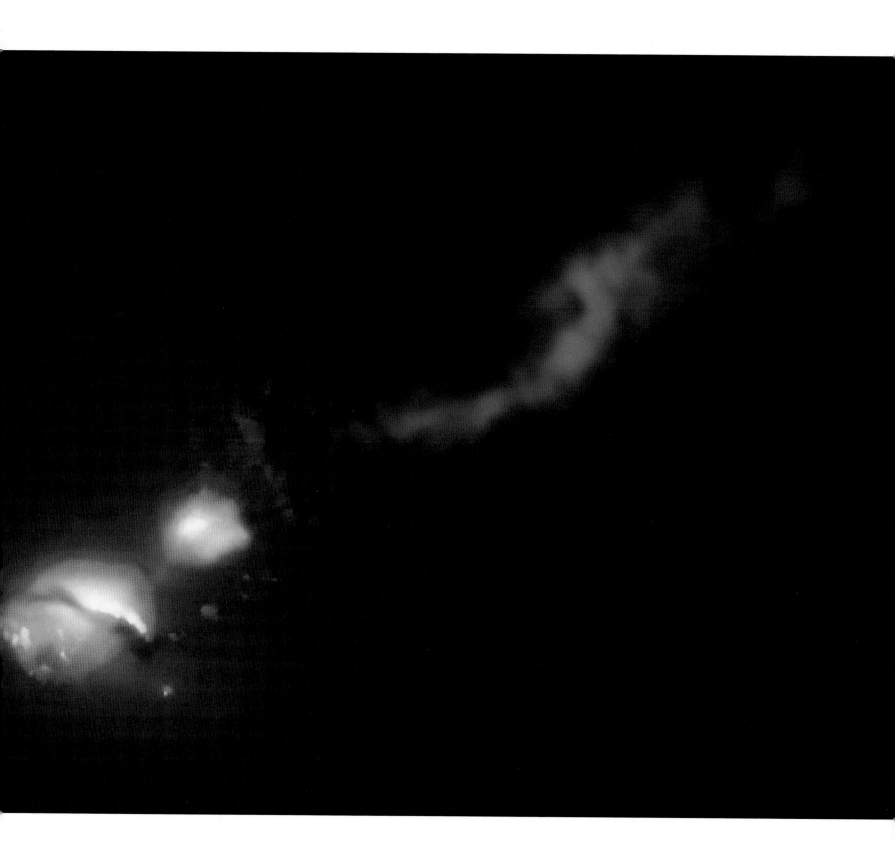

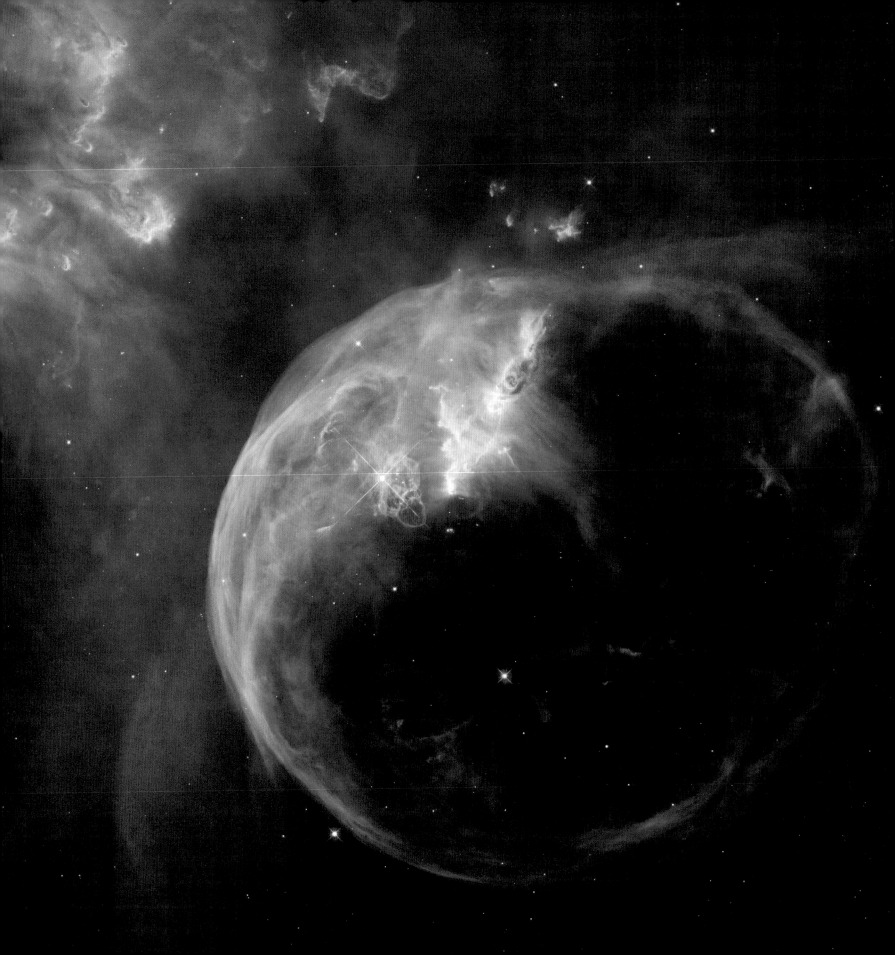

# HUBBLE UPDATE: 2012–2017

The last of the NASA space shuttle servicing missions to Hubble took place in May 2009. When the crew of seven astronauts completed repairing and upgrading the space telescope, Hubble reached the apex of its scientific performance.

During the final servicing mission in May 2009, astronauts installed two new instruments on the Hubble Space Telescope: the Wide Field Camera 3 (WFC3) and the Cosmic Origins Spectrograph (COS). The WFC3 has better resolution and a wider field of view than any of the previous Hubble cameras and greatly expanded Hubble's vision at near-infrared wavelengths. Its predecessor, the Near Infrared Camera and Multi-Object Spectrometer (NICMOS), installed on Hubble's second servicing mission in 1997, first pioneered Hubble's foray into the infrared universe.

NICMOS opened up an "undiscovered country" of fledgling young galaxies that are so far away, they can be seen only in infrared light because the expansion of the universe has stretched their light to longer wavelengths. WFC3 offered astronomers much better images than NICMOS. Its enhanced sensitivity pushed back the frontiers for a peek at primeval galaxies that existed just 400 million years after the Big Bang.

The astronauts performing the 2009 servicing mission accomplished a feat never envisioned by the telescope builders: on-site repairs in orbit. Two instruments, the Advanced Camera for Surveys (ACS) and the Space Telescope Imaging Spectrograph (STIS), had both stopped working. The ACS had an electrical short in 2007, and STIS had a power failure in 2004. To perform the repairs, the astronauts had to access the interior of the instruments, switch out circuit boards and reroute power. The successful completion of this task, along with the addition of the two new instruments, gave Hubble a full complement of five functioning instruments for its future observations. The five are still fully operational as of 2017.

Among the new discoveries, STIS photographed water plumes jetting into space off the surface of the Jovian moon Europa. The icy moon has been considered a prime candidate to search for life in its subterranean ocean. This new finding opens the exciting possibility that material from the subsurface ocean may rise to the surface to be analyzed by a future robotic sample-collection mission to Europa.

This chapter displays the Hubble Space Telescope at its peak imaging performance as it adds to its impressive legacy of discoveries. Because Hubble is in such robust operating condition, it will conduct complementary observations along with NASA's next big leap into space, the James Webb Space Telescope, scheduled for launch in 2018. While Webb's keen focus will unveil the universe in invisible infrared colors of light, Hubble will continue making spectacular observations in visible light.

Resembling a giant runaway balloon adrift among the stars, the Bubble Nebula, also known as NGC7635, lies 8,000 light-years away, in the constellation Cassiopeia the queen. Like a balloon, the Bubble Nebula is inflating, at a speed of 40,000 kilometers per hour. Its present diameter is 7,100 light-years across—almost twice the distance between our Sun and the next nearest star.

Supernova remnants can make a bubble in space, but the process here is different. An ultrahot young star several hundred thousand times brighter than our Sun and 45 times more massive (just beyond the upper left corner) is doing the heavy lifting. It produces a fierce stellar wind and intense radiation, which have inflated a structure of glowing gas that pushes against denser material outside of it.

## CHAPTER 11

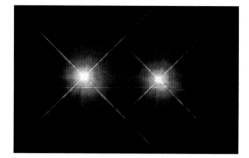

## ALPHA CENTAURI PAIR

At first glance, this image looks like a pair of automobile headlights. It's actually a portrait of the closest star system to Earth, the Alpha Centauri group. This system is made up of Alpha Centauri A (left) and Alpha Centauri B (right). A third member, the faint red dwarf Proxima Centauri, is not in this picture because it is 1.6 trillion kilometers from the two other stars in the system.

Alpha Centauri A is only slightly bigger than our Sun and is a similar temperature. Alpha Centauri B is somewhat smaller and cooler. Like a pair of skaters, the two stars pivot around each other, completing an orbit once every 80 years. They get as close to each other as roughly the distance between our Sun and the planet Saturn. By contrast, Proxima Centauri takes a whopping 500,000 years to make one orbit about its companion stars. Located 4.3 light-years from Earth, the Alpha Centauri group is approximately 270,000 times the distance from Earth to the Sun. An interstellar probe traveling at one-tenth the speed of light could reach the system in a little over 40 years.

## SPIRAL GALAXY M106

Messier 106, located a little over 20 million light-years away, is one of the brightest and nearest spiral galaxies to our own. This is a view of the inner region of the galaxy. Black dust clouds are silhouetted against the bright glow from stars at the galaxy's core. The interplay of light and shadow gives the center the appearance of a swirling witch's cauldron. Two brilliant blue spiral arms of hot young stars wrap symmetrically around this cauldron. They are punctuated by bright pinkish knots of glowing hydrogen gas in nebulas where new stars are bursting to life.

Unlike the normal arms, there are two pink arms made up of hot gas rather than stars. These extra arms appear to be an indirect result of jets of material produced by the violent churning of matter around the massive central black hole. As these jets travel through space, they disrupt and heat up the surrounding gas.

Amateur astronomer Robert Gendler retrieved archival Hubble images of M106 to assemble a mosaic of the center of the galaxy. He then used his own and fellow amateur astronomer Jay GaBany's observations of M106 to combine with the Hubble data in areas where there was less coverage and to fill in the holes and gaps where no Hubble imaging data existed.

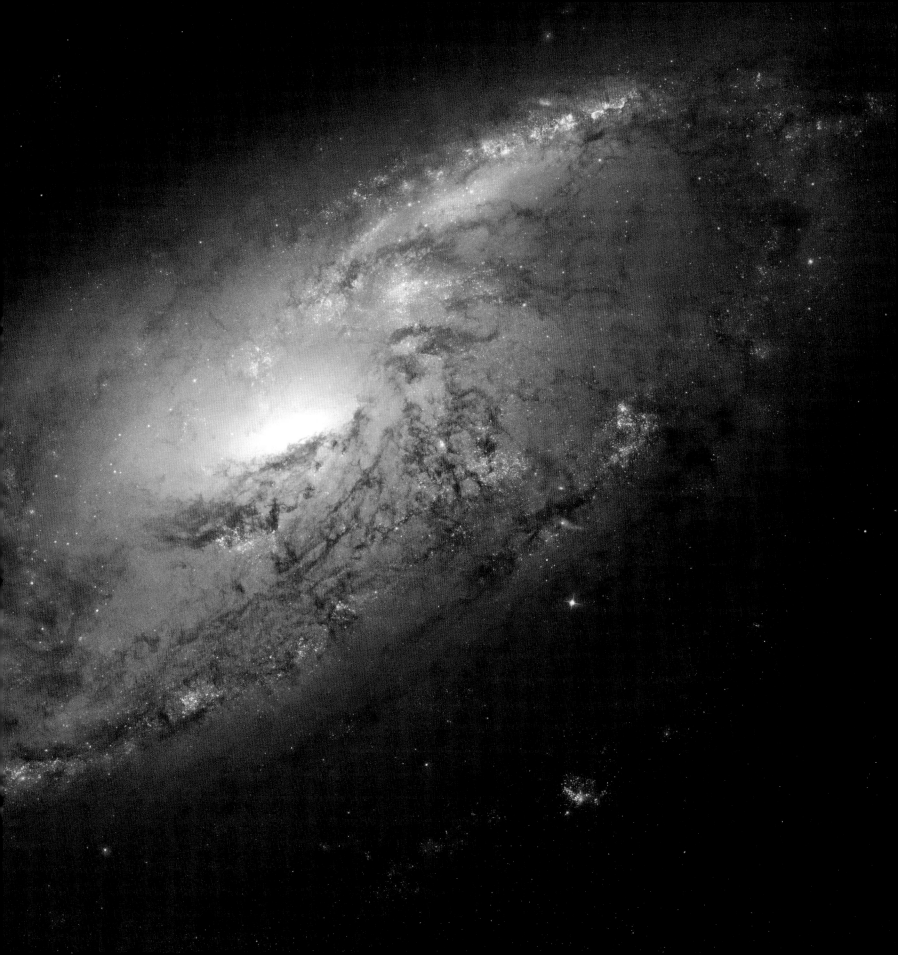

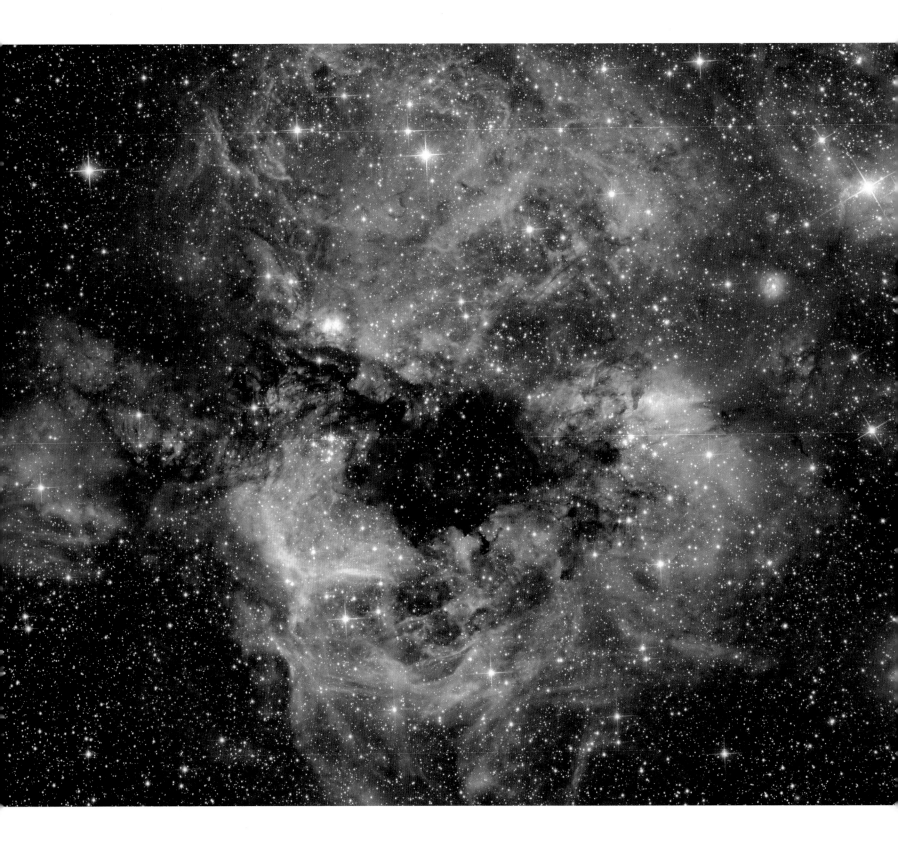

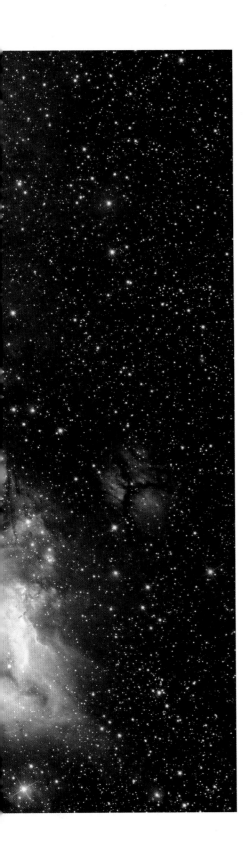

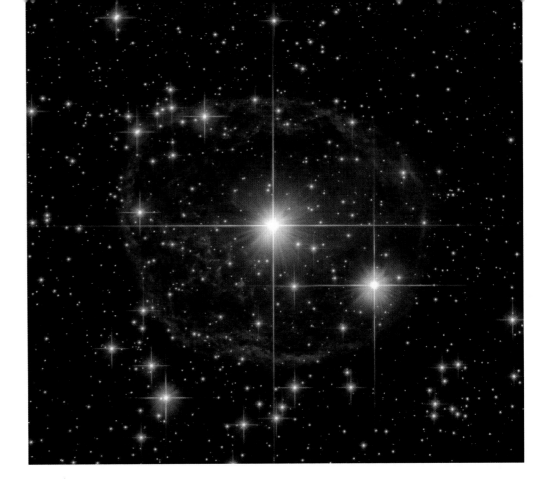

## BUBBLE AROUND A WOLF-RAYET STAR

A delicate-looking blue bubble of glowing hydrogen and helium encircles a Wolf–Rayet star known as WR31a, located 30,000 light-years away, in the constellation Carina the keel. The bubble was created when gas was ejected by the very hot Wolf–Rayet star approximately 20,000 years ago. Eight light-years across, the bubble is roughly twice the distance between our Sun and its nearest stellar neighbor, Alpha Centauri. A Wolf-Rayet star is more than 20 times the mass of our Sun and millions of times brighter, shedding about 2,000 billion billion tons of material every year. These stars live fast and die young, often in less than 100,000 years, ending their lives as supernovas.

## STAR-FORMING REGION IN THE LARGE MAGELLANIC CLOUD

This maelstrom of glowing gas and dark dust within one of the Milky Way's small satellite galaxies is the Large Magellanic Cloud (LMC; pages 114-115). The stormy scene shows a stellar nursery known as N159, over 150 light-years across. N159 contains many hot young stars emitting intense ultraviolet light, causing vast clouds of hydrogen gas to glow. The stars blow out torrential stellar winds of hot material that carve out ridges, arcs and filaments from the surrounding interstellar medium. N159 resides near the LMC's Tarantula Nebula, the largest, most violent star-forming region known in our Local Group of galaxies. Tidal forces caused by gravitational interactions with the Milky Way Galaxy have led to a relatively recent and ongoing burst of vigorous star birth throughout the LMC.

## CYCLOPS JUPITER

Jupiter's four large moons routinely cast shadows across the colorful face of the giant planet as they follow the celestial clockwork of their orbits. In this unique lineup with a moon, it looks as if Jupiter is staring at us like some malevolent space Cyclops with a 15,000-kilometer-diameter "eye." The "eye" is Jupiter's signature Great Red Spot, and its dark "pupil" is the pitch-black shadow of Jupiter's largest moon, Ganymede. The moon itself was outside the frame to the right as Hubble's camera captured the scene.

The Red Spot is an immense Earth-sized high-pressure storm akin to a hurricane. The clouds associated with the spot appear to rise about eight kilometers above the neighboring cloudtops. An observer located at the cloudtops would see the Sun blink out of the sky as the moon Ganymede passes in front of it.

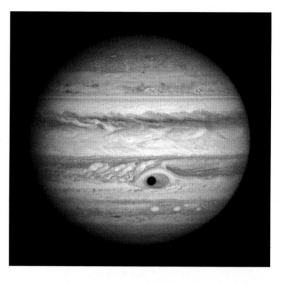

## JUPITER AURORA

Earth isn't the only planet where eerie glowing auroras light up the far northern and southern skies. One billion kilometers away, the giant planet Jupiter has a bright tiara of nonstop auroras.

As on Earth, auroras on Jupiter are generated by the planet's powerful magnetic field trapping electrically charged particles blasted into space by our Sun. Auroras appear when these charged particles from the solar wind slam into the planet's magnetosphere and are accelerated to high energies along magnetic field lines. When the particles cascade into the atmosphere near the north and south magnetic poles, they cause atmospheric gases to glow, like gases in a fluorescent light fixture. Jupiter's magnetosphere is 20,000 times stronger than the Earth's. It occupies an enormous volume of space, extending outward to 100 times the planet's width. The solar wind, a stream of electrically charged particles from the Sun, pushes on and stretches Jupiter's magnetosphere into a wind-sock shape. The wind sock's "tail" extends as far out as the orbit of Saturn, almost a billion kilometers beyond Jupiter. As part of a Hubble program to assemble global maps of the outer planets, this full-color disk of Jupiter was imaged at a different time as the auroras. The brightly glowing bluish auroras were photographed in ultraviolet wavelengths.

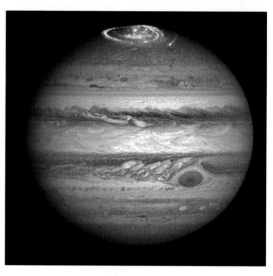

## MULTIPLE GALAXY-CLUSTER COLLISION

Hubble and its companion X-ray and radio observatories recorded a true clash of the Titans happening 5.4 billion light-years away. And the results look pretty messy. In this artificially colored view, four galaxy clusters are colliding. The blue X-ray features trace bright pockets of scorching gas heated to millions of degrees by being compressed in the collision. The pink structures are from radio data that identify enormous shock waves and turbulence. Similar to sonic booms, the shock waves are generated by the mergers of the clusters. Hubble resolves numerous galaxies in the pileup.

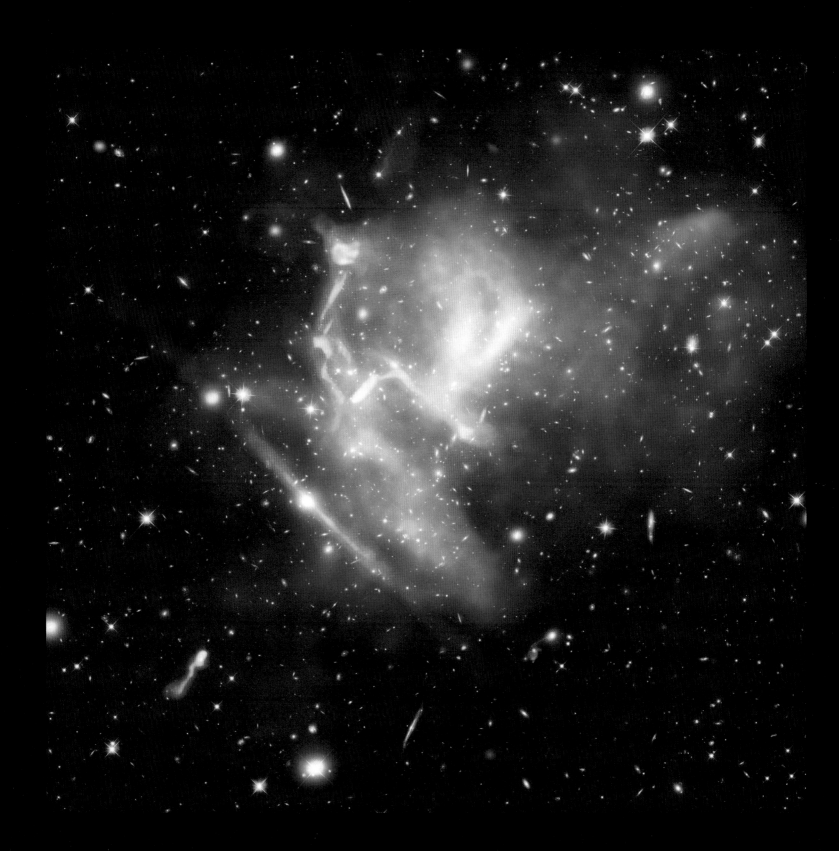

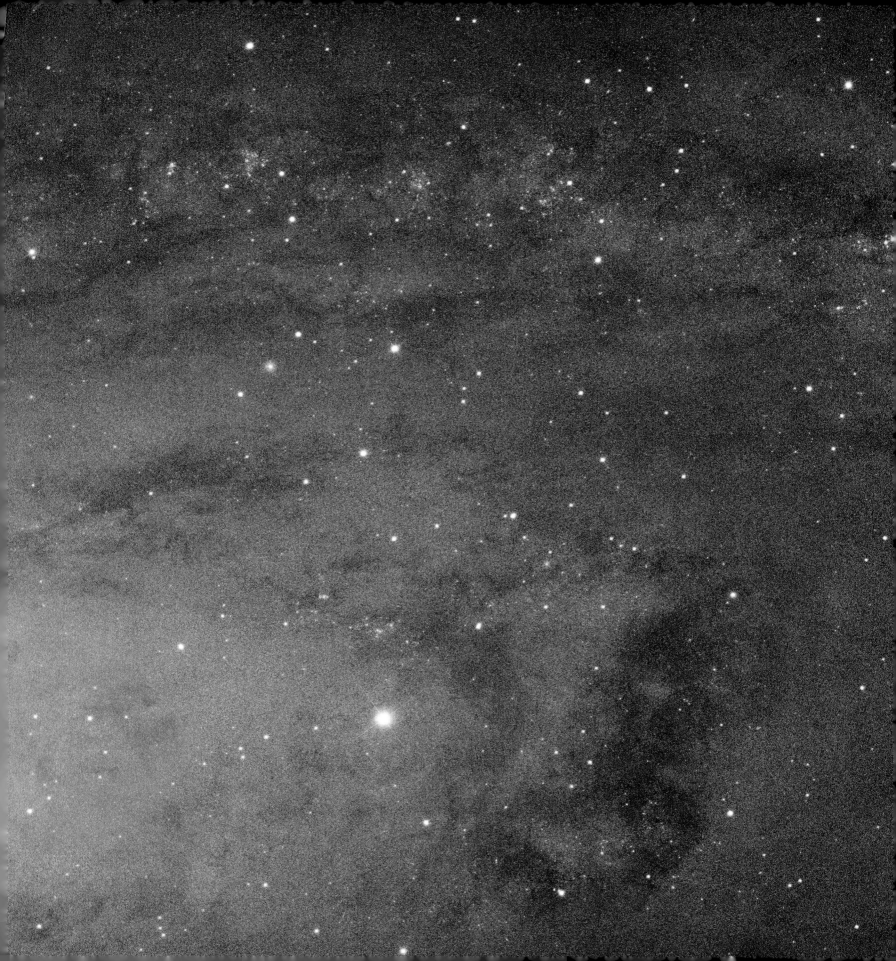

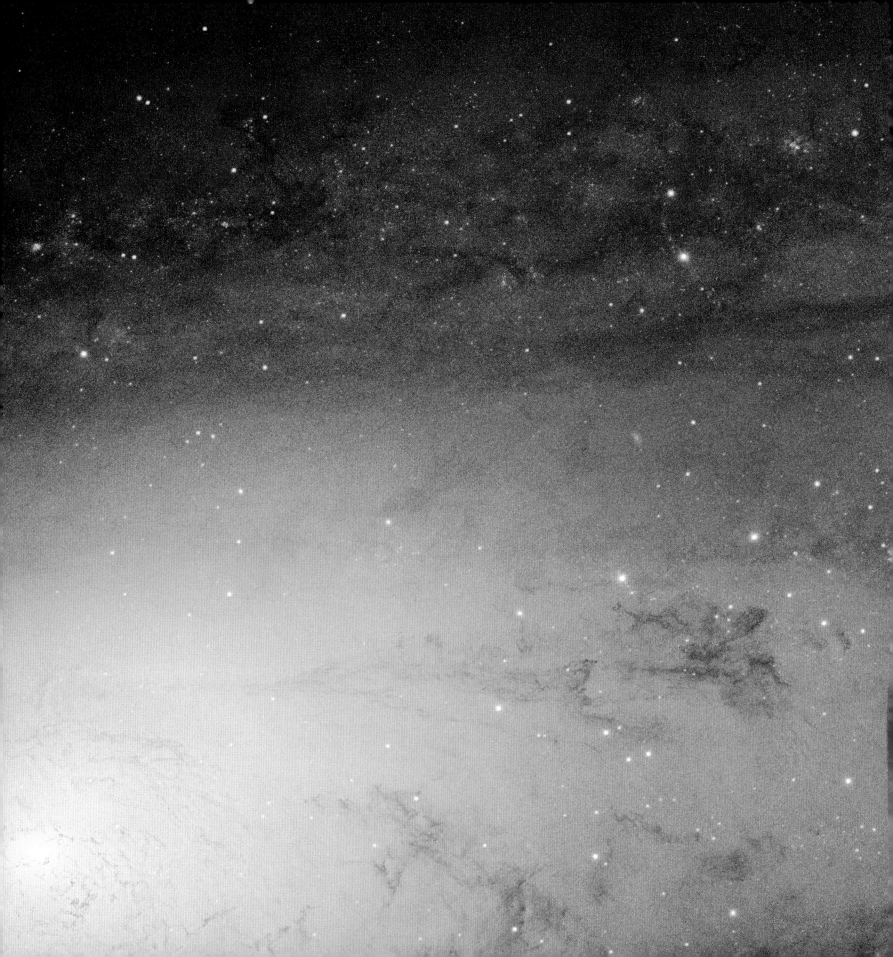

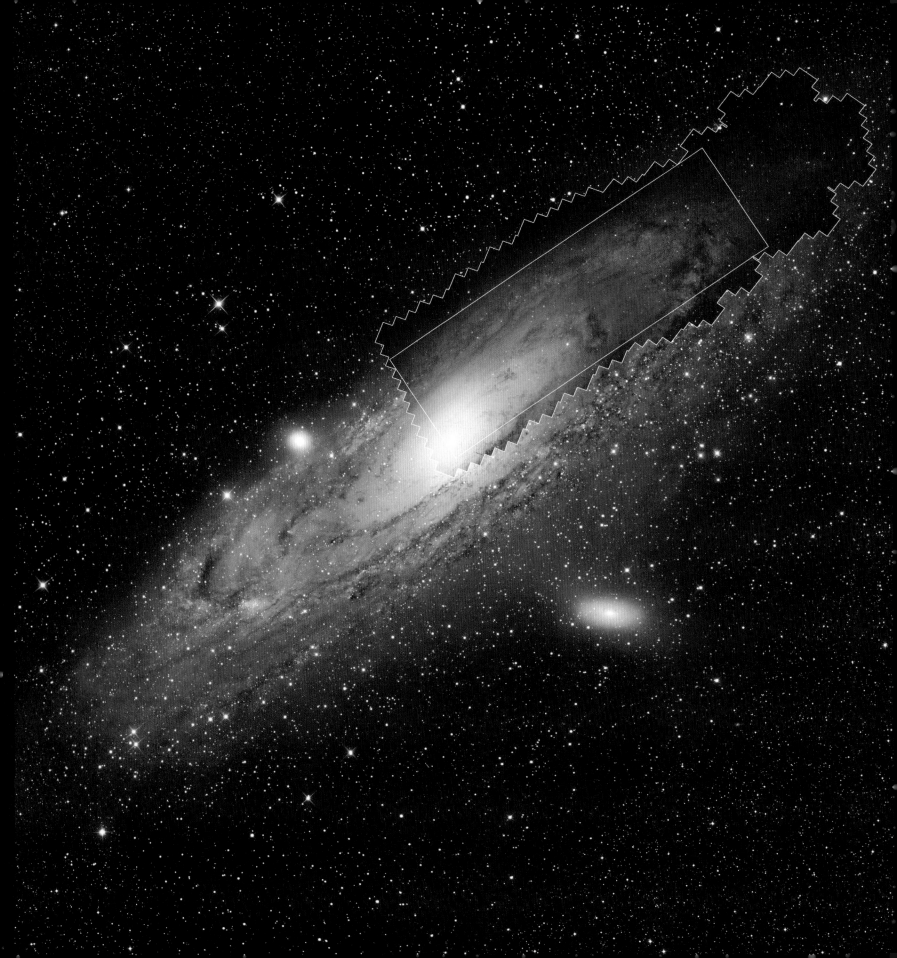

# ANDROMEDA GALAXY

When astronomers decided to use Hubble for a closer look at the Milky Way Galaxy's next-door neighbor, the Andromeda Galaxy, facing page, they faced a herculean task. Even though the magnificent spiral galaxy is 2.5 million light-years away, it stretches across a piece of sky as wide as six full Moons side by side. To capture a detailed slice of its splendor required a mosaic of more than 400 Hubble images digitally stitched together (foldout). The 61,000-light-year-long section of the galaxy's pancake-shaped disk resolves over 100 million stars, like grains of sand on a beach. Never before have astronomers been able to pinpoint individual stars over such a vast portion of a major spiral galaxy beyond our Milky Way. To see the detailed close-up of the region within the rectangle, fold out this page and the facing page.

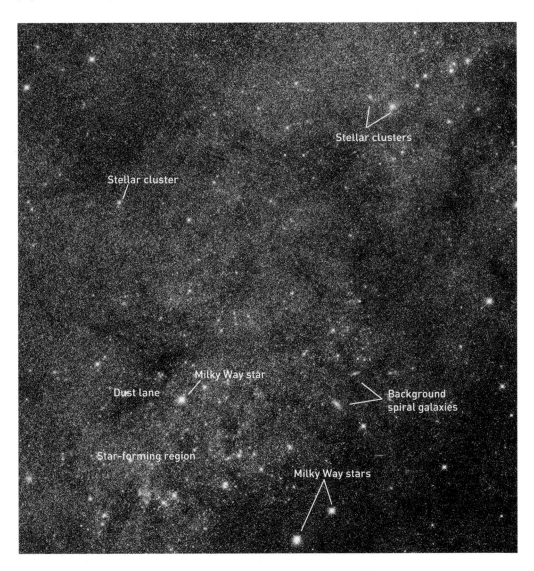

Left: The Hubble Space Telescope peers through the arms of the Andromeda Galaxy to clearly show background spiral galaxies more than a million light-years beyond. Stars in the Andromeda Galaxy itself are awash in the image.

307

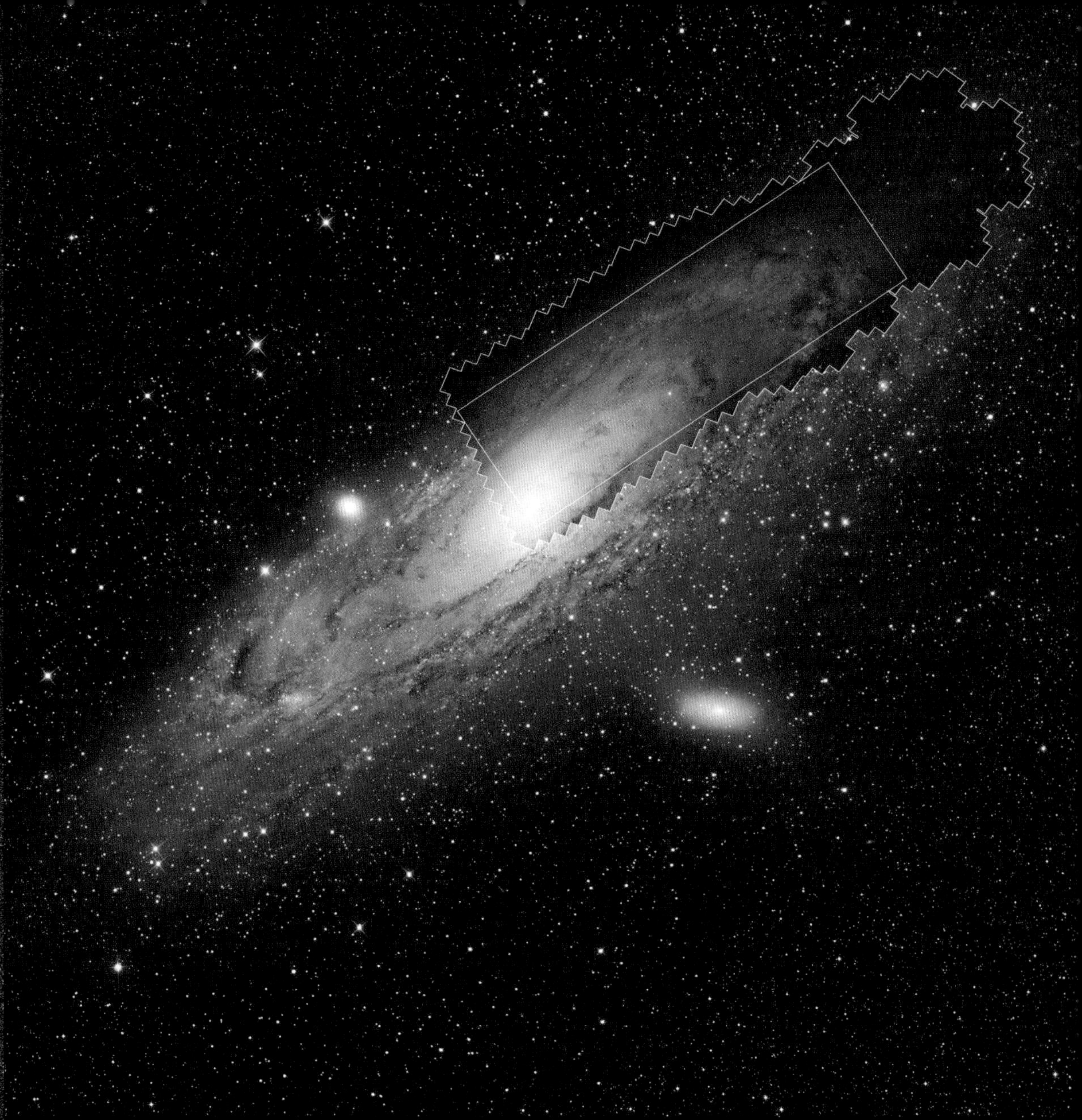

# ANDROMEDA GALAXY

The rectangle of the upper right sector of the Andromeda Galaxy, above, was selected for high-resolution inspection by the Hubble Space Telescope and is shown in detail at left. More than 400 Hubble images were acquired and seamlessly digitally assembled to complete this mapping project. The saw-toothed edges of the many Hubble frames needed for this celestial jigsaw are shown above. The apparent fine granular structure of the Andromeda Galaxy's spiral arms seen at left is, in fact, millions upon millions of individual stars in the big galaxy. The noticeably brighter stars are much closer to us, typically just a few thousand light-years away in our home galaxy, the Milky Way.

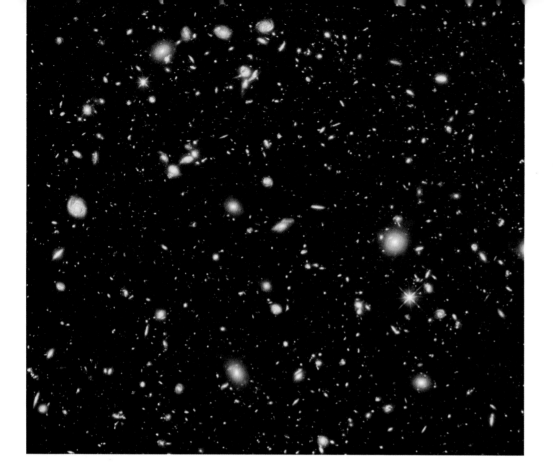

## TECHNICOLOR UNIVERSE

The richest and deepest view ever made of the universe was assembled from a decade's worth of long-duration exposures of a very small patch of sky in the southern-hemisphere constellation Fornax. A composite of separate exposures, this image was taken from 2002 to 2012 using two of Hubble's premier cameras.

This view, with seemingly infinite depth, combines a full range of colors stretching all the way from ultraviolet to near-infrared light. The colorful tapestry reveals a sea of 10,000 galaxies of myriad shapes and at various stages of evolution across a time span of over 13 billion years. This view is typical of what Hubble would see if it was pointed in almost any random direction on the sky. When astronomers extrapolate across the entire heavens, they estimate that the universe contains at least 200 billion galaxies.

This ultradeep probe reveals that significant galaxy evolution has taken place ever since the Big Bang, more than 13 billion years ago. Based on a statistical analysis of the masses of galaxies in the deep field, astronomers estimate that 90 percent of galaxies in the observable universe are too faint and too far away to be seen even with Hubble. This would boost the universe's estimated galaxy population to two trillion. These unseen small, faint galaxies from the early universe merged over time into the larger galaxies recorded by Hubble.

## MASSIVE GALAXY CLUSTER ABELL S1063

Besides being visually stunning, this megalopolis of galaxies, called Abell S1063, is a scientific gold mine for astronomers. The cluster contains the equivalent mass of 40 billion billion Suns, distributed among a staggering population of 1,000 galaxies that are largely disk-shaped like our Milky Way. Inside this mosh pit, galaxies zip around the cluster like bees around a hive. Unlike bees, they collide and merge to build massive elliptical galaxies containing monster black holes.

The huge mass of the cluster deforms space, like rolling a bowling ball across a rubber sheet. This distortion of space magnifies and brightens the light from galaxies that lie far behind the cluster, deep into the early universe. This effect is called gravitational lensing. Arcs of light seen throughout the cluster are evidence of the gravitational-lensing phenomenon and resemble distortions in a fun-house mirror.

For astronomers, the cluster becomes a huge zoom lens in space that allows them to see far distant galaxies that would otherwise be too faint to observe. The magnified galaxies give a peek at what could be the very first generation of objects that appeared after the Big Bang. Astronomers identified distorted images of 16 background galaxies. The foreground cluster is four billion light-years away, but the magnified galaxies are many billions of light-years farther away.

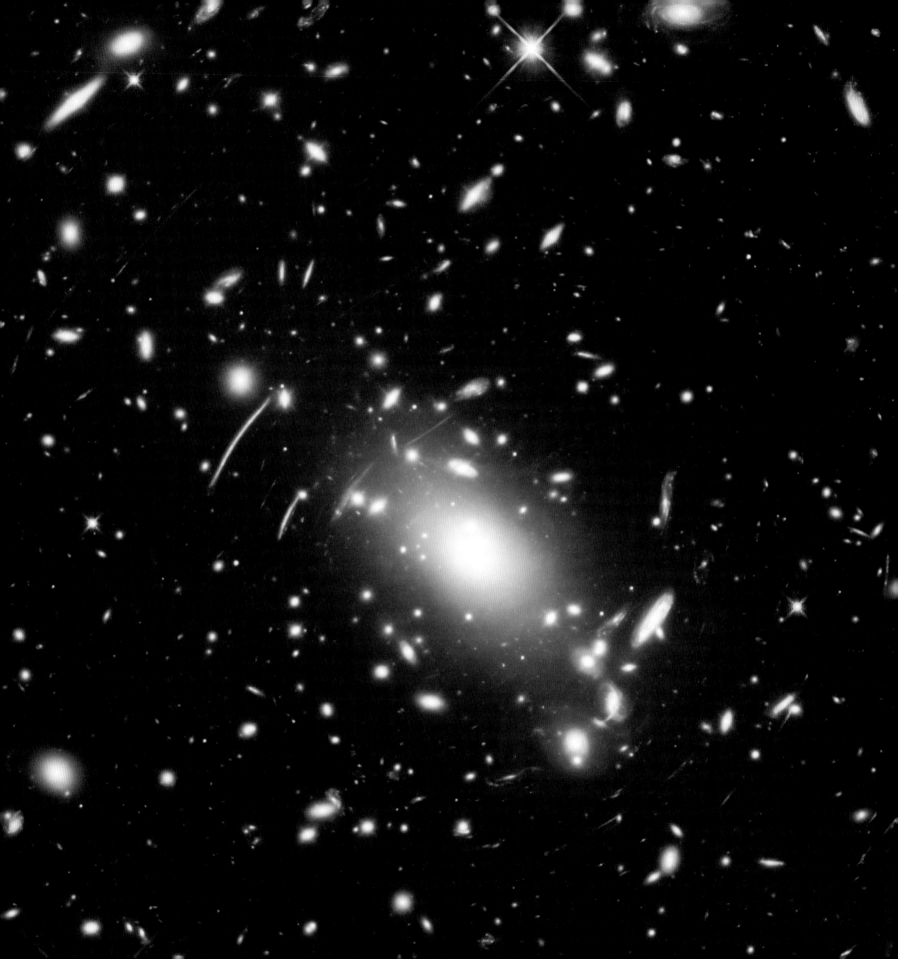

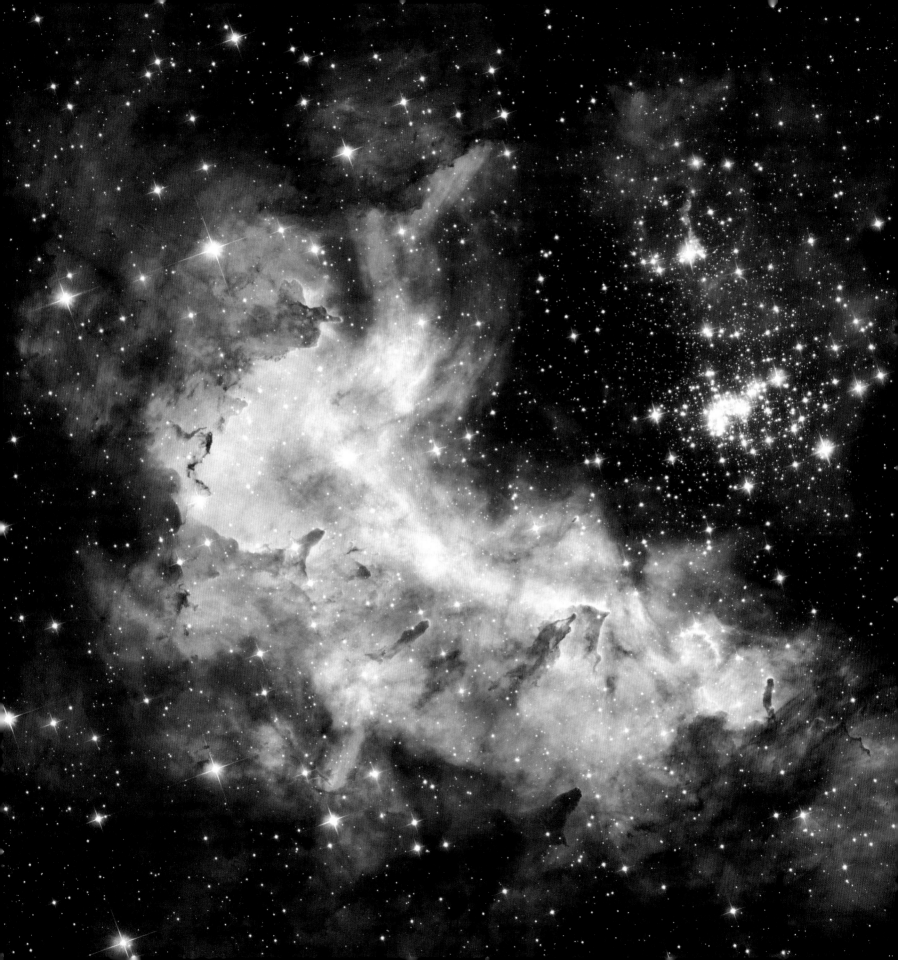

## STELLAR FIREWORKS IN WESTERLUND 2

Hubble presents a brilliant tapestry of young stars flaring to life in this panorama. The sparkling centerpiece is a cluster of 3,000 stars called Westerlund 2. The cluster resides in a raucous stellar breeding ground known as Gum 29, located 20,000 light-years from Earth in the constellation Carina the keel. Roughly 10 light-years across, the giant star cluster is only about two million years old. It contains some of our galaxy's hottest, brightest and most massive stars. Torrents of ultraviolet radiation from these heavyweights and hurricane-force winds of charged particles etch at the enveloping hydrogen gas cloud. This creates a fantasy landscape of pillars, ridges and valleys. The pillars, composed of dense gas, point to the central star cluster as the source of the blistering radiation. Other dense regions surround the pillars, including reddish brown filaments of gas and dust.

## MARS AT OPPOSITION 2016

Every two years, Earth makes its closest approach to Mars. Aside from its ruddy color and frosty polar caps, the red planet never looks quite the same on each visit due to its seasonal changes and wispy weather patterns.

This view was taken on May 12, 2016, when Mars was 80 million kilometers from Earth. The Hubble image reveals details as small as 40 kilometers across. The large dark region at far right is Syrtis Major Planum, an ancient, inactive shield volcano. Late-afternoon clouds surround its summit.

The dark areas are bedrock and fine-grained sand deposits ground down from ancient lava flows.

An extended blanket of clouds can be seen over the southern polar cap. The icy northern polar cap has receded to a comparatively small size because it is now late summer in the northern hemisphere.

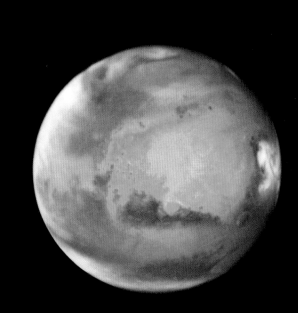

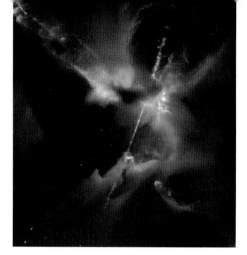

## LIGHT SABER FROM FORMING STAR

Some celestial objects are so arresting in their appearance, they can compete with the imaginings of science fiction writers. At first glance, this photo looks like a double-bladed light saber straight out of a *Star Wars* film. In the center of the image, partially obscured by a cloak of dust, a newborn star shoots twin jets out into space as a sort of birth announcement to the universe. The baby star resides in a turbulent birthing ground for new stars in the Milky Way Galaxy known as the Orion B Molecular Cloud Complex, located 1,350 light-years away.

The jets are created when gas from a disk surrounding the newly forming star rains down onto its surface. The material is super-heated and shoots outward from the star in opposite directions along the star's rotation axis. Much more energetic than a science fiction light saber, each jet collides with dense gas and dust along its path, clearing vast spaces, like a stream of water plowing into a sandpile.

## CRUCIBLE OF CHAOS (below)

This peek into the center of an ancient supernova explosion presents an eerie look at extreme physics gone wild. The object is the Crab Nebula, which lies 6,500 light-years away in the constellation Taurus the bull. A star was recorded exploding here in the year 1054 A.D. by ancient skywatchers. Nearly a millennium later, we see the expanding filaments of debris as a supernova remnant.

Looking deep into its heart, a region just three light-years across, Hubble easily spots the crushed core of the doomed star. It is the rightmost of the pair of bright stars at the center of the image (the left star is unrelated). The core is a neutron star, an ultradense ball of matter that furiously spins 30 times a second. It's the equivalent of a dense atomic nucleus, but the size of a city. The neutron star sends out machine-gun-like bursts of radiation and streams of antimatter particles. The whirling dynamo creates a mysterious halo-like magnetic loop expanding at half the speed of light. It sends ripples across the region, like waves from a stone tossed into a pond.

The orange filaments from the explosion glow with residual radiation. But to add to the weirdness, they are bathed in a blue light not seen in everyday experience. This is called synchrotron radiation and is produced by electrons spiraling along the neutron star's intense magnetic field lines.

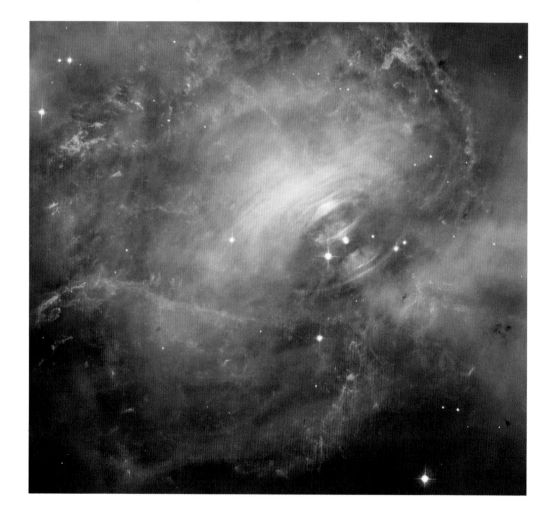

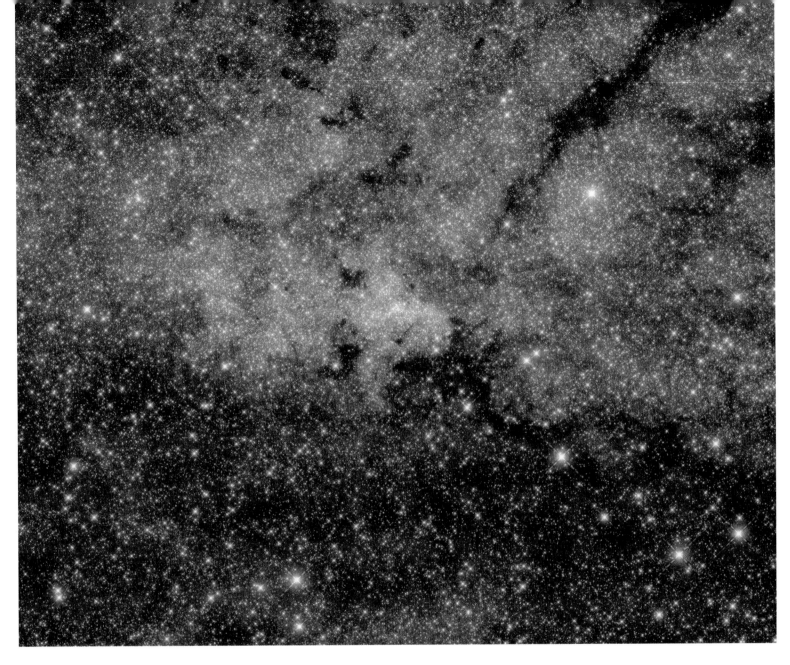

## JOURNEY TO THE CENTER OF THE GALAXY

Peering across 27,000 light-years, Hubble's near-infrared vision penetrates lanes of interstellar dust
to photograph the very center of our Milky Way Galaxy. Like the downtown of a bustling city, the core is
very crowded. The region is so tightly packed that it is equivalent to having one million stars crammed
into the volume of space between our Sun and its nearest stellar neighbor, the Alpha Centauri system,
located 4.3 light-years away. The bright blue stars in the image are foreground stars. The red stars
are either behind intervening dust or are embedded in dust themselves. Dense clouds of gas and dust
appear in silhouette against the bright background stars. The Milky Way's supermassive black hole,
about four million times the mass of the Sun, is located in the center of the image, obscured by interstellar gas and dust. A dense star cluster of 10 million stars surrounds the black hole.

## NGC2174

Seething radiation from a hot young star sculpts this fantasy landscape into the wall of a cold cloud of hydrogen laced with dust. It lies on the edge of NGC2174, a star-forming region 6,400 light-years away in the constellation Orion the hunter. Although star formation continues within these dusty clouds, they will likely be dispersed by the energetic young star within a few million years. This small portion of the large star-forming region spans six light-years. Dark brown and rust-colored dust clouds billow outward, framed against a background of bright blue gas. Young white stars sprinkle the glowing clouds, pushing away the dark stellar nurseries in which they formed.

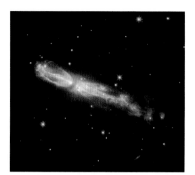

## CALABASH NEBULA

This very bizarre-looking interstellar object may, at first glance, resemble some kind of science fiction starship. The Calabash Nebula is a striking example of a brief stage in the death of a star like our Sun. The aging star is going through a rapid transformation, from a bloated red giant to a planetary nebula. At this stage of a star's existence, outer layers of its atmosphere are so hot that they escape deep into space along the doomed star's spin axis. Astronomers think that much of the gas flow in this photo stems from a sudden acceleration that took place only about 800 years ago. The jet-like exhaust is being propelled at a million kilometers per hour.

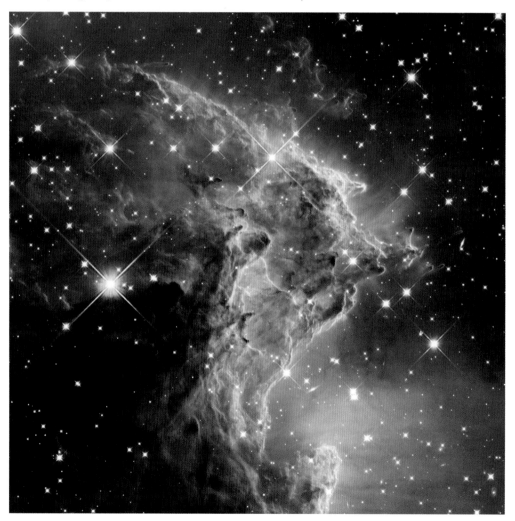

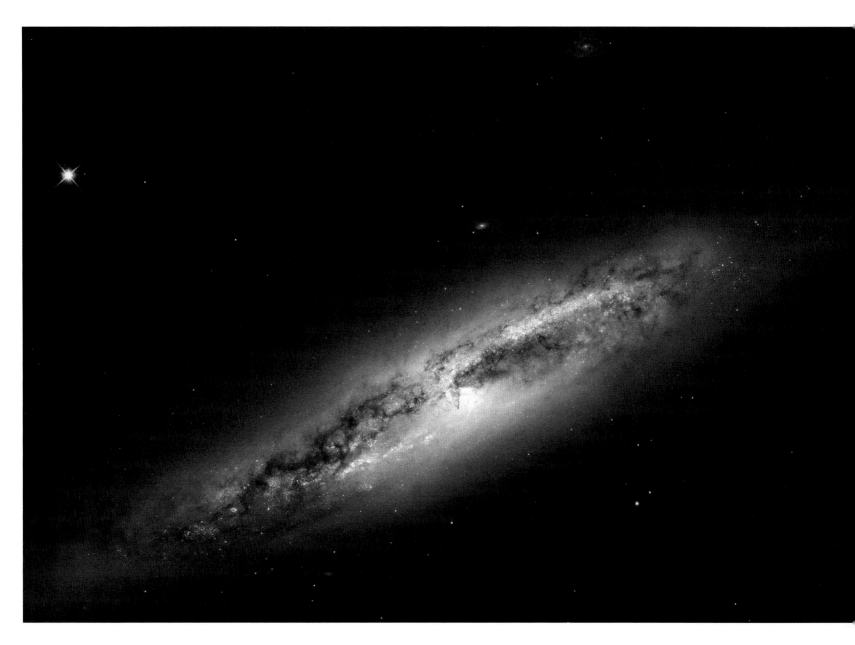

**BRIGHT BLUE ARMS IN GALAXY NGC4388**

Located some 60 million light-years away, NGC4388 looks like a typical spiral galaxy, with a dusty disk of stars and a bright central core. What's striking about this galaxy is the two brilliant arms ablaze with hot young blue stars winding out from the core, evidence that NGC4388 has hosted recent bursts of star formation. Due to its brilliant nucleus, this is one of the brightest galaxies in the neighboring huge Virgo cluster of galaxies.

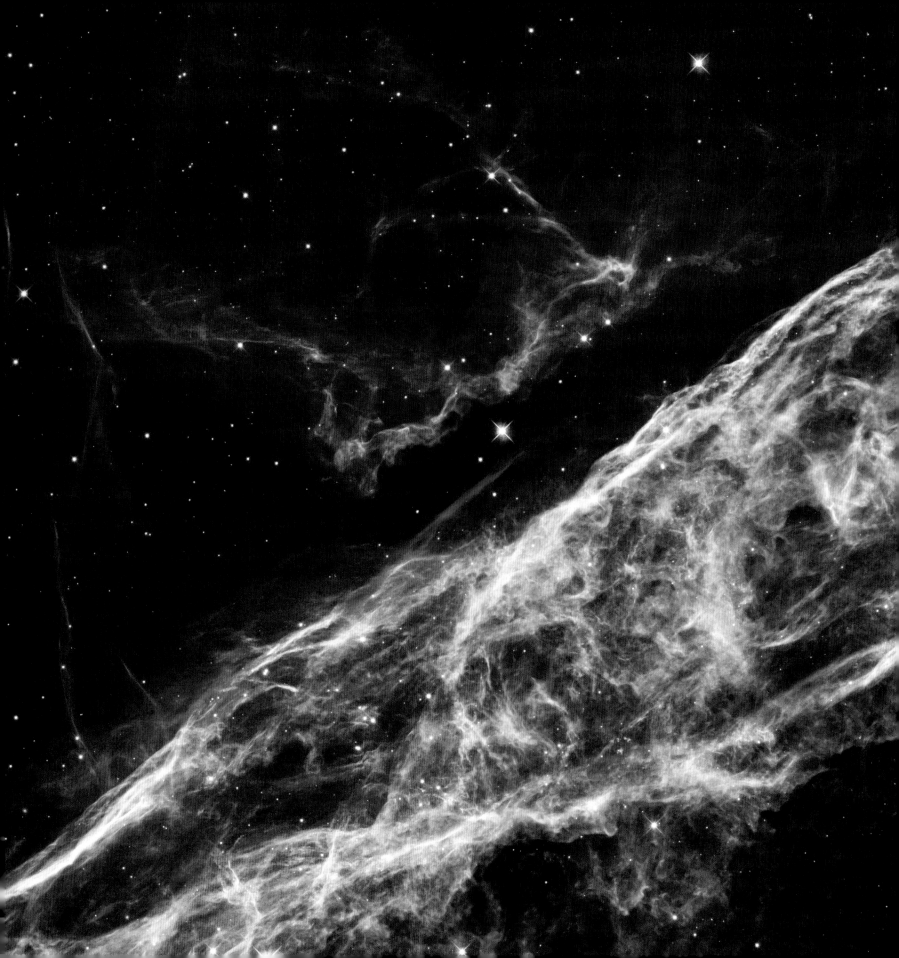

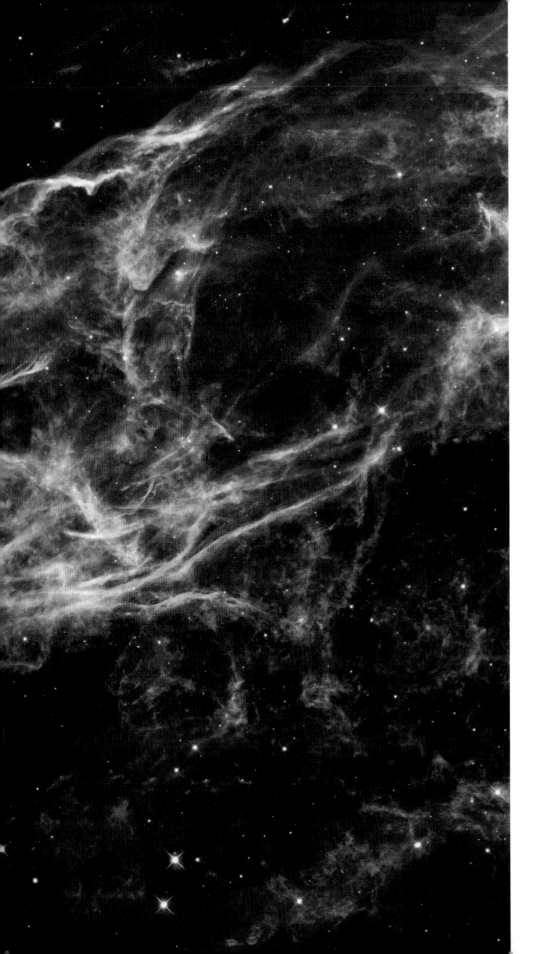

## VEIL NEBULA

This delicate tangle of glowing gas filaments is all that's left of a star that exploded about 8,000 years ago. It's just a small portion of the Veil Nebula, one of the best-known supernova remnants in the heavens. Located 2,100 light-years from Earth, in the constellation Cygnus the swan, the entire bubble-like structure—also known as the Cygnus Loop—spans approximately 110 light-years. As seen from Earth, it covers an area of sky over seven times the diameter of the full Moon.

The Hubble image zooms into a tiny fraction of the nebula's outer limb, located on the west side of the supernova remnant. This section of the outer shell is in a region known as NGC6960, or the Witch's Broom Nebula.

Before the supernova exploded, the bloated doomed star puffed off a superhot outflow of gas. Like an expanding soap bubble, it blew a large cavity into the surrounding denser interstellar gas.

As the shock wave from the supernova expands outward, it slams into the walls of this cavity. The so-called shock front forms the nebula's intricate structures. Bright filaments are produced as the shock wave interacts with a relatively dense cavity wall. Regions nearly devoid of material generate fainter structures. Variations in the temperatures and densities of the chemical elements produce the Veil Nebula's colorful appearance.

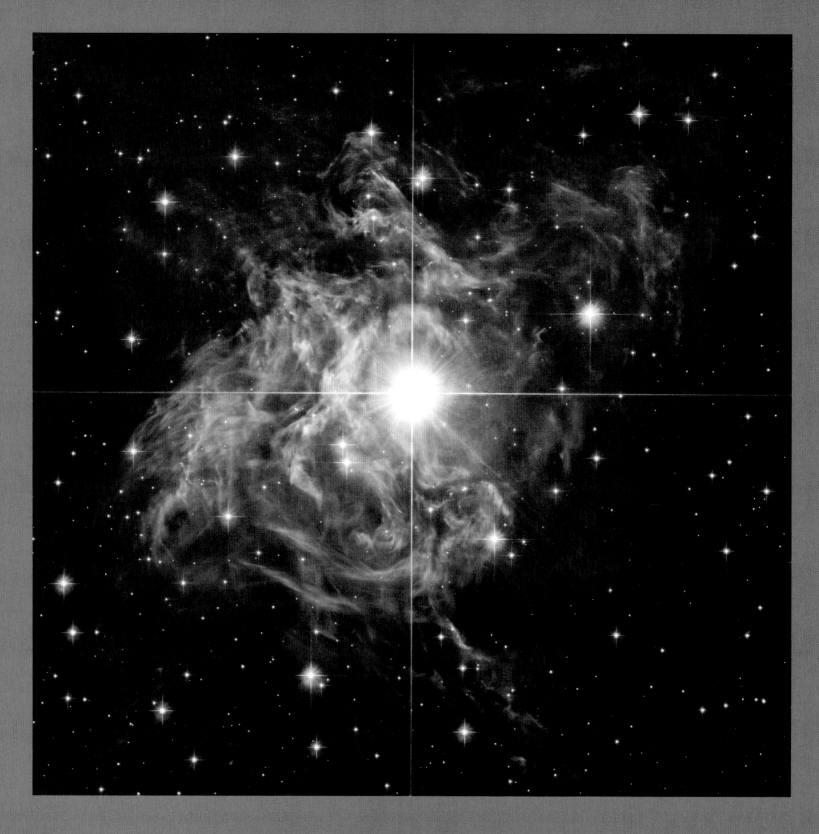

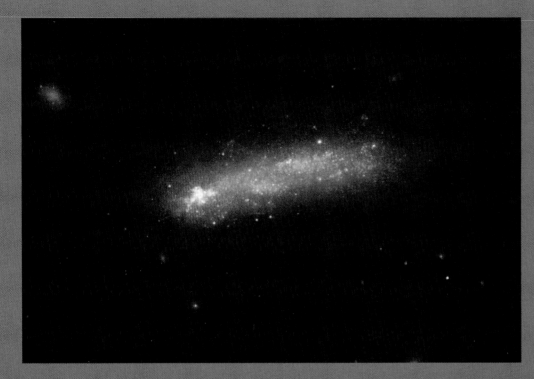

## FIREWORKS GALAXY

A firestorm of star birth is lighting up one end of the diminutive galaxy Kiso 5639. The dwarf galaxy is shaped like a flattened pancake, but because it is tilted edge-on to Hubble's view, it resembles a sky-rocket, with a brilliant blazing head and a long star-studded tail. Astronomers think that the frenzied star birth is sparked by intergalactic gas raining on one end of the galaxy as it drifts through space. Kiso 5639 offers a striking up-close example of what must have been common in the universe long ago, when young galaxies rapidly grew by pulling in gas from the surrounding neighborhood. Observations of the early universe with Hubble's Ultra Deep Field reveal that about 10 percent of all galaxies have these elongated shapes and are collectively called "tadpoles."

## BLACK HOLE'S NEXT BIG MEAL IN NGC4696

Eerie tentacles of dark dust and reddish glowing gas wrap around the bright center of the giant elliptical galaxy NGC4696. These filaments loop and curl inward, swirling around the supermassive black hole. They will be dragged into and eventually consumed by this cosmic monster. The host galaxy is the largest inhabitant of the huge Centaurus galaxy cluster 150 million light-years distant.

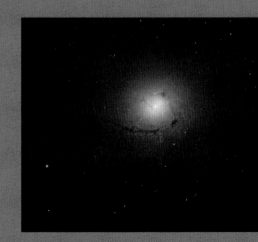

## CEPHEID VARIABLE STAR RS PUPPIS (left)

Resembling a holiday wreath sparkling with lights, the bright southern-hemisphere star RS Puppis is swaddled in a gossamer cocoon of reflective dust illuminated by the glittering star. RS Puppis rhythmically brightens and dims over a six-week cycle. The star burned steadily for most of its life, slowly consuming the fuel at its core to keep it shining brightly. However, once most of the hydrogen fuel was depleted, the star became unstable. It now expands and shrinks over 40 days, cyclically growing brighter and dimmer. This type of star is called a Cepheid variable, and RS Puppis is unusual in that it is one of the brightest stars in this class. It is also unusual because of a shroudlike nebula around it that is made up of shells of dust and gas puffed off into space by the dying star. The surrounding nebula flickers in brightness as pulses of light from the Cepheid propagate outward.

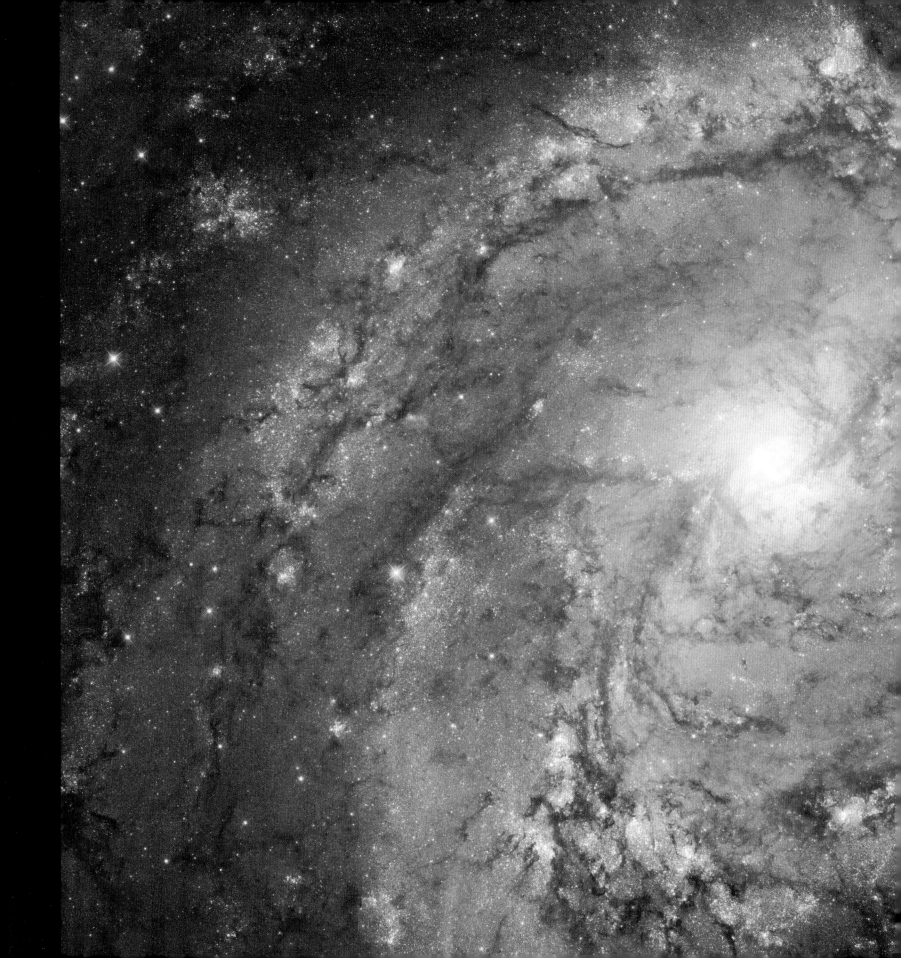

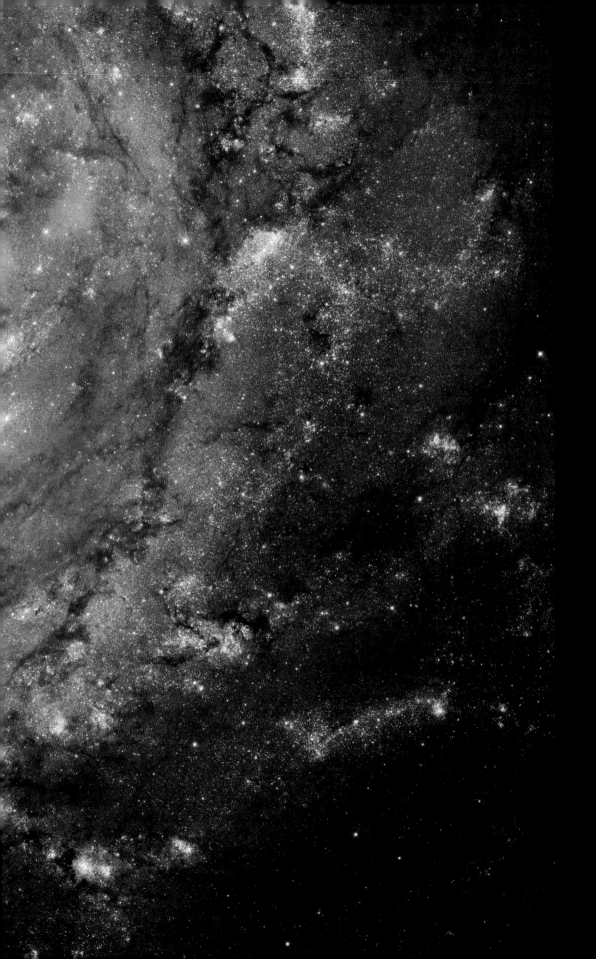

## SOUTHERN PINWHEEL GALAXY

Galaxy M83 is one of the closest barred spirals to us. Nicknamed the Southern Pinwheel, it presents a dramatic and colorful bird's-eye view of the architecture of a barred spiral galaxy. It features very prominent spiral arms traced by dark dust lanes silhouetted against a carpet of several thousand brilliant blue star clusters. A chain of reddish pink star-forming regions follows the blue lanes like garnets embedded in a necklace. The rich colors are evidence that this galaxy is much more ablaze with star formation than our own Milky Way. The effervescent star birth gives way to a fireworks barrage of supernova explosions as very massive stars self-destruct just a few million years after their birth.

The pair of prominent spiral arms form a barlike structure at the central hub. They funnel gas inward toward the galaxy's core. The gas is then used to create new stars and also to feed the galaxy's central black hole. This explains why many barred spirals, including Messier 83, have very active and bright centers.

The Pinwheel lies 12 million light-years away, near the southeastern tip of the constellation Hydra the water snake. Roughly 40,000 light-years across, the galaxy is half the diameter of our Milky Way.

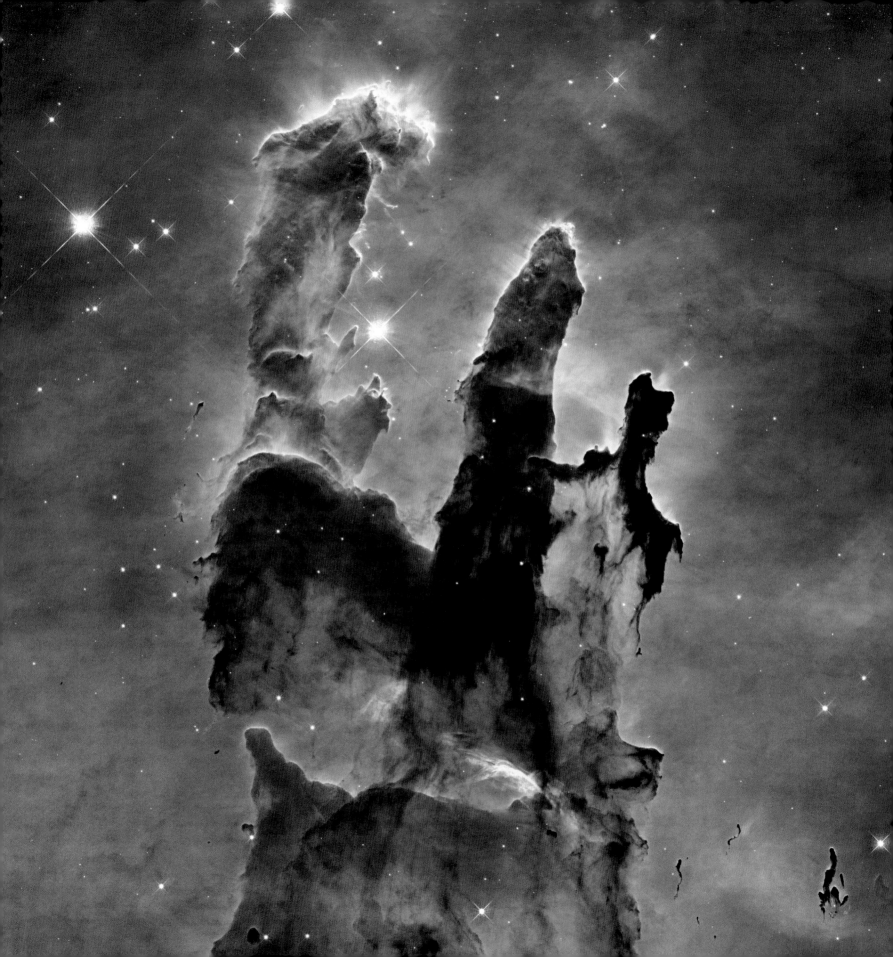

## HIGH-DEFINITION PILLARS OF CREATION

The most iconic celestial target ever imaged by Hubble was so awe-inspiring, it was worth revisiting with Hubble's new camera, the WFC3. Astronomers wanted a bigger and better image of the so-called pillars of creation. Compared with Hubble's 1995 snapshot (page 85), the new mosaic image stretches down to the base of the pillars and is somewhat sharper.

Having an eerie organic feel, three giant columns of cold gas are bathed in the scorching ultraviolet light from a cluster of young massive stars in a small sector of the Eagle Nebula (M16). Although such elephant-trunk-like features are common in star-forming regions, the M16 structures are, by far, the most photogenic and evocative.

Contrary to their nickname, these stalagmitelike structures are actively being ablated by fierce radiation from hot young stars embedded in the region. The ghostly bluish haze around the dense edges of the pillars is material heating up and evaporating into space. The pillars are so far away, Hubble recorded the light that left them 6,500 years ago. They may have already been swept away, like an incoming tide destroying a sand castle. The "tide" approaching the pillars is a blast wave from supernovas photographed in the region. But the light showing us the destruction will not reach Earth for another millennium.

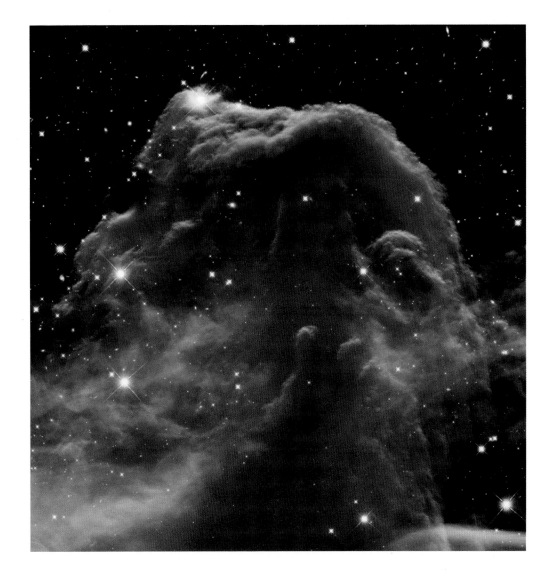

## HORSE OF A DIFFERENT COLOR

One of the most recognizable celestial objects in the history of astronomical imaging, the Horsehead Nebula was first photographed in 1888. The stalklike nebula looks a bit more like a giant seahorse rising from turbulent waves. Like the Eagle Nebula structures, it is a pillar of hydrogen laced with cold dust that resists being eroded by intense starlight in a young star-birth region. Astronomers estimate that the Horsehead Nebula has about five million years left before it is obliterated by blistering radiation.

Although the Horsehead has been photographed many times, Hubble shows us the horse in a different light. Hubble photographed the nebula in infrared, which can pierce through the dusty material that usually obscures the nebula's interior. The result is a rather ethereal and fragile-looking structure made of delicate folds of gas—very different to the nebula's appearance in visible light. The nebula lies 1,600 light-years away in the constellation Orion the hunter.

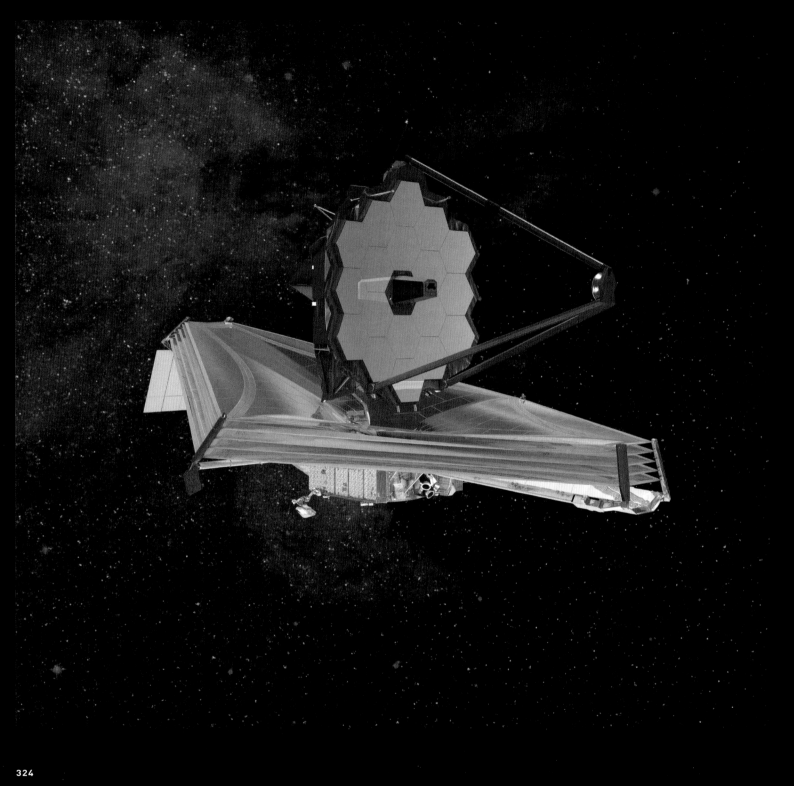

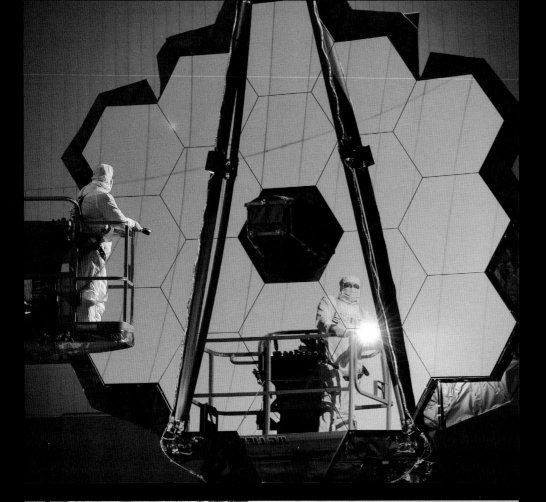

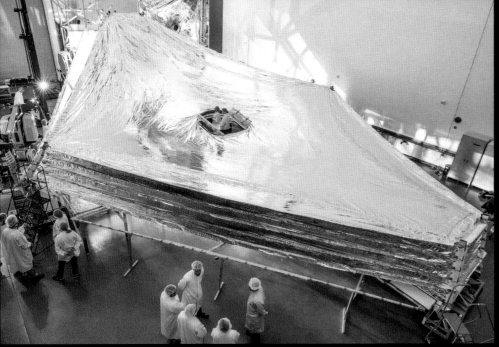

## JAMES WEBB SPACE TELESCOPE, HUBBLE'S SUCCESSOR

Scheduled for launch in 2018, NASA's James Webb Space Telescope will be located a million kilometers beyond the Moon in a gravitationally stable, dark and very cold region of the Earth-Moon system. Webb's giant segmented mirror will have seven times the light-collecting power of Hubble. Webb will pick up where Hubble left off, peering even deeper into the universe and further back in time.

Webb's capabilities are needed because the visible light from the first stars and galaxies to form more than 13 billion years ago has been stretched into infrared wavelengths by the universe's expansion. Hubble cannot detect this form of invisible light. Chilled to –225°C in the shadow of a giant sun shade, Webb will have infrared vision for peering into the undiscovered country of the primeval universe. Astronomers predict that roughly 200 million years after the Big Bang, stars began to form and, with them, embryonic galaxies. In capturing the ancient light from these objects, Webb will unveil the opening chapter in the history of the origin of the universe.

Looking much closer to home, Webb will search for the biochemical signatures of life on planets orbiting stars that are relatively close to our Sun. Webb's infrared capabilities will allow it to search for signs of oxygen, ozone, water vapor, methane and carbon dioxide in Earth-sized planets in the habitable zones around stars.

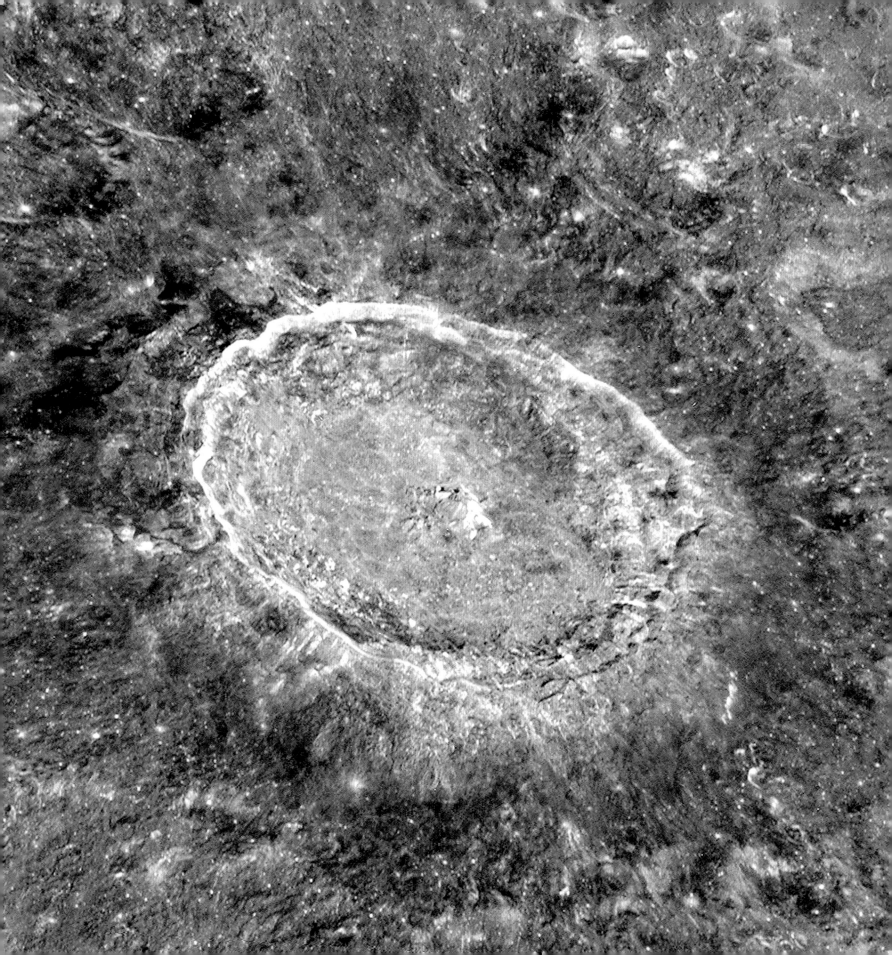

## HUBBLE DVDS

*Hubble: 15 Years of Discovery*. DVD. Directed by Lars Lindberg Christensen. European Space Agency, 2005.

*Hubble's Amazing Rescue*. DVD. Directed by Rushmore DeNooyer. Boston: Nova/WGBH, 2010.

*IMAX: Hubble*. DVD. Produced and directed by Toni Myers. Warner Bros. Picture Production in cooperation with NASA, 2010.

*Hubble: Mission Critical*. DVD. Directed by Tom Whitter. Discovery Channel International, 2010.

*Hubble's Amazing Universe*. DVD. Directed by Dana Berry. National Geographic Television, 2010.

## HUBBLE WEBSITES

http://amazing-space.stsci.edu/

http://archive.seds.org/hst/hst.html

http://asd.gsfc.nasa.gov/archive/hubble/

http://heritage.stsci.edu/

http://quest.nasa.gov/hst/

www.spacetelescope.org/

www.spacetelescope.org/projects/anniversary/

http://starchild.gsfc.nasa.gov/docs/StarChild/space_level2/hubble.html

## NOTABLE BOOKS AND REPORTS

Ad Hoc Committee on the Large Space Telescope, Space Science Board, National Academy of Sciences, National Research Council. *Scientific Uses of the Large Space Telescope*. Washington, D.C.: National Academy of Sciences, 1969. First comprehensive NASA study on the astronomical potential of an orbiting space telescope.

Smith, Robert W. *The Space Telescope: A Study of NASA, Science, Technology, and Politics*. New York: Cambridge University Press, 1989. Well-written history of the concept of a space telescope and an in-depth account of the design and manufacture of the Hubble Space Telescope. Published a few months before the telescope was launched into orbit.

Capers, Robert S. and Eric S. Lipton. "The Looking Glass: How a flaw reflects cracks in space science." Hartford, Connecticut: *The Hartford Courant*, March-April 1991. This four-part series of articles about Hubble's initially flawed optics won a Pulitzer Prize.

Livio, Mario, Keith Noll and Massimo Stiavelli, eds. *A Decade of Hubble Space Telescope Science*. Proceedings of the Space Telescope Science Institute Symposium, held in Baltimore, Maryland, 11-14 May 2000. New York: Cambridge University Press, 2003.

Christensen, Lars Lindberg and Bob Fosbury. *Hubble: 15 Years of Discovery*. New York: Springer, 2006.

*Hubble Space Telescope: Window to the Universe*. Washington, D.C.: National Aeronautics and Space Administration, 2010 (NP-2010-04-648-HQ).

Weiler, Edward J. *Hubble: A Journey Through Space and Time*. Foreword by Charles F. Bolden, Jr. Robert Jacobs, Dwayne Brown, J. D. Harrington, Constance Moore and Bertram Ulrich, eds. New York: Abrams Books, 2010.

Zimmerman, Robert. *The Universe in a Mirror: The Saga of the Hubble Space Telescope and the Visionaries Who Built It*. Princeton, New Jersey: Princeton University Press, 2010.

The Hubble Space Telescope captured this image of the 86-kilometer-wide lunar impact crater Tycho. Because the Moon has been mapped in great detail by lunar orbiting spacecraft, there is relatively little call for Hubble's intense gaze to be turned toward the Earth's natural satellite.

# INDEX

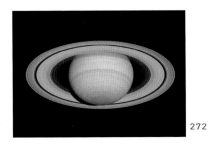

272

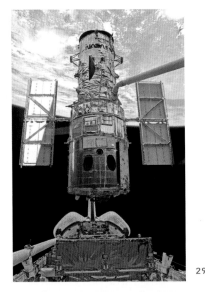

29

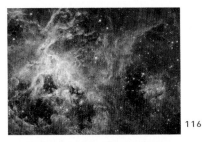

116

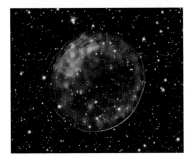

8

# PHOTO CREDITS

3 NASA, ESA, SAO, CXC, JPL-Caltech

6 NASA, ESA

8 X-ray: NASA, CXC, SAO, J. Hughes et al. Optical: NASA, ESA, Hubble Heritage Team (STScI/AURA)

10 NASA, ESA, F. Paresce, R. O'Connell

11 NASA

12-13 NASA, Hubble Heritage Team, A. Riess

14 NASA, ESA, M. Buie

15 NASA, ESA, Hubble Heritage Team (STScI/AURA), J. Bell, M. Wolff

16 NASA, ESA, Hubble Heritage Team (STScI/AURA)

17 NASA, H.E. Bond, E. Nelan, M. Barstow, M. Burleigh, J.B. Holberg

19 NASA

20-21 NASA, ESA, M. Davis

22 NASA, ESA, ESO

23 NASA, ESA, M. Livio, Hubble 20th Anniversary Team

24-29 NASA

30 NASA, ESA, H. Ford, G. Illingworth, M. Clampin, G. Hartig

31 NASA

32 NASA

33 NASA (both)

34 Left: NASA, ESA, Robert Williams, Hubble Deep Field Team. Right: NASA, ESA, Mark Dickinson

35-39 NASA, ESA, Robert Williams, Hubble Deep Field Team

40 NASA, ESA, G. Illingworth, R. Bouwens, HUDF09 Team

41 NASA, ESA, M. Trenti, L. Bradley, BoRG Team

42 G. Bacon/STScI

43 NASA, ESA, Roeland P. van der Marel, Frank C. van den Bosch

44 Left (both): NASA, ESA, Gary Bower, Richard Green. Top: NASA, ESA, D. Batcheldor, E. Perlman, Hubble Heritage Team, J. Biretta, W. Sparks, F.D. Macchetto. Bottom, right: NSA, ESA, John Bahcall, Mike Disney

45 NASA, ESA, S. Farrell

46 Top: NASA. Bottom: NASA/WMAP

47 Top: NASA. Bottom: NASA, ESA, A. Riess

48 NASA, ESA, A. Riess, S. Rodney

49 A Field/STScI

50 A Field/STScI

51 Left: NASA, HST, W. Freedman, R. Kennicutt, J. Mould. Right: Wendy L. Freedman, NASA

52 NASA, ESA, A. Riess, L. Macri, Hubble Heritage Team (STScI/AURA)

53 G. Bacon/STScI (both)

54 G. Bacon/STScI (both)

55: NASA, ESA, CXC, A. Mahdavi

56 NASA, ESA, CXC, M. Markevitch, D.Clowe

57 NASA, ESA, L. Bradley, R. Bouwens, H. Ford, G. Illingworth

58 NASA, ESA, M. Robberto, Hubble Space Telescope Orion Treasury Project Team

59 NOAO

60 NOAO (all)

61 Palomar Digital Sky Survey

62-63 NASA, ESA, Steve Lee, Jim Bell, Mike Wolff

64 Top: NASA, ESA. Ground Image: Canada-France-Hawaii Telescope, Hawaii. Bottom: NASA, ESA, D. Lafrenière

65 NASA, ESA, Hubble Heritage Team (STScI/AURA), R. Gendler

66 NASA/JPL

67 Alan Dyer

68 NASA (all)

69 NASA (all)

70-71 NASA, ESA, M. Robberto, Hubble Space Telescope Orion Treasury Project Team

72 NASA

73 NASA, ESA, SAO, CXC, JPL-Caltech

74 Right: NASA, ESA, R. O'Connell, B. Whitmore, M. Dopita. Left: ESO

75 NASA, ESA, Alan Stern, Marc Buie

76 NASA, ESA, Hubble Heritage Team (AURA/STScI)

77 NASA, ESA, F. Paresce, R. O'Connell (both)

78 NASA, ESA, D. Maoz

79 Top: NASA, ESA, Hubble Heritage Team. Bottom: NASA, ESA, Hubble Heritage Team (AURA/STScI)

80 Top: NASA, ESA, M. Regan, B. Whitmore, R. Chandar. Bottom: NASA, ESA, J. Garvin

81 Top: NASA, ESA, Michael S. Vogeley. Bottom: NASA, ESA, J. Madrid

82 NASA, ESA, Hubble Heritage Team (STScI/AURA)

83 NASA, ESA, N. Smith, Hubble Heritage Team

84 NASA, ESA, Hubble Heritage Team (STScI/AURA) (both)

85 NASA, ESA, J. Hester, P. Scowen

86-87 NASA, ESA, M. Robberto, Hubble Space Telescope Orion Treasury Project Team

88-89 NASA, ESA, M. Robberto, Hubble Space Telescope Orion Treasury Project Team

90 Left: NASA, ESA, C.R. O'Dell. Right NASA, ESA, K.L. Luhma, G. Schneider, E. Young, G. Rieke, A. Cotera, H. Chen, M. Rieke, R. Thompson

91 NASA, ESA

92 Left: NASA, ESA. Right: NASA, ESA, J. Hester

93 T. A. Rector, WIYN/AURA/NSF

94-95 NASA, ESA, Hubble Heritage Team (STScI/AURA)

96 NASA, ESA, Orsola De Marco

97 NASA, ESA, J. Hester

98 Left: NASA, ESA, R. Sahai. Right: NASA, ESA, A. Caulet

99 NASA, ESA

100 NASA, ESA

101 NASA, ESA, Hubble Heritage (STScI/AURA)

102-103 NASA, ESA, N. Smith, Hubble Heritage Team (STScI/AURA)

104-105 (STScI/AURA)

106 NASA, ESA, N. Smith (University of California, Berkeley), Hubble Heritage Team (STScI/AURA)

107 NASA, ESA, N. Smith (University of California, Berkeley), Hubble Heritage Team (STScI/AURA) (both)

108 NASA, ESA, N. Smith (University of California, Berkeley), Hubble Heritage Team (STScI/AURA)

109 NASA, ESA, and the Hubble SMR ERO Team

110 NASA, ESA, Hubble Heritage Team (STScI/AURA

111 NASA, ESA, Hubble Heritage Team (STScI/AURA)

112-113 NASA, ESA

114 Terence Dickinson

115 Alan Dyer

116-121 NASA, ESA, D. Lennon, E. Sabbi

122 NASA, ESA

123 NASA, ESA, Mohammad Heydari-Malayer

124 NASA, ESA, Jesús Maiz Apellániz

125 NASA, ESA, A. Nota

126 NASA, ESA, Hubble Heritage Team (STScI/AURA)

127 NASA, ESA, Hubble Heritage Team (STScI/AURA)

128 Top and Bottom: NASA, ESA, Mohammad Heydari-Malayeri

129 NASA, ESA, Q. D. Wang, S. Stolovy

130 NASA, ESA, Hubble Heritage Team (STScI/AURA)

131 Top: Rémi Lacasse. Bottom: NASA, ESA, Hui Yang

132 NASA, ESA

133 NASA, ESA

134 NASA, ESA, Hubble SM4 ERO Team

135 NASA, ESA, Hubble Heritage Team (STScI/AURA)

136-137 NASA, ESA

138 NASA, ESA

139 NASA, ESA, Hubble Heritage Team (STScI/AURA)

140 NASA, ESA (both)

141 Top: NASA, ESA, J. Bally, H. Throop , C. O'Dell. Center: NASA, ESA, C. Burrows, WFPC 2 Investigation Definition Team. Bottom: NASA, ESA, P. Hartigan

142 Left: NASA, ESA, K. Luhman. Top: NASA. ESA, J. Walsh. Bottom: NASA, ESA, D. Golimowski, D. Ardila, J. Krist, M. Clampin, H. Ford, G. Illingworth, ACS Science Team

143 Top: NASA, ESA, K. Sahu. Bottom: NASA, ESA, P. Kalas, J. Graham, E. Chiang, E. Kite, M. Clampin, M. Fitzgerald, K. Stapelfeldt, J. Krist

144 NASA, ESA

145 Top: NASA, ESA, Hubble Heritage Team (STScI/AURA). Bottom: NASA, ESA

146 NASA, ESA

147 NASA, ESA

148 NASA, ESA

149 Top: NASA, ESA, Martino Romaniello. Bottom: NASA, Hubble Heritage Team (STScI/AURA)

150 Top: NASA, ESA, Yves Grosdidier. Bottom: NASA, ESA, D Feiger

151 NASA, ESA, D. A Gouliermis

152-153 NASA, ESA, Jesús Maíz Apellániz

154-155 NASA, ESA

156 NASA, ESA, M. Robberto, Hubble Space Telescope Orion Treasury Project Team

157 NASA, ESA

158 NASA, ESA, Hubble Heritage Team (STScI/AURA)

160 NASA

161 Left: NASA, ESA. Right: NASA, ESA

162 NASA, ESA

163 NASA, ESA

164 NASA, ESA, D. Lennon, E. Sabbi

165 NASA, ESA, D. Lennon, E. Sabbi

166 Left: NASA, ESA, R. Humphreys. Right: NASA, ESA

167 Left: NASA, ESA, Hubble Heritage Team (STScI/AURA). Right: NASA, ESA, A. Dupree, R. Gilliland

168 Top: NASA, ESA. Bottom: NASA, ESA

169 Top: NASA, ESA, Hubble Heritage Team (STScI/AURA). Bottom: NASA; ESA; H. Van Winckel, M. Cohen

170 NASA, ESA, Hubble Heritage Team (STScI/AURA)

171 Nordic Optical Telescope

172 NASA, ESA, Hubble SM4 ERO Team

173 NASA, NOAO, ESA, Hubble Helix Nebula Team, M. Meixner, T.A. Rector

174 NASA, ESA

175 NASA, ESA

176 NASA, ESA, Hubble Heritage Team (STScI/AURA)

177 NASA, ESA, A. Fruchter

178 NASA, ESA, Hubble Heritage Team (STScI/AURA)

179 Top: NASA, ESA. Bottom: NASA, ESA

180 NASA, ESA (all)

181 NASA, ESA (both)

182 NASA, ESA, P. Challis, R. Kirshner, B. Sugerman

183 NASA, ESA, Hubble SM4 ERO Team

184 Left: NASA, ESA, Hubble Heritage (STScI/AURA) (all). Right: James Black

185 NASA, ESA, Hubble Heritage Team (STScI/AURA)

186 NASA, ESA, A. Loll, J. Hester

187 NASA, ESA, A. Loll, J. Hester

188 NASA, ESA, CXC, J. Hester

189 NASA, ESA, Hubble Heritage Team (STScI/AURA)

190 NCSA, M. Hall

192 NASA, ESA

193 ESA

194 NASA, ESA, J. Merten, D. Coe

196-197 ESA, NASA, J.-P. Kneib, R. Ellis

198 NASA, ESA, J. Rigby

199 Top: NASA, ESA. Bottom NASA, ESA, SLACS Survey Team

200-201 NASA, ESA, K. Sharon, E. Ofek

202 NASA, ESA

203 NASA, ESA, M. Postman, CLASH Team

204 NASA, ESA, H. Ford, N. Benitez, T. Broadhurst

205 NASA, ESA, Hubble SM4 ERO Team (all)

206 NASA, ESA, D. Coe, N. Benitez, T. Broadhurst, H. Ford

207 Left: NASA, ESA, R. Massey. Right: NASA, ESA, C. Faure, J. P. Kneib (all)

208 NASA, ESA, M.J. Jee, H. Ford

209 NASA, ESA, C. Heymans, M. Gray, M. Barden, STAGES collaboration

210 NASA, ESA, CXC, M. Bradac, S. Allen

211 NASA, ESA, CFHT, CXO, M. J. Jee, A. Mahdavi

212 NASA, ESA, Hubble Heritage Team (STSCI/AURA)

214-215 NASA, ESA, S. Beckwith, Hubble Heritage Team (STScI/AURA)

216 NASA, ESA

217 NASA, ESA, J. Lotz, M. Davis, A. Koekemoer (all)

218-227 NASA, ESA, M. Davis, A. Koekemoer

228 NASA, ESA, Hubble Heritage (STScI/AURA)

229 NASA, ESA, Y. Izotov, T. Thuan

230 NASA, ESA

231-233 NASA, ESA, Hubble Heritage Team (STScI/AURA)

234 NASA, ESA, K. Kuntz, F. Bresolin, J. Trauger, J. Mould, Y.-H. Chu

235 NASA, ESA, Hubble Heritage Team (STScI/AURA)

236-237 NASA, ESA, Hubble Heritage Team (STScI/AURA)

238-239 NASA, ESA, Hubble Heritage Team (STScI/AURA)

240 NASA, ESA, Hubble Heritage Team (STScI/AURA)

241 Top: NASA, ESA, Hubble Heritage Team (STScI/AURA). Bottom: NASA, ESA

242 NASA, ESA, Hubble Heritage Team (STScI/AURA)

243 NASA, ESA, A. Aloisi, Hubble Heritage Team (STScI/AURA)

244 NASA, ESA, Hubble Heritage Team (STScI/AURA)

245 ESO

246 NASA, ESA, S. Gallagher, J. English

247 NASA, ESA, Hubble Heritage Team (STScI/AURA)

248 NASA, ESA, Hubble Heritage Team (STScI/AURA)

249 NOAO

250 NASA, ESA, H. Ford, G. Illingworth, M.Clampin, G. Hartig, ACS Science Team

251 NASA, ESA, H. Ford, G. Illingworth, M.Clampin, G. Hartig, ACS Science Team

252-255 NASA, ESA, Hubble Heritage Team (STScI/AURA)

256 NASA, ESA, Hubble SM4 ERO Team

257 NASA, ESA, Hubble Heritage Team (STScI/AURA)

258 NASA, ESA

259 NASA, ESA

260-261 NASA, ESA, Hubble Heritage Team (STScI/AURA)

262 NASA, ESA, E. Karkoschka

264 NASA, ESA (all)

265 NASA, ESA, Hubble Heritage Team (STScI/AURA)

266 Left: NASA, ESA, H. Hammel (both). Right: NASA, ESA, J. Spencer

267 NASA, ESA, J. Spencer

268 NASA, ESA

269 NASA, ESA

270 NASA, ESA, P. James, S. Lee (both)

271 NASA, ESA, J. Bell, M. Wolff

272-273 NASA, ESA, Hubble Heritage Team (STScI/AURA)

274 NASA, ESA, Hubble Heritage Team (STScI/AURA)

275 Left: NASA, ESA, J. Clarke. Right: NASA, ESA, Hubble Heritage Team (STScI/AURA)

276 NASA, JPL

277 Top left: NASA, ESA, E. Karkoschka. Top right: NASA, ESA, L. Sromovsky, H. Hammel, K. Rages. Bottom left: NASA, ESA, Hubble Heritage Team (STScI/AURA). Bottom right: NASA, ESA, H. Weaver, A. Stern

278 NASA, ESA

280 Left: NASA, ESA, Hubble Heritage Team (STScI/AURA). Right NASA, ESA, H. Hammel

281 NASA, ESA, Hubble Heritage Team (STScI/AURA)

282 NASA, ESA, W. Keel, Galaxy Zoo Team

283 NASA, ESA, H.Bond, Hubble Heritage Team (STScI/AURA)

284 NASA, ESA, D. Jewitt

285 NASA, ESA, M. Wong, I. de Pater (all)

286 Left: NASA, ESA, H. Weaver, M. Mutchler. Right: NASA, ESA, S. Tereby

287 NASA, ESA, M. H. Wong, H.Hammel, I. de Pater, Jupiter Impact Team (all)

288 Left: NASA, ESA, J. Krist, K. Stapelfeldt, J. Hester, C. Burrows. Right: NASA, ESA, N. Tanvir, A. Fruchter

289 NASA, ESA, Hubble Heritage Team (STScI/AURA)

290 Left: NASA, ESA, J. Kenney, E. Yale. Right: NASA, ESA, R. Sahai

291 NASA, ESA, Hubble Heritage Team (STScI/AURA) (both)

292 NASA, ESA, K. Barbary, Supernova Cosmology Project (both)

293 NASA, ESA, D. Evans

294 NASA, ESA, the Hubble Heritage Team (STScI)

296 NASA, ESA

297 NASA, ESA, R. Gendler

298 NASA, ESA

299 NASA, ESA

300 Top: NASA, ESA, A. Simon (Goddard Space Flight Center).

Bottom: NASA, ESA, J. Nichols (University of Leicester), A. Simon (NASA), the OPAL Team

301 NASA, ESA, CXC, NRAO, R. van Weeren (Harvard-Smithsonian Center for Astrophysics), G. Ogrean (Stanford University)

302 R. Gendler

303-307 NASA, ESA, J. Dalcanton, B.F. Williams and L.C. Johnson (University of Washington), the PHAT Team, R. Gendler

308 NASA, ESA, H. Teplitz, M. Rafelski (Caltech), A. Koekemoer (STScI), R. Windhorst (Arizona State University)

309 NASA, ESA, J. Lotz (STScI)

310 NASA, ESA, the Hubble Heritage Team (STScI), A. Nota (ESA/STScI), the Westerlund 2 Science Team

311 NASA, ESA, J. Bell (ASU), M. Wolff (Space Science Institute)

312 Left: NASA, ESA, D. Padgett (GSFC), T. Megeath (University of Toledo), B. Reipurth (University of Hawaii). Right: NASA, ESA, J. Hester (ASU), M. Weisskopf (NASA)

313 NASA, ESA, the Hubble Heritage Team (STScI), T. Do and A. Ghez (UCLA), V. Bajaj (STScI)

314 Left: NASA/ESA. Right: NASA, ESA, Judy Schmidt

315 NASA, ESA

316 NASA, ESA, the Hubble Heritage Team (STScI)

318 NASA, ESA, the Hubble Heritage Team (STScI), H. Bond (Pennsylvania State University)

319 Top: NASA, ESA, D. Elmegreen (Vassar College). Bottom: NASA, ESA, A. Fabian (University of Cambridge)

320 NASA, ESA

322 NASA, ESA, the Hubble Heritage Team (STScI)

323 NASA, ESA, the Hubble Heritage Team (STScI)

324-325 NASA

326 NASA, ESA, H. Bond, Hubble Heritage Team (STScI/AURA)

# THE AUTHOR

Terence Dickinson is the author of 15 astronomy books. He is the former editor of *SkyNews*, the Canadian astronomy magazine. In the 1960s and 1970s, he was a staff astronomer at the McLaughlin Planetarium at the Royal Ontario Museum and the Strasenburgh Planetarium in Rochester, New York. In recognition of his contributions to public appreciation of astronomy, Mr. Dickinson has received numerous awards, including the New York Academy of Sciences' Book of the Year Award, the Astronomical Society of the Pacific's Klumpke-Roberts Award and, in 1995, the Order of Canada. Asteroid 5272 Dickinson is named after him.